T0295374

MATERIALS SCIENCE AND TECHNOLOGIES

ALUMINUM ALLOYS: PREPARATION, PROPERTIES AND APPLICATIONS

MATERIALS SCIENCE AND TECHNOLOGIES

Additional books in this series can be found on Nova's website
under the Series tab.

Additional E-books in this series can be found on Nova's website
under the E-books tab.

MATERIALS SCIENCE AND TECHNOLOGIES

ALUMINUM ALLOYS: PREPARATION, PROPERTIES AND APPLICATIONS

ERIK L. PERSSON

EDITOR

Nova Science Publishers, Inc.

New York

For permission to use material from this book please contact us:
Telephone 631-231-7269; Fax 631-231-8175
Web Site: http://www.novapublishers.com

Additional color graphics may be available in the e-book version of this book.

LIBRARY OF CONGRESS CATALOGING-IN-PUBLICATION DATA

Aluminum alloys : preparation, properties, and applications / editor, Erik
L. Persson.
p. cm.
Includes bibliographical references and index.
ISBN 978-1-61122-311-8 (hardcover)
1. Aluminum alloys. I. Persson, Erik L.
TN775.A4587 2010
620.1'86--dc22

2010041279

Published by Nova Science Publishers, Inc. ✝ New York

CONTENTS

PREFACE

Aluminium alloys are widely used in engineering structures and components where light weight or corrosion resistance is required. This book presents current research from across the globe in the study of aluminum alloys, including the casting methods for aluminum sheet and their effect on microstructural evolution; aluminum alloy anodes application for the removal of boron from drinking water by electrocoagulation; aluminum alloys used for corrosion resistance in structures submerged in marine environments; aluminum as an energy carrier; laser welding of aluminum alloys; and aluminum alloy heat treatments.

Chapter 1 - The studied material is a 5xxx aluminium alloy. It contains two kinds of intermetallic particles $Al_x(Fe,Mn)$ et Mg_2Si. This alloy is currently used in car industry as reinforcement pieces or in packaging industry as bottle liquids box lid.

The aluminium alloys sample is observed by X ray micro-tomography performed at the ESRF (European Synchrotron Radiation Facility). The obtained 3D images show complex shapes of intermetallic particles, generated during the alloy solidification process. Particles fill vacant spaces between aluminum grains. Therefore the final sheet properties depend on intermetallic particle shapes and notably on the matrix-particle interface properties. The goal of the current chapter is to classify intermetallic particles versus their shapes based on a morphological study and local curvature information.

The obtained three dimensional images are segmented, and then intermetallic particles are identified in a data base, each particle being stored as a map of voxels.

A morphological study is first performed; the particles are described by a set of measurements: volume, surface, shape indexes, geodesic elongation, and inertial parameters. In a second step, the boundary of each particle is meshed by a marching cubes triangular meshing with the Amira® software. A simplification of the surface is performed by a contract edge algorithm. At last principal curvatures (k_{min} and k_{max}) are estimated on each facet centre of the mesh. The bivariate distribution of k_{min} and k_{max} is estimated on every particle from data obtained on its boundary, providing a geometric signature of interface portions of the surface of particles. A statistical analysis is performed to classify particles according to their shape. This statistical study proceeds in two steps: the statistical analysis of curvature measures which provides a first classification of the particles according their interface geometrical properties; a final statistical analysis based on all the measurements to classify particles according to their shapes.

This approach enables the authors to follow the evolution of the populations of particles during the high temperature forming of the aluminium alloy.

Finally a prospective approach to model the complex shape of particles by a probabilistic model is proposed.

Chapter 2 - The chemistry of aluminium complexes covers quite bulk fields, as shown by a survey covering the crystallographic and structural data of over one thousand and two hundred examples. About nine percent of those complexes exist as isomers and are summarized in this paper. Included are distortion (92.2 %), ligand (4.3 %), *cis* – *trans* (2.5 %) and polymerisation (1 %) isomerism. These are discussed in terms of coordination about aluminium atom, and correlations are drawn between donor atoms, bond lengths and interbond angles. Distortion isomers differ only by degree of distortion in Al-L bond distances and L-Al-L bond angles are the most common.

Chapter 3 - Aluminum sheet is widely preferred for its light weight, strength and corrosion resistance. It is used in numerous applications such as automotive, transport, packaging, construction and the printing industry to name but a few. The two most popular processes for its production are direct chill (DC) casting of ingots and twin roll casting (TRC) of coil slabs.

DC casting involves modest solidification rates resulting in a rather coarse cast grain size. Further downstream processing consists of homogenization of the ingot followed by a series of hot and cold rolling steps. The main advantage of this route is that the elevated temperature processing steps allow the elimination of casting segregation phenomena, consequently resulting in homogeneous properties at final gauge. When high performance service properties are required (e.g. formability), DC cast material is preferred.

In TRC, aluminum exits the caster rolls in the form of a coil slab and is then directly cold rolled to final gauge. The process offers low capital investments, savings in energy and low operational cost together with some metallurgical advantages (e.g. higher solidification rates/solute leves and finer constituent particles). However, if an intermediate homogenization treatment is not introduced, metal produced via this route can be rather unstable at final gauge.

With continuous pressure to reduce the industry's carbon footprint, TRC is gaining popularity and more work is now focused on the optimization of this processing route to produce material with performance similar to DC.

In this review, an overview of the two casting methods along with the associated downstream processing for the production of rolled aluminum sheet and their effect on the microstructural evolution is presented.

Chapter 4 - The present chapter provides an electrocoagulation process for the amputation of boron-contaminated water using aluminium alloy as the anode and stainless steel as cathode. The various parameters like effect of pH, concentration of boron, current density and temperature has been studied. Effect of co-existing anions such as silicate, fluoride, phosphate and carbonate were studied on the removal efficiency of boron. The results showed that the optimum removal efficiency of 90% was achieved at a current density of 0.2 A dm^{-2}, pH of 7.0. The study shows that boron adsorption process was relatively fast and the equilibrium time was 60 min. First and second-order rate equations were applied to study adsorption kinetics. The adsorption process follows second order kinetics model with good correlation. The Langmuir, Freundlich and D-R adsorption models were applied to describe the equilibrium isotherms and the isotherm constants were determined. The experimental adsorption data were fitted to the Langmuir adsorption model. The thermodynamic parameters such as free energy (ΔG^0), enthalpy (ΔH^0) and entropy changes

(ΔS^0) for the adsorption of arsenate were computed to predict the nature of adsorption process. Temperature studies showed that adsorption was endothermic and spontaneous in nature.

Chapter 5 - The present chapter provides a review on a variety of aluminium alloy sacrificial anodes used for protection against corrosion of structures submerged in a marine environment. The superiority of aluminium alloy anodes over other sacrificial anodes and their characterisation are explained in detail. The performance of anodes depends not only on the alloy composition but also on the treatment / homogenization technique that is to be used for their preparation. This aspect is reviewed critically as both these factors play an important role in enhancing the life of anodes significantly. The activation mechanism of aluminium alloy anodes based on the surface free energy concept is an advanced technique. This concept is explained after critically reviewing it and pointing out the disadvantages of mechanisms proposed by earlier investigators. It is also mentioned that the surface free energy phenomenon is a novel, simple and non-electrochemical tool for understanding their working mechanism, as well as for the development of efficient sacrificial anodes. Finally, the necessity of innovation of a smart anode to protect the structures against corrosion both in marine and fresh / river water environments, the critical steps involved in their development is emphasized.

Chapter 6 - In this chapter, a brief overview and perspectives of the development of aluminum-derived energy are given. Aluminum has received a lot of attention particularly in recent years for energy applications because of its outstanding properties including high electrochemical equivalent (2980Ah kg^{-1}), highly negative voltage (-1.66V versus SHE in neutral electrolyte; -2.33V versus SHE in alkaline electrolyte) as well as high energy density (29.5 MJ kg^{-1}). Other attractive features of aluminum are its abundance and recyclability.

All these features make aluminum and its alloys promising candidates to meet the urgent demand of advanced energy carrier arising from energy and environmental crises. Through proper activation, aluminum and its alloys are capable of reacting with water to generate hydrogen or being directly used as a "fuel" to produce electricity through aluminum-air semi-fuel cells.

The hydrogen produced from aluminum-water reaction can be delivered to fuel cells or internal combustion engine for further energy conversion.

The technical characteristics and development challenges of the above-mentioned two aluminum energy conversion routes, i.e. aluminum-assisted hydrogen production and aluminum-air semi-fuel cells, are briefly reviewed.

In spite of the existence of challenges, this study indicates that aluminum and its alloys may serve as an excellent energy carrier over the near to long-term.

Chapter 7 - Aluminium alloys are largely used in transportation industry, as automotive and aeronautics where the high strength-to-weight ratio is valuable, but also in corrosive environments due to the excellent corrosion resistance they exhibit, as in tanks, pressure vessels and packaging. Despite their wide acceptance in industry, they have limited weldability. Arc welding processes as tungsten inert gas (TIG) and metal inert gas (MIG) have been used as well as electron beam and laser welding. For high productivity and demanding manufacturing, laser welding is an alternative process to consider.

Laser welding is a precise high energy density process that reduces the fusion and heat affected zones, minimising some of the major weldability problems. However, the high reflectivity of aluminium to carbon dioxide laser has restricted its use. Infrared solid state lasers, as Nd:YAG, both disc and rod, and the recent high power fiber lasers showed to be more adequate for laser welding of Al alloys.

This chapter is structured as follows: Laser welding process is discussed focusing on the effect of operating parameters on the weld quality, welding modes (keyhole and conduction) and process control. An overview of the weldability problems of major groups of industrial aluminium alloys is given. Recent and emerging techniques are presented and discussed.

Chapter 8 - The objects of the author's paper are aluminum alloy samples (AASs) contained the different amount of Cu, Mn, Mg, Si and Li. The author are modeling the features of microstructure of potential relief of an Aluminum Alloy Sample (AAS) and studying its transformation under both imposed fatigue deformation and wetted by liquid metals (Ga; or Hg; Li ;In). Although fatigue in an AAS is characterized by permanent changes of its mechanical impedance, the author illustrate the main ideas by using only the «time series» allied with effective internal friction Q-1eff of an AAS.

There is not a common law permits one in advance to determine both what kind of micro - structures and how many of micro - structures will be simultaneously belong to aluminum alloy (AA) after the N- number of cycles are loaded.. AAs like B-95 or 7075 are heterogeneous materials for which the more energy can be absorbed by selected micro-regions of a tested sample. So micro-crack in the space of AAS and alarm state of AAS arises. Each micro-regions will to contribute (the Q-1 k belong to k-th micro - region) to the effective internal friction – Q-1 eff accordance with fit statistic g k. The author find a number of micro-regions - L and series g k and Q k from the experimental data like as the internal friction (Q-1) eff versus both, the number of cycles - N and the deformation - ε. Series g k and Q k (k = 1,2,3,…,L) present the microstructures of AASs. So the monitoring of AASs alarm states was made.

In this paper also is presented the original technology to forecast fatigue damage of an aluminum alloy sample. Here was used the fatigue sensitive element (FSE). The various impurities in Aluminum implemented the various resistances to fatigue of an Aluminum Alloy Sample. In the other hand, the author used multiphase heterogeneous mixtures (MHMs) which contents a variable volume of initial components. It is selected MHMs are using for produce FSEs. The correlation between output parameters of the MHMs and volume contents of initial components and form of their grains are given in the author's previous papers, The present paper is aimed to establish the correlation of the FSEs microstructures changes and corresponding changes of the aluminum alloy microstructures at imposing the same spectra deformation on both of them. A change of FSEs microstructure investigated by using their effective electrical resistance Reff= data.

Chapter 9 - The growing importance of hydrogen among the alternative energy sources has given rise to new studies concerning the effect of hydrogen on the technological properties of metallic alloys. Among the most commonly used metallic materials are cast and wrought aluminium alloys, by virtue of their light weight, fabricability, physical properties, corrosion resistance and low cost. However, the interaction of hydrogen with aluminium alloys has not yet been fully understood.

In this research investigation hydrogen was introduced into the surface layers of important technological and industrial cast and wrought aluminium alloys via cathodic hydrogen charging technique. The effect of hydrogen absorption on the structure, surface microhardness and mechanical behavior of aluminium alloys was studied. In addition, the absorption .of hydrogen was found to depend strongly on the charging conditions. The presence of aluminium hydride in the surface layers of hydrogen charged aluminium alloys, after intense charging conditions, was observed.

Hardening of the surface layers of aluminium alloys, due to hydrogen absorption and in some cases hydride formation, was also observed. Tensile experiments revealed that the ductility of aluminium alloys decreased with increasing hydrogen charging time, for a constant value of charging current density, and with increasing charging current density, for a constant value of charging time. However, their ultimate tensile strength was slightly affected by the hydrogen charging procedure. The cathodically charged aluminium alloys exhibited brittle transgranular fracture at the surface layers and ductile intergranular fracture at the deeper layers of the alloy. Finally, the effect of deformation treatment (cold rolling) on the hydrogen susceptibility of aluminium alloys was studied.

Chapter 10 - This chapter presents the results of an experimental study done on a wrought aluminum alloy (AlCu2,5Mg). This alloy has been heat treated in order to maximize its mechanical properties, especially its yield strength and its ductility. The experiment consisted in choosing 10 identical pieces of alloy which were to be differently heat treated regarding the quenching and artificial aging temperature. The methodology consists in the following steps: adopting the heat treatment technology for the proposed alloy; choosing the necessary hot working installations; choosing the equipments used for studying the yield strength; programming the experiment and the analytical interpretation of the results.

The present chapter presents the analysis of the yield strength and of the aging and quenching heating temperature. The equation for mechanical properties theoretical determination in standard or special heat treatment conditions has also been calculated. There have been carried out two nomograms for quick graphic determination of the necessary technology.

In: Aluminum Alloys
Editor: Erik L. Persson

ISBN: 978-1-61122-311-8

Chapter 1

MORPHOLOGICAL ANALYSIS OF COMPLEX THREE-DIMENSIONAL PARTICLES APPLICATION TO ALUMINIUM ALLOY 5XXX

Estelle Parra-Denis* and Dominique Jeulin

MINES Paristech, CMM- Centre de morphologie mathématique, Mathématiques et Systèmes, 35 rue Saint Honoré - 77305
Fontainebleau cedex, France

ABSTRACT

The studied material is a 5xxx aluminium alloy. It contains two kinds of intermetallic particles $Al_x(Fe,Mn)$ et Mg_2Si. This alloy is currently used in car industry as reinforcement pieces or in packaging industry as bottle liquids box lid.

The aluminium alloys sample is observed by X ray micro-tomography performed at the ESRF (European Synchrotron Radiation Facility). The obtained 3D images show complex shapes of intermetallic particles, generated during the alloy solidification process. Particles fill vacant spaces between aluminum grains. Therefore the final sheet properties depend on intermetallic particle shapes and notably on the matrix-particle interface properties. The goal of the current chapter is to classify intermetallic particles versus their shapes based on a morphological study and local curvature information.

The obtained three dimensional images are segmented, and then intermetallic particles are identified in a data base, each particle being stored as a map of voxels.

A morphological study is first performed; the particles are described by a set of measurements: volume, surface, shape indexes, geodesic elongation, and inertial parameters. In a second step, the boundary of each particle is meshed by a marching cubes triangular meshing with the Amira® software. A simplification of the surface is performed by a contract edge algorithm. At last principal curvatures (k_{min} and k_{max}) are estimated on each facet centre of the mesh. The bivariate distribution of k_{min} and k_{max} is estimated on every particle from data obtained on its boundary, providing a geometric signature of interface portions of the surface of particles. A statistical analysis is performed to classify particles according to their shape. This statistical study proceeds in

* Email: Estelle.Parra@ensmp.fr

two steps: the statistical analysis of curvature measures which provides a first classification of the particles according their interface geometrical properties; a final statistical analysis based on all the measurements to classify particles according to their shapes.

This approach enables us to follow the evolution of the populations of particles during the high temperature forming of the aluminium alloy.

Finally a prospective approach to model the complex shape of particles by a probabilistic model is proposed.

INTRODUCTION

The 5xxx aluminium alloys include two types of intermetallic particles: Al_x(Fe, Mn) and Mg_2Si. Those particles are broken up within the aluminium matrix, and take their form during the alloy solidification process. They are located in the vacant space of aluminium grains. Thus they show varying complex shapes. During the industrial rolling process, used to transform an ingot into a stampable sheet, the particles are broken and distributed within the sheet volume. The study of particle shape during the industrial rolling process leads to knowledge about the final mechanical properities of the aluminium alloy sheet (Feuerstein, 2006).

The 3D observation techniques used to study materials are: the successive polishing of the sample with an image capture at each successive step, and the X ray microtomography. The X ray microtomography technique provides the acquisition of three dimensional images in a non detstructive way with a high resolution (up to 0.7μm in the present case). The studied images are acquired on the ID 19 beam line at the ESRF "European Synchrotron Radiation Facility". The used electron beam energy is 120 KeV. On these images, one can observe the three dimensionnal alloy microstructure on a $1x1x10mm^3$ sample (Buffières et al, 1999).

The aim of this chapter is to characterize the three dimensional shape of intermetallic particles, and to classify them into a family of shape. Particles belonging to a family present the same mechanical ability versus their break-up during the rolling process. The obtained families are used to study by finite element methods the aluminium alloy formability (Moulin, 2008), (Moulin et al, 2009), (Moulin et al, 2010).

In the litterature, many 3D shape analysis methods exist. Nevertheless, they mostly deal with simple shapes, or star-shaped object (Gardner et al, 2003)(Holboth, 2003). For more complex shapes, studies based on deformable template using splines are used for example for morphometric studies (Bookstein, 1997). These technics consist in the comparison of shapes to a model, by measuring the distance between shapes or calculating the mathematical transformations needed to fit them.

This paper proposes a methodology to carry out 3D complex shape analysis based on two approaches (Parra et al, 2007)(Parra et al, 2008). This methodology uses 3D measurements to characterize the particle shapes. There are three different kinds of measurements: morphological measurements (Parra et al, 2008) such as volume, surface, elongation; mechanical measurements such as distribution mass index; and surface measurements based on local curvature (Parra et al, 2007). This article presents the use of these three dimensional measurements, and their synthesis by statistical data analysis to sum up the measured information in an adapted representation space. In this space we propose a classification of intermetallic particle according to their shape. Finally, a prospective approach to model the complex shape of particles by a probabilistic model is proposed based on morpholgical measurements performed on the sample.

The paper is organized as follows: in the first part we present the used morphological measurements. Firstly we quickly deal with classical measurements, and then we propose tridimensionnal measurements based on the geodesic distance. In the second part, we present inertia measurements. They are original methods using mass distribution information in the form of shape indexes (Parra et al, 2007). In the third part, we present local curvature measurements (Parra et al, 2008). We first define the principal curvatures, and then we present the bivariate distribution of k_{min}-k_{max} and its use. In the fourth part, the application of the different kinds of measurement to classify particles is presented. Finally, the last part is dedicated to a prospective approach to model intermetallic particles.

MORPHOLOGICAL MEASUREMENTS

In this paragraph a set of morphological mesurements used to characterize three dimensional shapes is given. The studied shapes are complex. They show various types of non convex and branched shapes, which cannot be modelled by a simple geometrical approach. The presented parameters are divided into three groups: classical measurements, shape indexes and geodesical measurements.

Classical Measurements

The most classical measurements on binary objects in image analysis are their volume and their area:

- The volume (V) is estimated by counting the number of voxels.
- The surface area (A) is estimated by the method of Crofton (1868). It is proportional to the average of the projected area on several planes of the space.

Shape Indexes

Shape indexes compare the studied shape to a reference one.

The sphericity index is presented by equation 1. It compares the shape to a sphere. It is equal to one for a sphere and its value is very sensitive to the object boundary noise.

The compacity index is obtained by changing the proportionality constant (equation 2), since it compares the studied shape to a cube.

$$I_S = 36\pi \frac{V^2}{S^3} \qquad (1)$$

$$I_c = 6\frac{I_s}{\pi} \tag{2}$$

Geodesic Measurements

In this section, the presented parameters are based on the geodesic distance introduced by Lantuejoul and Maisonneuve in 1984 (Lantuejoul and Maisonneuve, 1984). The geodesic distance is a mesure which allows apprehending connected objects, and particularly non convex objects.

The geodesic distance separating x_1 and x_2 belonging to a given shape X is the length of the shortest paths connecting x_1 to x_2 and remaining included in X. From this distance definition a set of useful measurements can be defined : the geodesic propagation function, the geodesic centre and radius, the geodesic length, and the geodesic elongation index.

The Geodesic Propagation Function

From this distance, the geodesic propagation function of a given point x_1 of X is defined by equation 3.

$$\forall x_1 \in X, P_x(x_1) = \max\left\{d_g(x_1, x_2) \mid x_2 \in X\right\} \tag{3}$$

Geodesic Centre and Geodesic Radius

The geodesic centre of X is defined as the regional minimum of its propagation function. The presence of noise on the object boundary does not affect the position of the geodesic centre. The value of the propagation function on the geodesic centre is the geodesic radius. It corresponds to the smallest ball included in an object (equation4). It provides information about the size of the core of the shape.

$$\mathrm{Rg}(\mathrm{X}) = \operatorname*{Inf}_{\mathrm{x_i} \in \mathrm{X}} \left[\operatorname*{Sup}_{\mathrm{x_j} \in \mathrm{X}} \left(\mathrm{d_g}(\mathrm{x_i}, \mathrm{x_j}) \right) \right] \tag{4}$$

The Geodesic Length

The geodesic length of X is defined as the length of the longest geodesic path included in X. It is the maximal value of the geodesic propagation function on X, as given by equation 5.

$$\mathrm{Lg}(\mathrm{x}) = \operatorname*{Sup}_{\mathrm{x_i}, \mathrm{x_j} \in \mathrm{X}} \left[\mathrm{dg}(\mathrm{x_i}, \mathrm{x_j}) \right] \tag{5}$$

It is not sensitive to the presence of noise on the object boundary. It gives an idea of the studied object complexity, if it is compared to its Feret diameter, and geodesic radius.

The Geodesic Elongation Index

The geodesic elongation index IG_g (equation 6) is a three dimensionnal extension of the two dimensionnal geodesic index (Lantuejoul and Maisonneuve, 1984). It characterizes the elongation or the spread of an object X. It is normalized as 1 for a ball. The normalisation by a ball seems natural since it corresponds to the more compact three dimensional objects.

$$IG_g = \pi \frac{L_g^3(X)}{6V(X)} \tag{6}$$

MECHANICAL MEASUREMENTS

In the following section, shape indexes are based on the solid mechanic theory, and more precisely on the object inertial properties.

The definition of the inertia matrix is first reminded. Then, two new parameters based on the inertia moment values are presented: their definition, and their notable properties.

Inertia Matrix

The inertia tensor of an object X, composed by n sub elementary volumes is defined by the equation 7. The variables x_i, y_i, and z_i are the X voxel coordinates in the direct space centered on its inertia center, and with axes parallel to the image one. The mass distribution of X is supposed to be uniform. The elementary volume of the obect is the voxel with an elementary mass equal to 1.

$$[J_0] = \begin{pmatrix} I_{ox} & -I_{xy} & -I_{xz} \\ -I_{xy} & I_{xy} & -I_{zy} \\ -I_{xz} & -I_{yz} & I_{oz} \end{pmatrix} \tag{7}$$

$$\text{with} \quad \underbrace{\begin{matrix} I_{xy} = \sum_n m_i(y_i.x_i) \\ I_{yz} = \sum_n m_i(y_i.z_i) \\ I_{xz} = \sum_n m_i(z_i.x_i) \end{matrix}}_{\text{inertia poducts}} \quad \underbrace{\begin{matrix} I_{ox} = \sum_n m_i(y_i^2 + z_i^2) \\ I_{oy} = \sum_n m_i(x_i^2 + z_i^2) \\ I_{oz} = \sum_n m_i(y_i^2 + x_i^2) \end{matrix}}_{\text{inertia moments}}$$

The inertia tensor is symmetric and positively defined. It admits a base composed by three orthogonal eigen vectors $\overrightarrow{u_1}$, $\overrightarrow{u_2}$, $\overrightarrow{u_3}$. Their components in relation to the initial base are given by the transition tensor T (equation 8). The name of this base is the principal inertia base, its axes being the principal inertia axes. In this base, the inertia tensor is diagonal. It is called the principal inertia tensor, its diagonal terms being called the principal inertia moments: I_1, I_2, I_3.

$$T = \begin{pmatrix} u_{11} & u_{12} & u_{13} \\ u_{21} & u_{22} & u_{23} \\ u_{31} & u_{32} & u_{33} \end{pmatrix} \qquad [J_0] = \begin{pmatrix} I_1 & 0 & 0 \\ 0 & I_2 & 0 \\ 0 & 0 & I_3 \end{pmatrix} \tag{8}$$

Mass Distribution Parameters or Normalized Inertia Moments

The mass distribution parameters are the value of principal inertia moments (I_1, I_2, I_3) of the object. Their values depend on the object shape, and are linked to its distribution mass (Parra-Denis et al, 2005).

The inertia moments are normalized to be free from the object volume; they are noted λ_1, λ_2, λ_3 and they are defined by:

$$\lambda_i = \frac{I_j}{I_1 + I_2 + I_3}, \text{ i=1,2,3} \tag{9}$$

They present particular mathematical properties: by definition their sum is equal to 1, and they are ordered (equations 10 and 11).

$$\lambda_1 + \lambda_2 + \lambda_3 = 1 \tag{10}$$

$$\lambda_1 \geq \lambda_2 \geq \lambda_3 \tag{11}$$

$$\forall i,\ \lambda_i \leq 0.5 \tag{12}$$

$$\lambda_2 \geq 0.5 \cdot (1 - \lambda_1) \tag{13}$$

From the equations 9, 10 and 11, and the inertia moments definition, the inequalities 12 and 13 can be deduced. From the equation 9 and 10, and the inequalities 11, 12, and 13, it is possible to design a graph representing the possible values of λ_1 versus λ_2 (figure 1).

The normalized inertia moments have a rich physical interpretation as shape parameters. Indeed, they permit to estimate the similarity of a given shape with reference mass distribution shapes: spherical mass distribution, and cylindrical mass distribution (flat type and needle type).

The graph corresponding to λ_1 versus λ_2 (figure 1) represents a triangular zone of allowed values for any objects or weigthed couple of points. At the triangle vertices, 3 types of mass distribussion are distinguished: spherical, flat and needle. Between those extremitiees, shapes vary continuously. Along the triangle edges, shapes are prolate ellipsoid type, oblate ellipsoid type or flat ellipse type. The projection in this graph of the normalized inertia moments of any object gives a prior knowledge about is shape.

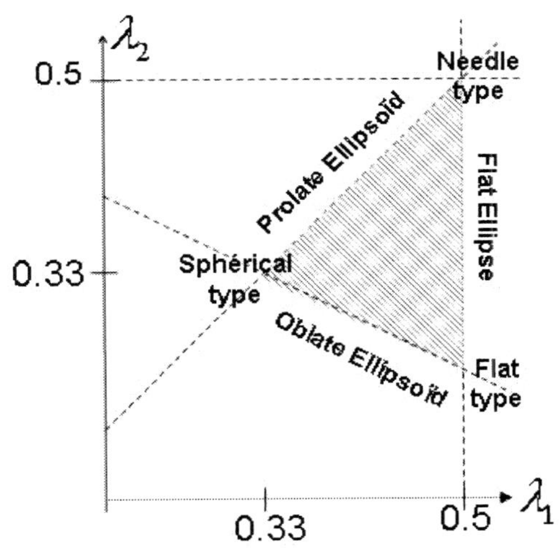

Figure 1. theorical graph of λ_2 *versus* λ_1.

Local Curvature Measurements at the Surface

The local curvature surface study gives informations about the relations betwen the matrix interface and the particles. Thus it gives an idea about the interactions of an object with its behavior.

The local curvature surface study is inspired from works developed in the metallurgy field by Alkemper and Voorhees (Alkemper et al 2001; Voorhees et al, 2002), who use the local curvature histogram to follow the evolution of aluminium dendrites during the solidification process.

In this section, the local curvature measurement on a three dimensional particle is presented. Firstly, the curvature at a surface in the three dimensional space is defined. Secondly, the study of principal curvature graph is explained: its geometrical properties, and its use for the proposed intermetallic particle study. Lastly, we apply the presented methodology to the study of aluminium alloy intermetallic particle.

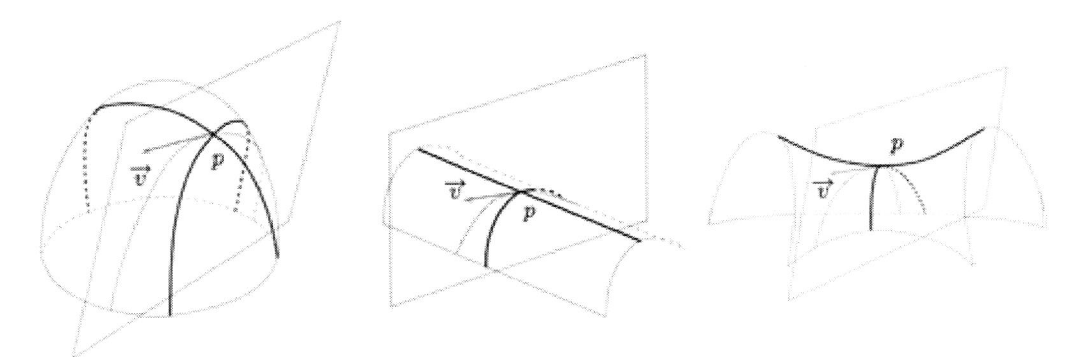

Figure 2. Example of typical interface geometry à a given point p of the surface.

The Three Dimensional Surface Curvatures

Let S be a smooth surface in the Euclidean space R^3, and x be a point of S. The normal vector n is defined at x on the surface S. Every plane P containing n intersects S with a curve C. For each plane P_i, the radius of curvature r_i of C_i at x is given. The minimum curvature radius R_{min} and the maximum one R_{max} are obtained for a set of two orthogonal planes. They define the maximum curvature k_{max} (inverse of the radius) and the minimum curvature k_{min}, called principal curvatures.

The k_{min}-k_{max} Graph

It is possible to quantify the shape of the object-matrix interface by looking at the distribution of the principal curvatures (k_{min}-k_{max}).

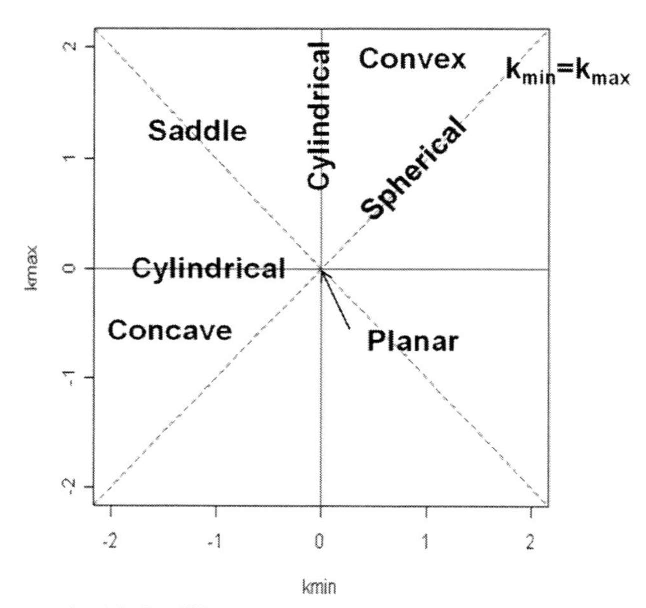

Figure 3. The k_{min}-k_{max} graph with the different types of portion intrefaces corresponding region.

Geometrical Properties of the k_{min}-k_{max} graph

The (k_{min}-k_{max}) graph presented in the reference article of Besl and Jain in 1986 (Besl, 1986) presents the geometrical properties of interface portions (figure 3). The shape of the interface can be determined from the values of k_{min} and k_{max}. Examples of different geometrical interface portion types are illustrated on the figure 3. We present the striking properties of this graph (see figure 2):

- All the points of the surface are located on the left of the line corresponding to $k_{min} = k_{max}$. All the interface portions belonging to this line have a spherical shape.
- The planar interface portions are located at the origin of the graph, where $k_{min} = k_{max} = 0$.
- The cylindrical interface portions have $k_{min} = 0$ or $k_{max} = 0$; if $k_{min} = 0$ the portion has a valley shape, if $k_{max} = 0$ it has a ridge shape.

- The hyperbolic interface portions or saddle shape have one positive and one negative curvature.
- If the two principal curvature values are positive, the interface portion is concave.
- If the two principal curvature values are negative, the interface portion is convex.

The Bivariate Distribution (k_{min}-k_{max})

On the (k_{min}-k_{max}) graph each point represents a portion of the object interface.

Implementation In the present chapter, the local curvature measurement is performed with Amira® software. The principal curvatures of intermetallic particles are measured on the surface meshing of the intermetallic particles produced by Amira®. The surface meshing of an object is usually used in materials field, to carry out a mechanical analysis of an object by finite elements calculation, directly from three dimensional images. The choice of the technique to mesh the surface is a difficult task (Frey and George, 1999).

The geometry of aluminium alloys containing intermetallic particles is complex (Figure 4). It is far from simple geometries and then the use of regular meshing of points is not possible. The chosen technique to mesh particles is a frontal non structuring method. The idea behind the frontal methods is to mesh the field by successive iterations, propagating a front inside the surface (starting on the surrounding area of the field), until its complete covering.

The non structuring meshing method used is based on the Marching Cubes algorithm (Lorensen and Cline, 1987). From the 3D voxel image, the surface of the object is digitized into triangles to mesh. Then from the discrete surface, the volume is meshed into linear tetrahedrons by using a frontal method implemented into Amira® software. This method allows varying the length of elements of the model without creating discontinuities. The number of triangular elements needed to cover the surface depends on its complexity. It is optimized to obtain a good definition of the surface with a low number of degrees of freedom.

After the meshing, a simplification algorithm of the surface by collapsing edges is applied (Hoppe et al, 1993). This algorithm consists in merging the two extremities of edge into an apex, with a collapsed criterion based on the minimization of an energy function. The energy function consists in two terms: the distance between the simplified and the original meshing, the spatial repartition of point improvement (which ensures the meshing does not present strong bump).

Figure 4. Example of intermetallic particle contained in AA 5xxx meshed by Amira®. The number of facets of this meshing is equal to 8044.

Measurement parameters used with Amira® The measurement of the curvature with Amira® software is based on the meshing obtained by the marching cubes algorithm. The principal curvatures are estimated either for each triangle of the surface, or for each points of the meshing. The algorithm works by a local fitting of the surface with a quadric form. The eigenvalues and eigenvectors of the quadric form correspond to the principal curvature values and to the directions of principal curvature.

The parameters of the algorithm are:

- The number of neighbours: it determines which triangles and which points are considered to be neighbours of a given triangle and of a given point. If the value of this input is one, only triangles sharing an edge with a given triangle, and only points directly connected to a given point are considered.
- The number of iterations: it determines how many times the initial curvature values computed for a triangle or for a point are being averaged with the curvature values of direct neighbour triangles or points. This parameter interferes only with the scalar curvature values, not with the directional curvature.

A preliminary study on different reference shapes has been performed, in order to validate the method of estimation: cylinder, sphere, torus, ellipsoid, with variable lengths and orientations. For each tested shapes, we introduce variations of all the meshing parameters : simplification of the meshing varying between {0;1,4;2,8}, number of neighbours varying between {1;6;16}. Finally the retained parameters for further measurements are a simplification of the meshing equal to 1.4, a number of layers equal to 6, and a number of average equal to zero (Figure 4 presents an example of meshing obtained with the presented methodology on an intermetallic particle).

Effects of the meshing refinements on the k_{min}-k_{max} graph Theorically, the k_{min}-k_{max} graph of a cylinder or a sphere is perfectly known. For the sphere, it presents a single peak having for value the inverse of its radius. For a cylinder, it presents two peaks, the first one is for the cylindrical surface, its k_{max} value is equal to the inverse of the cylinder radius, and its k_{min} value is equal to zero. The second peak is for the planar portions of the cylinder; its value is (0,0).

The figure 5 shows on the k_{min}-k_{max} graph the effects of the meshing reffinement on a 32 voxels radius cylinder (the number of neighbors is 6 and the number of iteration is 0).

When no refinement is applied, the surface milling due to the digitalization of the object is important; thus the measured k_{min}-k_{max} graph shows a lot of false values. When a 1.4 meshing refinement is used (ie the « *contract edges* » algorithm is applied when the distance between two edges is less or equal 1.4, i.e 2 times the size of a voxel in the present study), the k_{min}-k_{max} graph shows two point clouds centereds on the expected peaks. For a 2.8 meshing refinement, the k_{min}-k_{max} graph also shows two point clouds centereds on the expected peaks, there is less erroneous value in comparison with the one of the 1.4 refinement.

In the case of intermetallic particles, the obtained meshing is defined by a large number of facets, from one thousand up to at least ten thousands (Table 1). A bidimensional histogram is used to easily exploit the obtained information (Parra-Denis et al, 2007).

Nevertheless, additionaltests realized on intermetallic particles shows meshing refinements greater than 2.1 introduce a too strong smoothing of the surface particle, with an important loss of information about the particle shape.

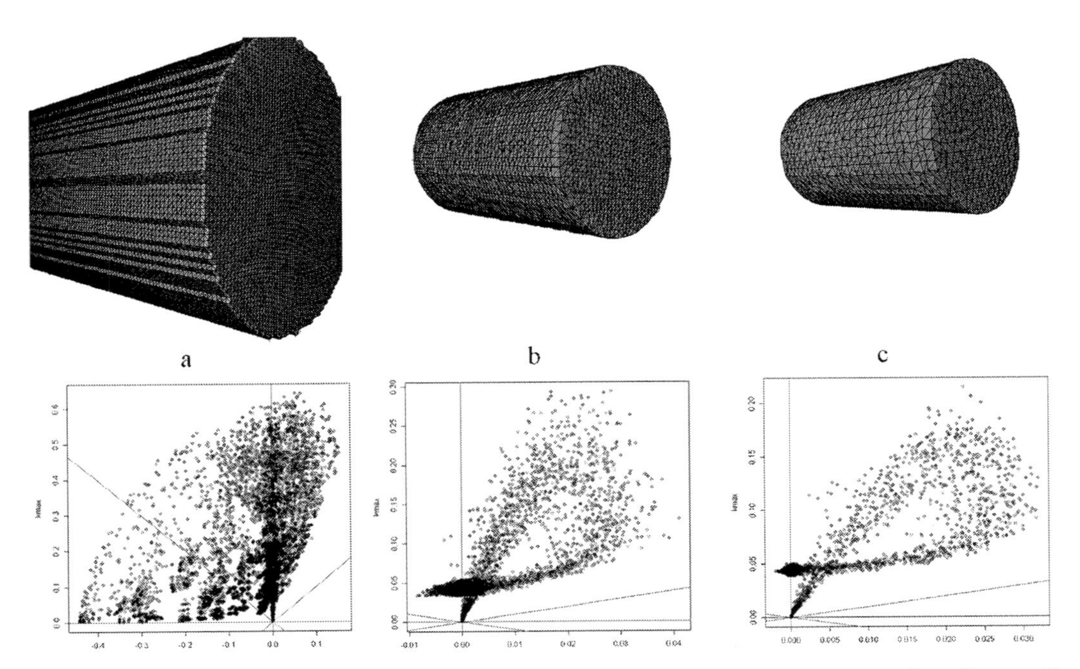

Figure 5. Illustration of meshing refinement effects on the surface curvature measurements of a cylinder with radius 32 voxels ; applied meshing refinements equal to : (a) 0 (b) 1,4 (c) 2,8.

From the results obtained by the Amira software, the (k_{min}-k_{max}) graph and the associated (k_{min}-k_{max}) bivariate distribution are obtained by a function writing with the R© software. The bivariate distribution (k_{min}-k_{max}) has been calculated for each intermetallic particle of the data base. They are weigthed, in the way each triangle of the meshed particles is counted according to the percentage of the surface it represents. The principal curvature histogram is used for comparing intermetallic particles belonging to different rolling process deformation steps. A prestudy about the histogram boundaries was performed. A histogram of the minimal curvature and a histogram of the maximal curvature were obtained on the entire set of intermetallic particles (figure 6).

Table 1. Meshing number elements in relation to the simplification of the meshing applied and to the shapes of the studied object

		Cylindre3D	Cylindre3D_tourné	Ellipsoide_oblate	Tore
contract edges : 2.8	Points	878	821	1766	2464
	Faces	1752	1638	3528	4928
	Edges	2628	2457	5292	7392
contract edges : 2.0	Points	1734	1642	3476	4986
	Faces	3462	3278	6948	9972
	Edges	5193	4917	10422	14958
contract edges : 1.4	Points	3994	3558	6805	9952
	Faces	7984	7112	13602	19886
	Edges	11976	10668	20403	29829
contract edges : 0.0	Points	32678	39816	76534	110656
	Faces	65352	79628	153064	221312
	Edges	98028	119442	229596	331968

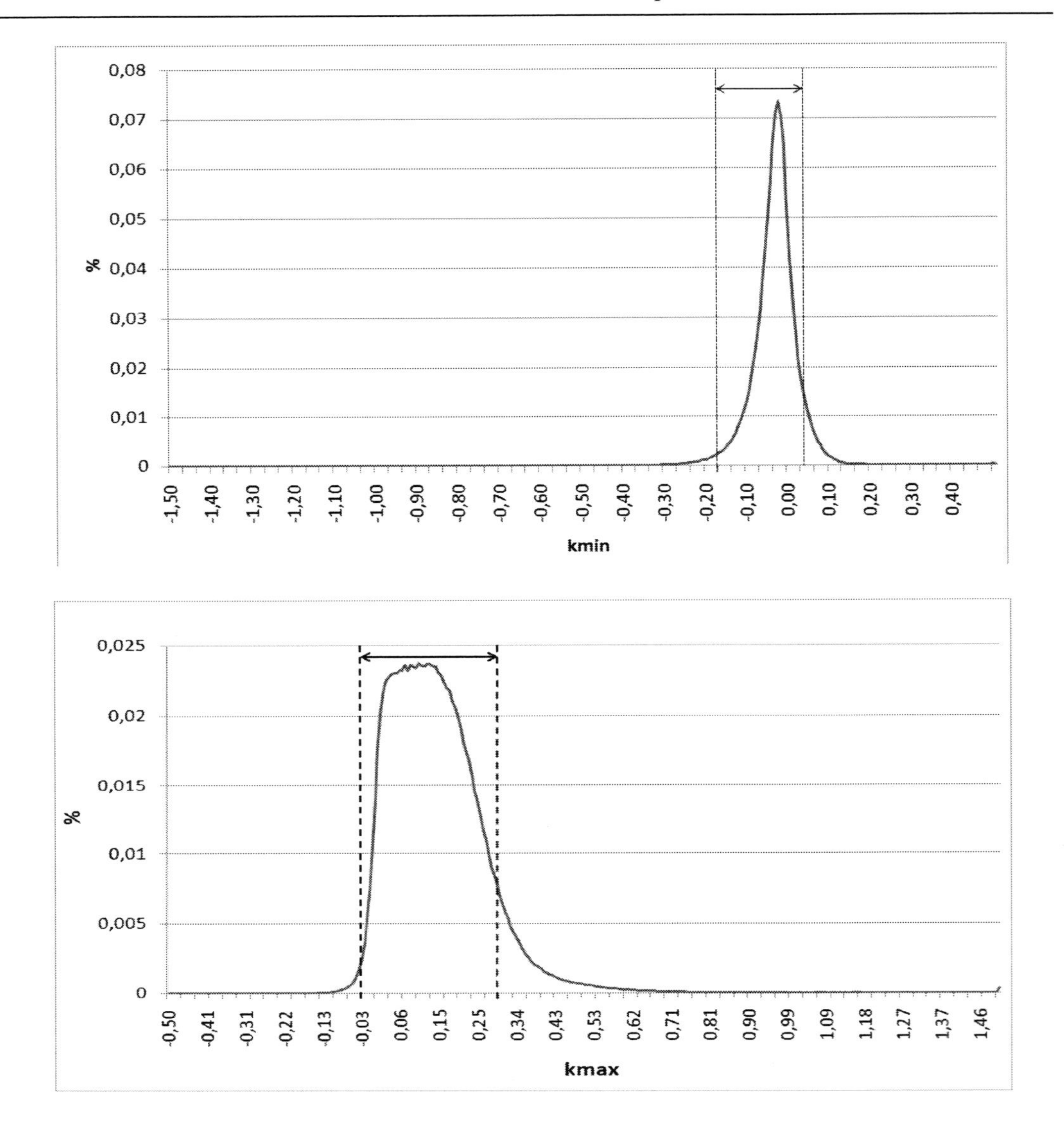

Figure 6. Study of the principal curvature values on the entire set of intermetallic particles (a) k_{min} (b) k_{max}.

Table 2. Boundaries of the principal curvature histogram for the entire data base of intermetallic particles

Min k_{min}	Max k_{min}	Min k_{max}	Max k_{max}
-0.1733	0.0333	-0.0133	0.3000

On these histograms, we chose the maximum and minimum boundaries for k_{min} and k_{max} (see tab table 2), so that almost 90% particles were contained in the chosen interval (89.89% particles for the k_{min} interval and 90.02% for the k_{max} interval). Five more classes were added. These classes correspond to the marginal curvature values in relation with the interface geometry: concave, convex, saddle, cylindrical (bridge and valley).

Application of the Presented Methodology to the Aluminium Alloy 5xxx Study

The studied intermetallic particles are observed on Aluminium alloy 5xxx three dimensional images acquired by X-ray microtomography at the European Synchrotron Radiation Facility (Parra-Denis, 2007). Samples corresponding to successive hot rolling process steps are studied: 0%, 2%, 10%, 12%, 19%, 41%, 82%, and 123.5%.

Pre Treatement

The X ray microtomography obtained images are studied by three dimensionnal image analysis to extract the different type of intermetallic particles: iron rich particles, voids and silicon-magnesium particles. A multi class thresholding is applied, its flow chart being presented on the figure 7 (Parra-Denis et al, 2005). Each particle is then extracted in an image corresponding to its bounding box (the new image facettes are parallel to the ones of the original image). A data base is created. It contains, for a deformation stage, the whole set of extracted intermetallic particles, classified according their constituants, and individualized in a bounding box. Each imagette of the data base contains its relative coordinates regarding the initial image sample it comes from. The three dimensional thresholded sample image can be reconstructed (Figure 8). The 5xxx alluminium alloy 10% deformed sample contains 4921 iron rich particles.

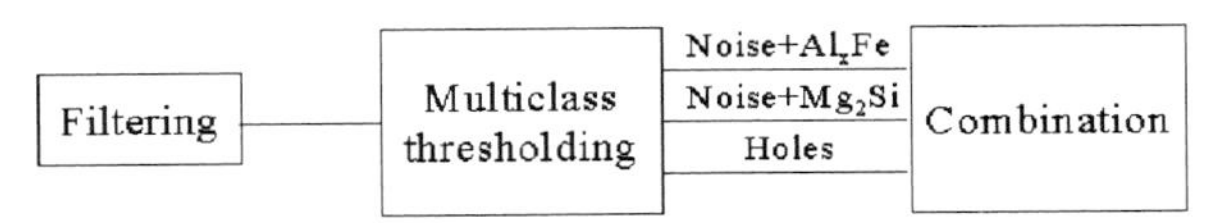

Figure 7. Schematic description of the multi class segmentation algorithm.

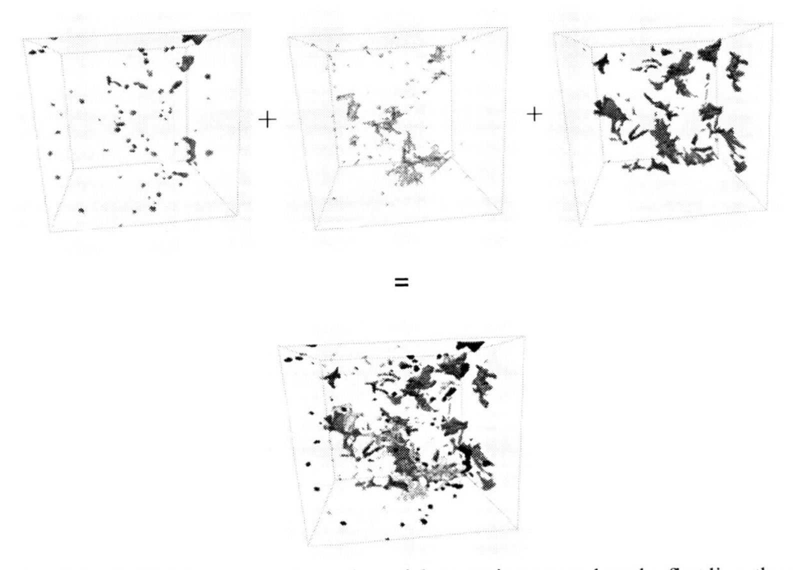

Figure 8. Example of thresholded image sample; each particles type is extracted on the first line, the entire sample particles being reconstructed below; black: voids; dark grey: Al_xFe; grey : Mg_2Si.

Illustration of the Presented Measurements for Three Particles Extracted from the Data Base

For each intermetallic particle of the data base, the previous mentionned parameters were been measured. The AA5xxx intermetallic particles show various complex shapes (figure 9).

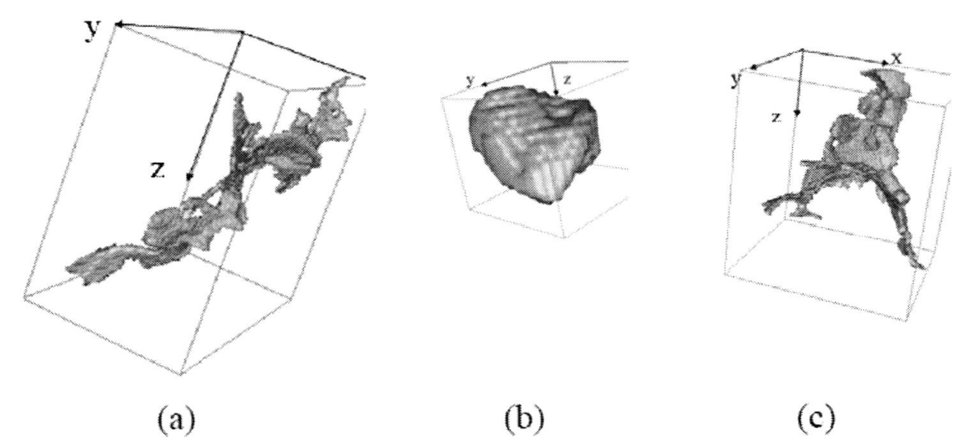

Figure 9. AA5xxx Iron rich particles form a 10% deformed sample (a) particle n°22, (b) particle n°1448, (c) particle n°4249.

Table 3. Results of the parameter measurements in voxels size on iron rich particles of the 5xxx data

Measurements	V	S	I_s	I_c	IG_g	λ_1	λ_2
n°22	17397	15238,9	0,01	0,005	101,6	0,49	0,44
n°1448	2121	954,7	0,585	0,321	1,2	0,47	0,36
n°4249	28619	21858,3	0,009	0,005	33,2	0,4	0,35

Nevertheless, some kinds of shapes seem distinguishable: compact particle shape, sheet particle shape, and more complex particles formed with branches and sheets. In this paragraph, parameters measured on iron rich particles $Al_x(Fe,Mn)$, extracted from a 10% deformed AA5xxx sample, and presenting *a priori* a certain variety of shapes are presented. Their measured parameters are summarized in the table 3:

- The particle n° 22 presents a sheet shape.
- The particle n°1448 is a compact shape particle.
- The particle n°4240 shows a complex branched shape.

Particle N°22

The low shape indexes values I_c and I_s indicate that the 22 particle shape is far from a cube or a sphere. The values of the shape indexes are very close to zero showing the 22 particle is essencialy constituted by surface zones. The important value of IG_g, higher than 100 indicates the object is elongated. The normalized inertia moments λ_1 and λ_2 equal to 0.5 indicates thar the particle 22 have a needle type mass distribution.

The local curvature bivariate distribution in figure 10 (a), presents a concentrated cloud of interface portions. This cloud is located in the saddle type region, near from the central point (k_{min}=k_{max}), and also from the lines k_{min}=0 and k_{max}=0. That suggests the 22 particle has a great number of interface portions with cylindrical bridge type and also an important number of planar type. The 22 particle is formed with branches and planes which present intersections.

Particle N°1448

The low values of shape indexes indicate that the 1448 particle presents a shape far from the reference ones (sphere and cube). The shape indexes are close to 0.5 indicating the 1448 particle is quite compact. The elongation value is close to one. The normalized inertia moments, by projection in the λ_1-λ_2 space (Figure 1), show a mass distribution with oblate ellipsoïd type.

The local curvature bivariate distribution indicates that the 1448 particle has some cylindrical valley, and saddle type, together with an important number of concave interface portions. Thus the particle is mostly concave.

Particle N°4249

The low values of shape indexes indicate the 4249 particle presents a shape far from the reference ones (sphere and cube). Their very low values indicate, as well as for the 22 particle, that the 4249 particle is essentially constituted on surface. The important value of the geodesic elongation index shows that the particle is elongated. The normalized inertia moments are located in the centre of the triangle in the λ_1-λ_2 space (figure 1), showing that the 4249 particle has a mass districbution far from the reference mass distributions. The curvature bivariate distribution in figure 10 (c) presents, for the 4249 particle, cynlindrical valley and bridge interface, with a majority of saddle type interface, indicating a complex geometry of the particle with branches and planes layout.

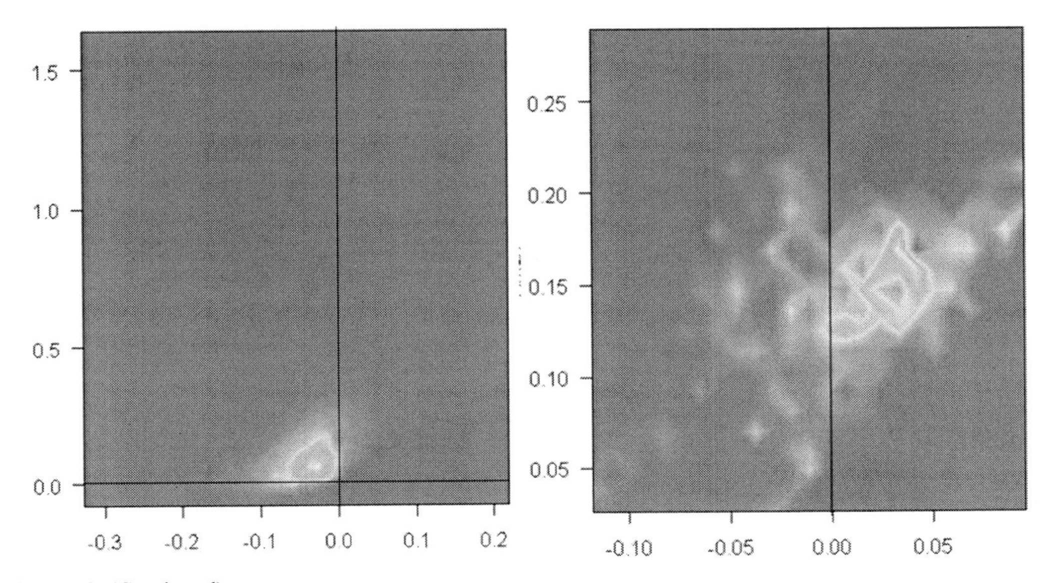

Figure 10. (Continued).

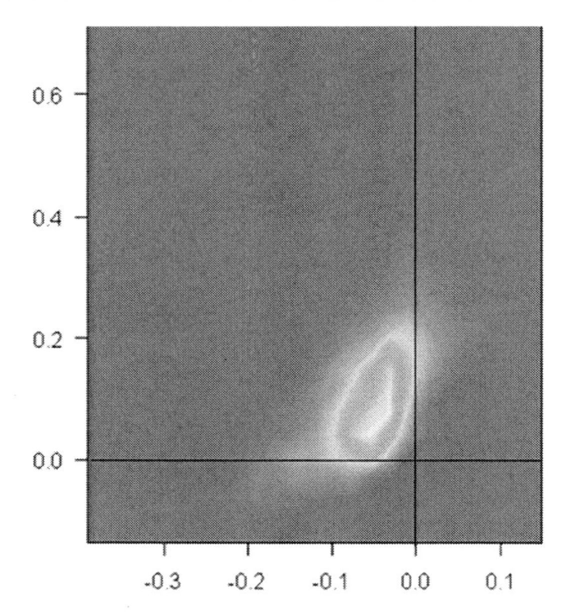

Figure 10. Bi dimensionnal k_{min}-k_{max} histogram of iron rich particles of the AA5xxx data base (a) particle n°22 (b) particle n°1448 (c) particle n° 4249.

CONCLUSION

The particle measurements highlight the complex shape caracteristics of intermetallic particle.

The parametrical study based on the selected morphological parameters (chosen for their physical interpretation), allows us to describe the particle shape characteristics. The matrix particle interface curvature study shows the singular geometrical zones of the particles. These zones correspond to a special localization for mechanical stresses during the rolling process. The study of the results on the entire population of intermetallic particles is presented in the next section.

Statistical Analysis

For a given deformation, the number of intermetallic particles contained in the aluminium sample is of the order of several thousands. To perform an analysis on the intermetallic particle shape from the presented parameters, a statistical study is needed. It permits to summarize the acquired information in an adapted space, where the particle will be classified according their shape peculiarities, characteristic of their ability to break up during the rolling process.

Several types of measurement were previously presented: morphological measurements, inertia based measurements, and local curvature measurements. The two first measurements type are individual, that is to say that a particle have a parameter measured. The local curvature measurements for a particle are summarized in the bivariate distribution of k_{min}-k_{max}.

To use the entire set of parameters measured on intermetallic particles, we proceed in two steps:

- The statistical analysis of curvature measurements on the entire set of intermetallic particles: the aim is to extract reduced information about curvature corresponding to a particle classification in the curvature characteristic space.
- The global statistical analysis on the entire set of measurements, to classify intermetallic particles according their shape. A correlation study is first performed to only keep the uncorrelated measurements versus intermetallic shape.

Classification of Intermetallic Particles in the Factorial Correspondance Analysis (FCA) Space of Principal Curvature Histogram

The bivariate distribution of k_{min}-k_{max} can be considered as a set of parameters, each parameter being a cell of the histogram. The distribution being normalized versus the particle surface, each cell value represents the particle interface proportion which has the geometrical properties defined by its principal curvatures.

Each particle of the data base is defined by a 405 length vector. It corresponds to the 400 cells of the k_{min}-k_{max} bivariate distribution, completed by the 5 margin curvature cells. The k_{min}-k_{max} bivariate distribution is sparse even if its bounds were optimized; the zero values represent curvatures which are not present on the particle surface.

A statistical analysis is needed to summarize the information from the entire set of descriptor obtained by the bivariate distribution k_{min}-k_{max} analysis. A Factorial Correspondence Analysis (FCA) is used. The FCA has been introduced in the 60's by JP Benzécri (Benzécri, 1980). It allows going through quantitative and qualitative datas on a set of individual. The aim of the FCA is to look for the best simultaneaous representation of the two sets line and column of a data table. The FCA corresponds to a general analysis of a weigthed cloud of points in a space with a χ^2 metric.

In the data analysis of the bivariate distribution of k_{min}-k_{max}, each intermetallic particle corresponds to a line of the contingency table, the columns being the values of the bivariate distribution. The k_{min}-k_{max} graph being normalised, the sum of all the value of the graph is equal to one. The intermetallic particles form a cloud of points in the projection plane corresponding to by the factorial axes. In this space, it is possible to classify the particles according to their interface curvature properties. In the literature, many classification family methods exist (Benzécri, 1973): the non hierarchical classification (it produces a fixed number of clusters), the Hierarchical Ascendant. Classification (HAC), merging the elements two by two according to their similarity leading to a unique cluster.

In the case of intermetallic particle classification, we have no prior knowledge about the number of cluster to constitute; a HAC is the more appropriate method. The Ward algorithm (Ward, 1963) is used. It merges the two nearest clusters, taking as the distance between the two clusters the loss of inertia encountered by the merging. The distance between the two merged clusters is called the inter cluster distance. The lower the cluster distance is, the more similar are the merged clusters of particles in their curvature sense. Respectively, the larger the cluster distance is, the less similar are the merged clusters of particles.

The HAC leads to a hierarchical partition in n clusters, which can be represents by a tree, the dendrogram (figure 11). The dendrogram is a useful tool to represent the n-1 steps of the particles classification leading to one cluster. A cut of the tree at a given level gives a partition of the particle set. The aim of the HAC is to indicate the presence of groups. If at a level of the dendrogram the intra cluster inertia suddenly grows, the partition into clusters at this level is considered as relevant.

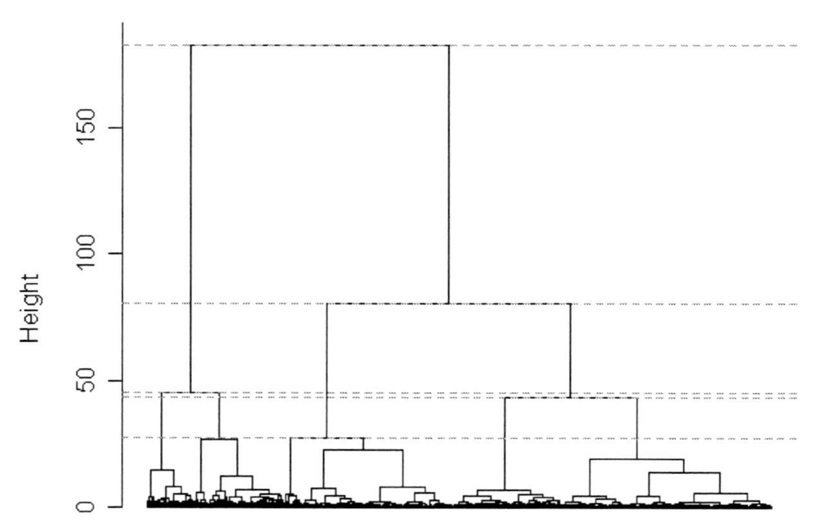

Figure 11. AA5xxx iron rich particle dendrogram in the FCA space of the bivariate distribution of k_{min}-k_{max} according the Ward criterion.

The figure 11 presents the dendrogram obtained by aplying the Ward criterion on the iron rich particles contained in the AA5xxx sample deformed at 10%. The number of cluster is obtained by looking at the dendrogram. On the Figure 11 AA5xxx iron rich particle dendrogram in the FCA space of the bivariate distribution of k_{min}-k_{max} according the Ward criterion. The inter cluster distance varies between 0 and 160. The last nodes, marked by red dotted lines on the Figure 11 represent a high inter cluster distance. A partition obtained by cutting the dendrogram between the nodes having the values 35 and 45 is relevant. This pruning leads to a partition of the intermetallic particles population in 5 clusters. The 5 identified clusters have specific geometrical properties:

- The cluster 1 contains particles with a lot of saddle interface portions. They correspond to branched particles (figure 12).
- The cluster 2 contains unbranched particles with a considerable number of planar interface portions, as well as bridge cylindrical type portions (figure 13). The intermetallic particles constituting this group are flat and present an important elongation.
- The cluster 3 contains essentially particles with nappe shape showing slight curves with a 120° angle (figure 14).
- The cluster 4 contains slender shape particles, with a majority of bridge cylindrical type cylindrical interface portion as well as planar (figure 15).
- The cluster 5 contains compact particles showing spherical interface portion (figure 16).

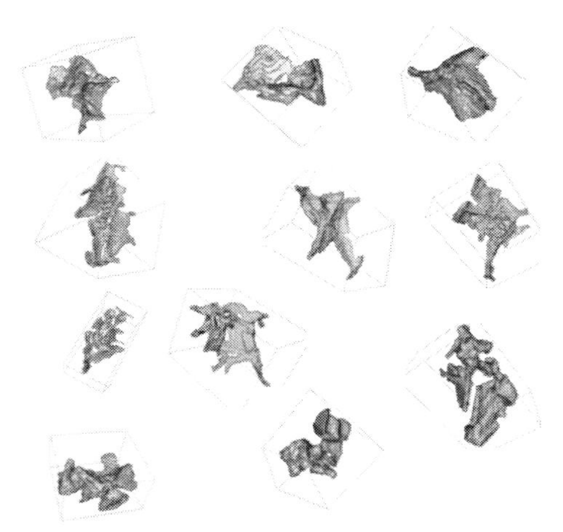

Figure 12. Example of particles contained in the cluster 1 obtains by HAC on the FCA of the bivariate distribution of k_{min}-k_{max}.

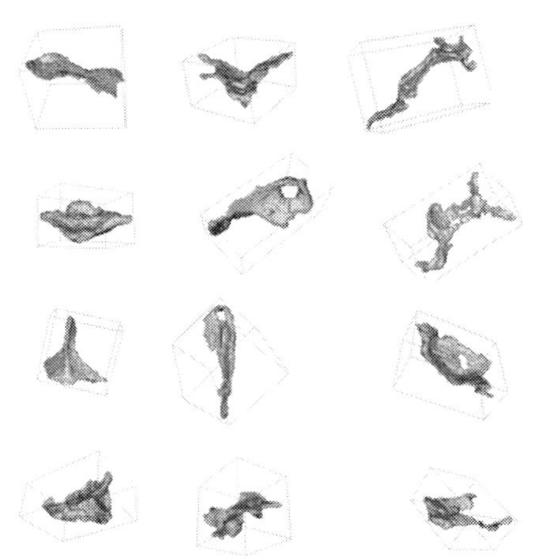

Figure 13. Example of particles contained in the cluster 2 obtains by HAC on the FCA of the bivariate distribution of k_{min}-k_{max}.

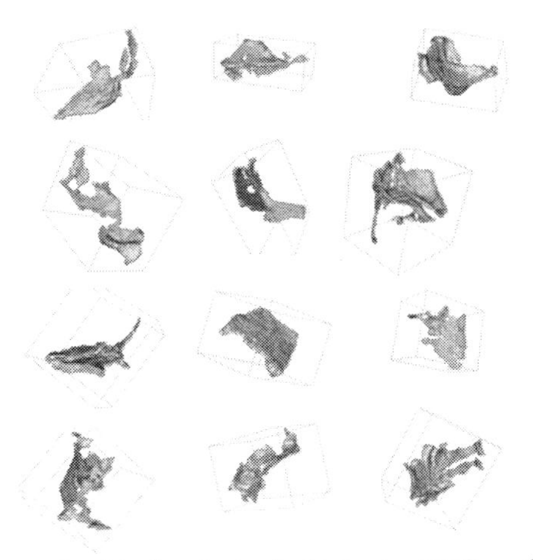

Figure 14. Example of particles contained in the cluster 3 obtains by HAC on the FCA of the bivariate distribution of k_{min}-k_{max}.

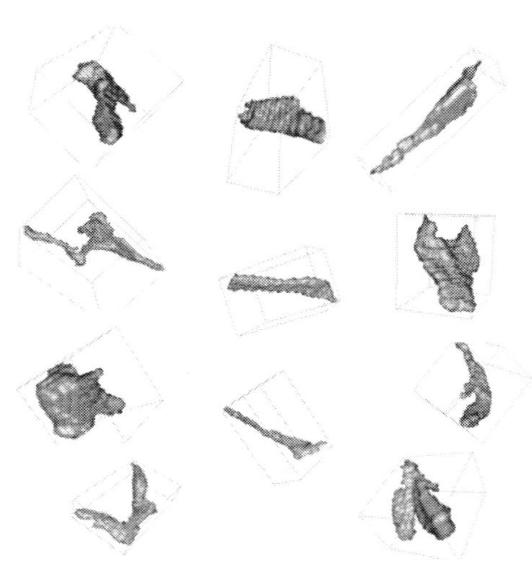

Figure 15. Example of particles contained in the cluster 4 obtains by HAC on the FCA of the bivariate distribution of k_{min}-k_{max}.

The caracterisation of three dimensional shapes using local curvature type information is efficient. This method is original and gives information about the complex nature of the three dimensional interface between intermetallic particles and aluminium matrix.

A suitable classification of complex shape objects according to their geometrical interface properties is thus allowed.

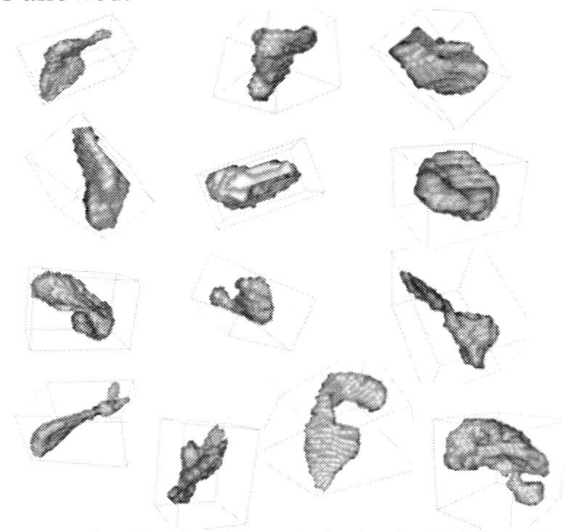

Figure 16. Example of particles contained in the cluster 5 obtains by HAC on the FCA of the bivariate distribution of k_{min}-k_{max}.

Principal Component Analysis (PCA) of the Parameters

The Principal Component Analysis (PCA) is a data analysis method which principal idea have been proposed by Pearson in 1901 (Pearson, 1901), and was developed by Hotelling in 1933 (Hotelling, 1933).

From a population of individual characterized by a set of quantitative variables, the PCA structures and summarizes the data by using a linear combination of the variables. They generate new variables (the principal components, or eigen functions), with decreasing variances (or eigenvalues μ_i). In the case of intermetallic particle study, a study about the correlation of the variables was performed. It permits to identify the uncorrelated variables, ie the variables carrying different type of information versus intermetallic particles. They are: the volume, the sphericity index, the geodesic elongation index, the curvature indexes (i.e. the cluster label obtained by HCA on the bivariate distribution of k_{min}-k_{max} FCA), and the normalized inertia moments λ_1 and λ_2. These measurements have not the same dimensions. It is then necessary to apply a centered normalized PCA (to not favour the measurements with a large variance).

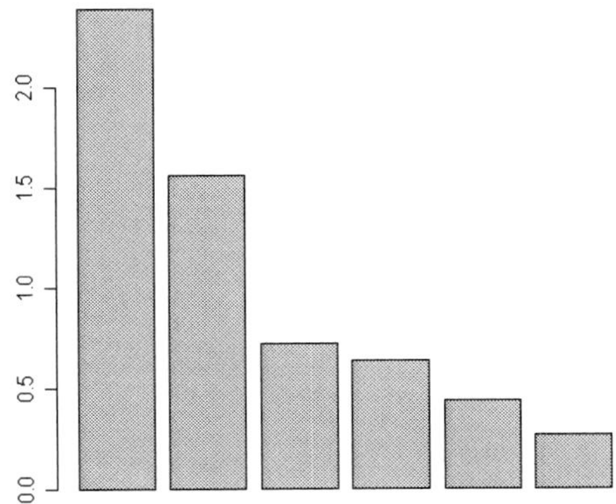

Figure 17. Graph of the eigen values of iron rich intermetallic particles cointaining in the AA 5xxx deformed at 10% μ_1 =2,38, μ_2 =1.56, μ_3 =0,72, μ_4 =0,44, μ_5 =0,27.

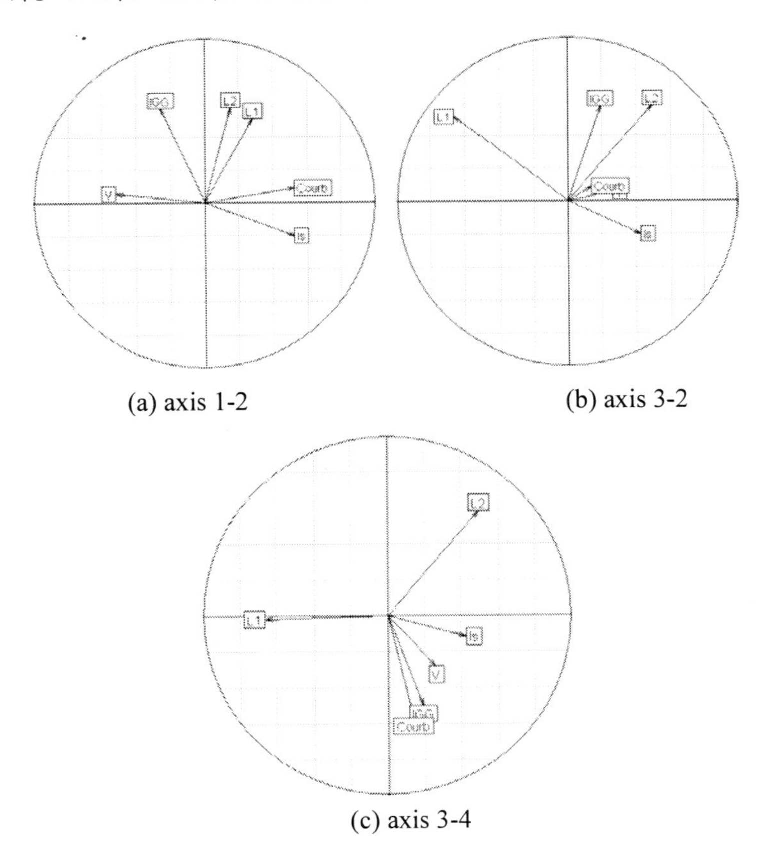

Figure 18. Correlation circle obtained for the PCA of the intermetallic particles contained in the AA5xxx 10% deformed sample.

The PCA on the presented mesures on interemetallic particles contained in AA5xxx deformed at 10% shows eigen values with the following contributions (figure 17): e_1= 39.69%, e_2 = 25.99%, e_3= 12,00%, e_4= 10,58%, e_5= 7.31%, e_6= 4,44%. Considering the obtained percentages, the fisrst 4 axes represent 88.25% of the data variance, and are sufficient to well describe the data.

Each PCA axis gives the coefficient of correlation of the factor with the parameters, illustrated by the correlation circles, giving the projection of the parameter space axes into the new space obtained by the PCA (Figure 18). According to the correlation with the parameters, it is possible to interpret the shape trends carried by this new representation space.

The axis 1 is strongly positively correlated with the curvature parameters as well as with the sphericity index, and strongly negatively correlated with the volume. According to this axis, the more a particle is positive the more its shape is spherical, its volume small and its interface shows a lot of spherical portions. Conversely, if the particle is located on the left side of the axis 1, the largerits volume, and its shape is complex (in relation with the first group defined by the FCA).

The axis 2 is strongly positively correlated with the geodesic elongation index, as well as with the two first normalized inertia moments. According to this axis, the more a particle is positive, the more it presents an elongated shape, independently from its volume and its classification in the local curvature space.

The axis 3 is strongly negatively correlated with the first normalised inertia moment and positively correlated with the second normalized inertia moment. It suggests the more negative is the particle according to this axis, the more its mass distribution is planar. Conversely, the more positive is the particle according this axe, the more its mass distribution is spherical and its sphericity index high.

The axis 4 is strongly positively correlated with the second normalized inertia moment and strongly negatively correlated with the geodesic elongation index and the curvature indexes. It suggests that the more positively is positioned a particle according to this axis, the more its mass distribution is of needle type and it shows a membership to the first cluster of the curvature space (corresponding to branched particles), but with a low geodesic elongation oindex.

Hierarchical *Ascendant* Classification *(HAC) of Intermetallic Particles in the CPA Space*

The physical interpretation of the PCA axes allow one to analyse the cloud of points obtained by projecting the particles into this new representation space, and thus to identify the clusters of particles presenting shape similarities.

In the CPA space, the intermetallic particles are classified. As previously explained we apply a HCA with the Ward method. The aim is to merge particles according to their shapes, to obtain homogenous clusters regarding the break up ability of the particles. The Figure 19 presents the dendrogram obtained by applying the Ward method on the iron rich intermetallic particles population extracted from a 10% deformed AA5xxx projected in the parameters space.

From the inter cluster distance a pruning of the dendrogram between the nodes 165 and 200 leads to a particles partition into seven clusters. The obtained seven clusters of intermetallics particles present the following properties:

- Cluster 1 represents 13.6% of the particles. It contains elongated particles, cutted out, and with a planar zone.
- Cluster 2 (19.1% of the particles) contains elongated particles with an orthogonal plan layout.
- Cluster 3 (19.0% of the particles) contains particles with needle type mass distribution, characterized by a large elongation but with no junction.
- Cluster 4 (8.7% of the particles) contains planar particles with nappe shape.
- Cluster 5 (19.9% of the iron rich particles) contains elongated particles.
- Cluster 6 (12.1% of the iron rich particles) contains particles which presents a cuppel shape.
- Cluster 7 (7.5% of the iron rich particles) contains particles with a high sphericity index, as well as a small volume.

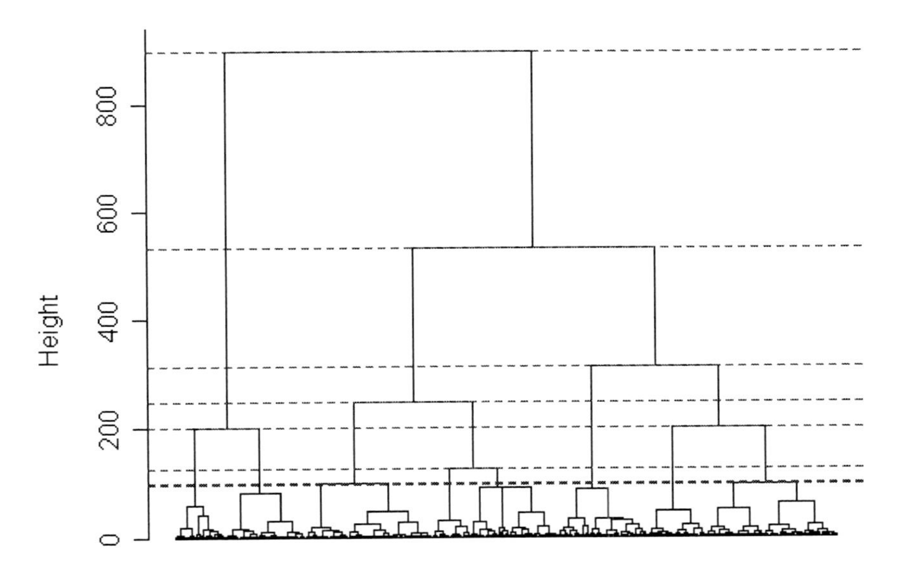

Figure 19. Dendrogram of iron rich intermetallic particles containes in AA5xxx in the space of PCA obtained by HAC with the Ward criterion.

SIMULATION OF THE MORPHOLOGY OF INTERMETALLIC PARTICLES IN ALUMINIUM ALLOYS

The microstructure of a material determines its mechanical abilties during the applied industrial process.

From this statement, it is easily understable that the numerical simulation of a real material is a key step to grasp its micro structure and thus its performance along strain for further mechanical studies.

In the following we first present a reminder about the probabilistic model of Boolean type. Second, we show the results obtained by image analysis on AA5xxxdeformed at 10%. Finally, we show the advantages of this method and its limits.

Reminder on Probabilistic Model of Structures of Boolean Model Type

The Boolean Model

Definition The Boolean model is a flexible model coming from the random set theory (Matheron, 1967, 1972). It permits to describe two phase structures such as aluminium alloys, which contains breaked-up grains in an aluminium matrix. It is obtained by the implantation of random primary grains A' on a Poisson points process x_k with the intensity θ_d. The Boolean model A is a random closed set obtained by the union of the A'_{x_k} (equation 14). It results a p proportion of the A phase, and a proportion q=(1-p) for its complement.

Figure 20. Clustering of intermetallic particles contained in AA5xxx deformed at 10%.

$$A = \bigcup_{x_k \in (\theta_d)} A'_{x_k} \tag{14}$$

Properties A Boolean model which only depends on the θ_d point density and on the gains formation laws presents the 3 following properties:

- The union of two independent Boolean models is a Boolean model.
- The dilation of a Boolean model by a fixed compact is a Boolean model.
- The intersection of a Boolean model with a i-plan of infinite dimension is a Boolean model in a lower dimensional space.

Models derived from the theory of random closed sets are characterized by their Choquet capacity. For a compact set $\check{K}$, the Choquet capacity T(K) is given by the equation 15.

$$T(K) = 1 - Q(K) = P\left\{K \cap A \neq 0\right\} = 1 - P\left\{K \subset A^c\right\} \tag{15}$$

$$Q(K) = V_v\left(A \ominus \check{K}\right) \tag{16}$$

$$T(K) = V_v\left(A \oplus \check{K}\right) \tag{17}$$

$$Q(h) = Q(x, x+h) = P\left\{x \in A^c, x+h \in A^c\right\} \tag{18}$$

where $A' \oplus K = \bigcup_{x \in K} A_x$ is the result of the dilation of A' by K and $A \ominus K = \bigcap_{x \in K} A_x$.

The functional $Q(K)$ can be written as the volume fraction of the morphological erosion of the complementary phase of A by K(equation 16), and $T(K)$ as the volume fraction of the dilation of A by K (equation16).

The equations 16 and 17 correspond to notions direcly mesurable by image analysis:

- If K is a singleton x (x being at the origin of coordinates or at any point), the Choquet capacity $T(x)$ is the volume fraction of the random set A.
- If $K = \{x, x + h\}$, K is a bi-point and Q(x, x+h) is the covariance of the complementary set A^c written Q(h) (equation 18). It depends only on h for a stationary random set.

For a Boolean model the Choquet capacity can be written by the equation 19, μ_d is the Lebesque measure in $\mathbb{R}^d$, and $\overline{\mu_d}$ is mean on the realization set A'.

$$T(K) = 1 - e^{-\theta_d \overline{\mu_d}\left(A' \oplus \check{K}\right)} = 1 - q^{-\frac{\overline{\mu_d}\left(A' \oplus \check{K}\right)}{\overline{\mu_d}(A')}} \tag{19}$$

The Choquet capacity is thus easily estimated by image analysis.

Random primary grains The description of models fitting a physical process simulation, can be made by using different typeS of primary grains A', in relation with the phase to describe. For instance, one can use a population of spheres, Poisson polyedra, sphero-cylinders...

In our case, the simulation of aluminium alloy, we want to simulate the aluminium grains growth to be able to describe the shape of intermetallic particles; as intermetallic particles are formed in the vacant spaces between aluminium grains during the alloy solidification. The geometrical shape of alumium grains presents dendrites; in a simplified approach, they can be modelled by the union of spherical primary grains.

Geometrical Covariogram of a Population of Spheres

The geometrical covariogram of a set X is given by the Lebesgue measure of the intersection of X with its translated by a vector $\vec{h}$ (equation 20). If $h = 0$, thus $K(0)$ is the Lebesgue measure of X (volume in 3D).

$$K(h) = \mu_d\left(A \cap A_{-h}\right) = \mu_d\left(A \ominus h\right) \tag{20}$$

For a population of spheres with random diameters following a probability density function f(x), the geometrical covariogram is defined by equation 21. The reduced covariogram is defined by equation 22.

$$\overline{K(h)} = \frac{\pi}{6}\left(\int_h^\infty x^3 f(x)dx - \frac{3}{2}\int_h^\infty x^2 f(x)dx + \frac{h^3}{2}\int_h^\infty f(x)dx\right) \tag{21}$$

$$\overline{r(h)} = \frac{K(h)}{K(0)} \tag{22}$$

Covariogram for different usual distribution laws In this section, we present the geometric covariogram obtained from known probability density functions used to model aluminium alloys with spherical random primary grains.

Spheres with diameter D For a single sphere with diameter D, the covariogram is given by:

$$\overline{r(h)} = 1 - \frac{3h}{2D} + \frac{1}{2}\left(\frac{h}{D}\right)^3 \text{ for } 0 \le h \le D\text{, else } r(h) = 0$$

Population with 2 sphere diameters For a population of spheres with 2 diameters D_1 and D_2 with probability respectively p_1 and p_2, as $p_1+p_2=1$, the covariogram is given by:

$$\overline{r(h)} = 1 - \left(\frac{3h}{2}\left(p_1 D_1^2 - (1-p_1) D_2^2 \right) + \frac{1}{2} h^3 \right) \times \frac{1}{\left(p_1 D_1^3 - (1-p_1) D_2^3 \right)}$$

for $0 \le h \le D_1 \le D_2$

Gamma Law For a population of spheres with diameters following a gamma distribution function, the covariogram is given by:

$$\overline{r(h)} = (1 - F(h, a+3, b)) - \frac{3h}{2b} \frac{1}{a+2} (1 - F(h, a+2, b)) + \frac{1}{2} \left(\frac{h}{b} \right)^3 \frac{1}{(a+2)(a+1)a} (1 - F(h, a, b)) \quad \text{for } 0 \le h$$

with $F(x,a,b) = \frac{1}{b^a \Gamma(a)} \int_0^x u^{a-1} e^{-\frac{u}{b}} du$ and $f(x,a,b) = \frac{1}{b^a \Gamma(a)} x^{a-1} e^{-\frac{x}{b}}$.

Exponential law It presents a specific case of the gamma law with a = 1, its covariogram is:

$$\overline{r(h)} = \left(1 - \frac{h}{2D} \right) e^{-\frac{h}{D}} \text{ for } 0 \le h < 2D \text{, else } 0$$

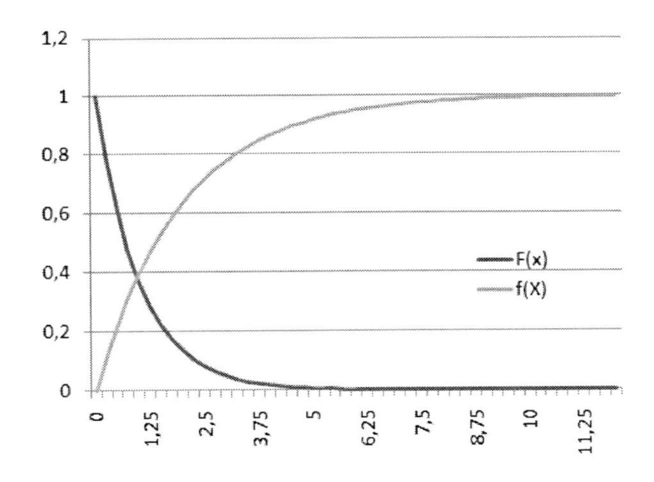

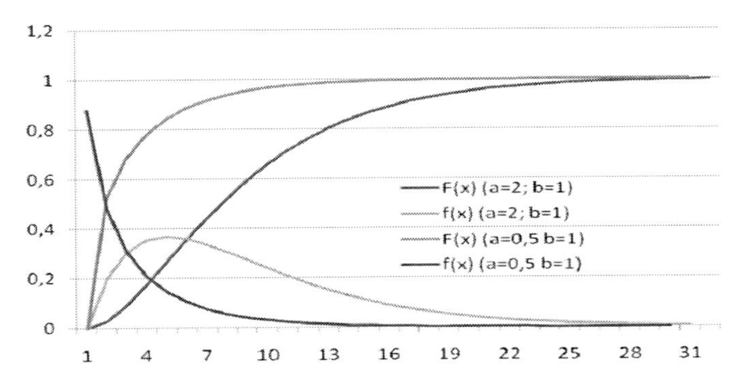

Figure 21. Illustration of the distribution law (a) Gamma law (b) Exponential Law (with λ=0,5).

Covariance of a Boolean Model of Spheres

For a Boolean model of spheres, the covariance of the complementary phase Q(h) is expressed by equation 23, where q is the volume fraction of the complement (here the intermetllic particles). The distribution function parameters are then set, by fitting the measured covariance by image analysis on interemetallic particle to the theorical covariance of the Boolean model which grain diameters following a given probability distribution function.

$$Q(h) = q^{2-\bar{r}(h)} \tag{23}$$

Experimental Results

Measurement of the Covariance

Principle In image analysis, the covariance corresponds to the correlation of the image with its translated in a given direction of the space (Matheron, 1967, Delarue and Jeulin 2001). Given a set X in the image, its covariance, in a given direction, is written $C(h)$. It defines the probability for this set translated by $\vec{h}$ to intersect the initial set (equation 24).

$$C(h) = P\left\{x \in X \bigcap X_{\overrightarrow{-h}}\right\} \tag{24}$$

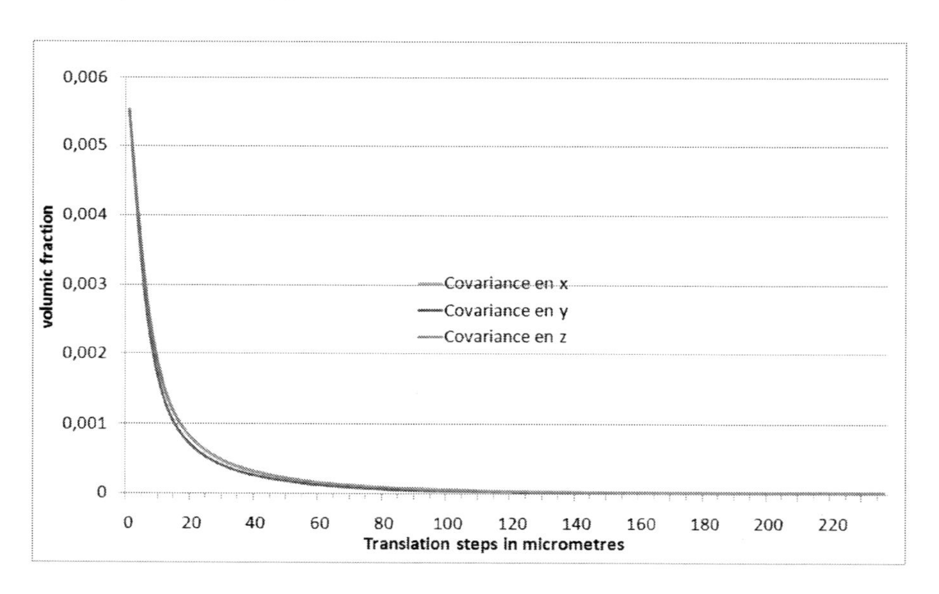

Figure 22. Iron rich particle covariance of AA 5xxx deformed at 10% along x y and z image axes.

In image analysis, the covariance presents rich information. For instance, it gives information about the anisotropy and the periodicity of material organization in the considered h direction. Furthermore, the covariance is used to control the sampling. Lastly, it gives information about the microstructure of the sample (volume fraction and characteristic length).

Experimental covariance The covariance is measured on the whole binary image of the AA5xxx deformed at 10% along the 3 orthonormal axes of the image (x, y, z) which are in relation with the rolling axes. For each axis, we translate the image with a step h, and we measure the volume of the intersection of the translated image with the initial one. The Figure 22 shows the result of the covariance for the $Al_x(Fe,Mn)$ particles dispersed in the AA5xxx deformed at 10%. At this deformation, we consider that the aluminium presents few microstructure changes in comparison with the undeformed alloy. The covariance on the Figure 22 shows that the intermetallic distribution in the sample is perfectly isotropic, the measurements along each axis being identical.

The value of the covariance at the origin corresponds to the volume fraction Vv of intermetallic particles in the alloy; it is equal to 0.55%. We can also determine the characteristic scale of the material defined from the plateau of the covariance theorically obtained at $(Vv)^2$. Here it is equal to 3.059 10^{-5}%. This value is quickly reached for a translations step equal to 52.5μm. It corresponds to the maximum diameters of aluminium grains for the Boolean model.

Estimation of the Model Parameters from the Covariance

We compare the measured to the theorical covariance Q(h) obtained for a Boolean model of spheres with different probability density functions (pdf) of the diameters according to the ones previously presented. The equation 23 is used: q represents in the model the volume fraction of the intermetallic particles; it is equal to 0.55%.

The geometrical covariogram $\overline{r(h)}$ of a sphere population depends on different parameters, according to the used pdf of the diameters. For each tested function, the covariances are fitted by minimizing the mean of the absolute relative deviations. The retained parameters are presented in the Tablet Table 4, and the obtained covariances are presented on the Figure 23.

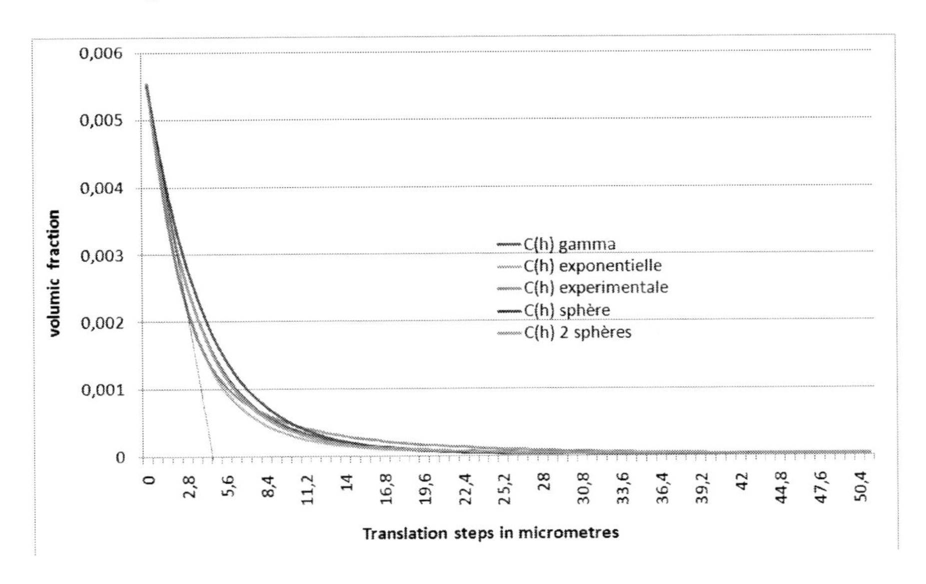

Figure 23. Experimental and theorical covariances: red $Al_x(Fe,Mn)$ particles in the AA5xxx deformed at 10%, dark blue Gamma density function, pink Exponential density function, blue population of spheres, Violet 2 population of spheres.

The pdf giving the best fit of the covariance of the AA5xxx alloy from a Boolean model of spheres is a gamma pdf with parameters equal to a = 1.52 and b = 11.

Table 4. Retained parameters used to model the AA 5xxx deformed at 10% according to a Boolean model with spherical primary grains with different diameters pdf

Density function of the diameters	parameters
Population of spheres with diameters D	D=45 ie 31,5μm
Population of 2 Spheres	D_1=30 ; D2=45 ; P1=0,8 ie D_1=21 μm ; D2=31,5μm
Gamma pdf	a= 1,52 ; b= 11 ie 7,7 μm
Exponential pdf	b=11 ie 7,7 μm

From this pdf, points are randomly dispersed in an image (of volume equal the one of the sample 1280x1280x2048 voxels) according a Poisson point process, to obtain a volume fraction equal to 99.45%. The number of points to dispersed depends on the volume fraction of the spheres (depending on their mean volume equal to $\pi b^3 a(a+1)(a+2)4/3$ for a gamma pdf) implanted on each point of the Poisson processs according a soft core model (Moreaud and Jeulin, 2005).

Spheres with random diameters following the estimated gamma pdf are thus implanted on the set of drawn points (Figure 24). The simulation of intermetallic particles is obtained by taking the complement of the simulated Boolean model (Figure 25).

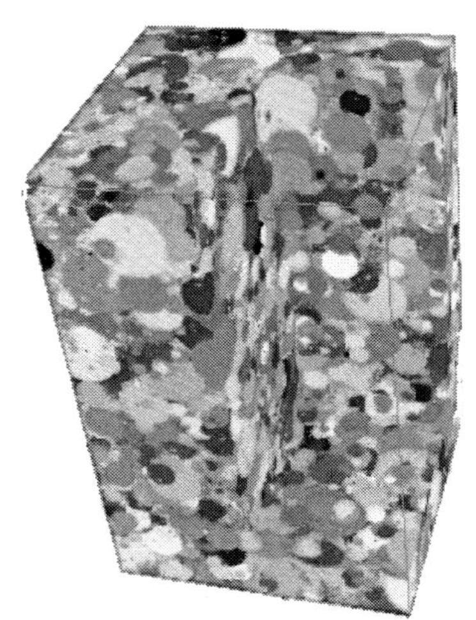

Figure 24. Three dimensional simulation of a Boolean model of spheres according to a gamma distribution of the radius (a= 0.1737 and b=34) and with volume fraction equal to 99.45%.

Considering the size of the image to simulate (1280x1280x2048), we reconstruct the image by section. At each step of the reconstruction the particle are labelled and extracted in a sub volume into two data bases.

The first data base contains the entire particles, the second one the particles hitting the edges of the successive sections. After a step of reconstruction of the particles, an entire data base of simulated particles is created.

For each particle of this data base the set of previous measurements is performed, except for the measurement of the surface curvature. Indeed the simulated particle are obtained by a Boolean model of spheres, and therefore their curvatures satisfy $k_{min} = k_{max}$ and the radii of curvatures must follow the used Gamma pdf.

They cannot reproduce the variablility observed in the real sample, where in general $k_{min} \neq k_{max}$.

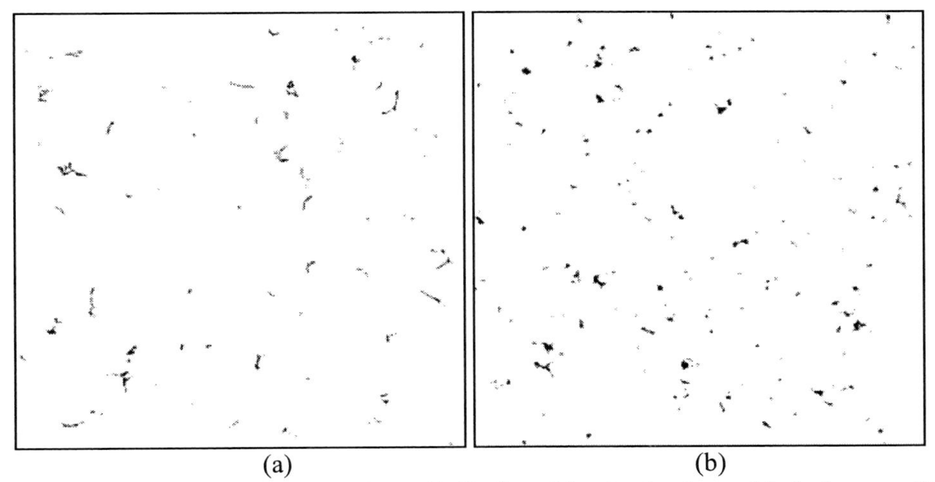

Figure 25. (a) slice of the AA5xxx deformed at 10% (b) slice of the simulated material; the intermetallic particles are in white.

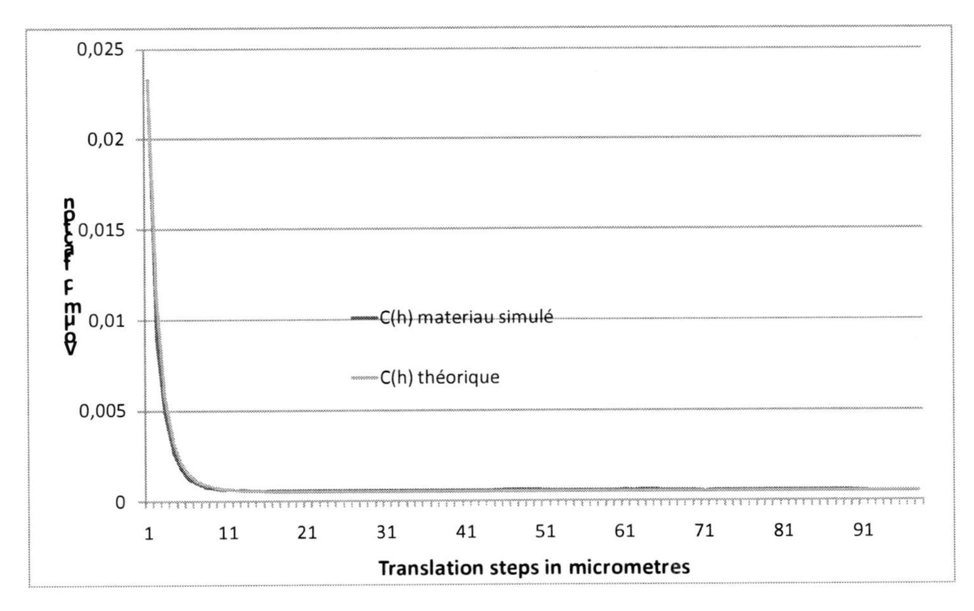

Figure 26. Covariance of the simulated material compared with the theorical one obtained with a Gamma function (b=11and a=1.52).

The volume fraction of the particles cointained in the simulated material is equal to 2%. It is thus more important than the measured volume fraction on the AA5xxx equal to less than 0.5%. The number of extracted particles in the sample of real material is 4922, and it is equal to 14800 for the simulated one. It is because the simulated matrixwas been fitted on a part of the intermetallic population (the Mg_2Si particles and the voids were not been taken into account, based on mechanical studies (Feuerstein, 2006)). Nevertheless, the Figure 26 shows the simulated material with the same covariance as the fitted one. The model is therefore validated.

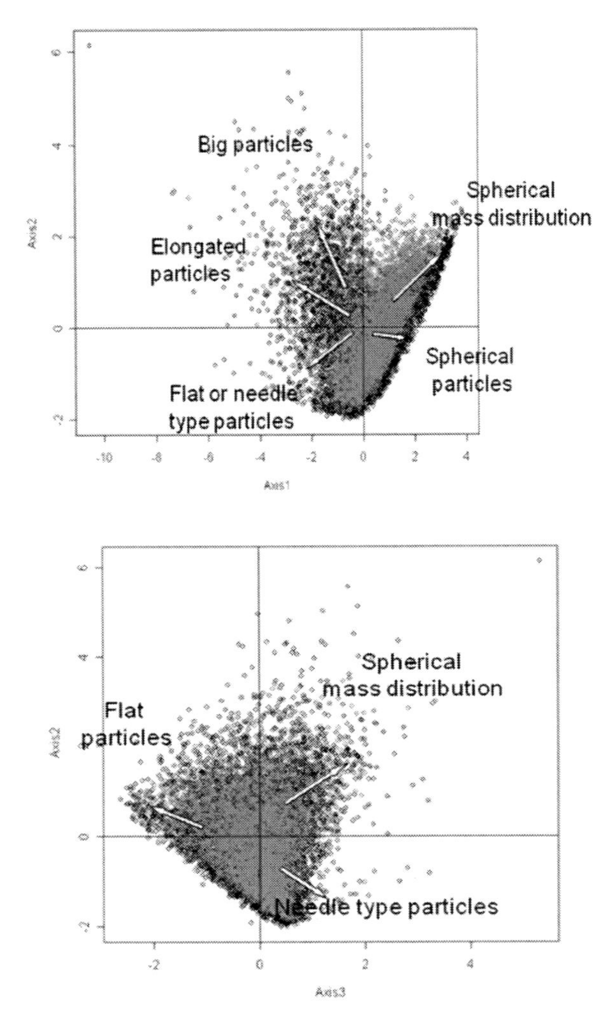

Figure 27. First Plane of the PCA of the intermetallic particles contained in AA 5xxx deformed at 10: in black real particles, in red simulated particles.

Projection of the Simulated Population of Particles into the Parameter Space of the Real Particles

In the case of intermetallic particles analysis, we have shows in the previous section the use of statistical analysis. Here the statistical analysis is performed on the same parameters, except for the curvature one, as this parameter does not carry useful information for the simulated particles. In this space, the first axis is dedicated to changes of elongation; on this axis, the more a particle is elongated the more important is its volume. The second axis links the normalized inertia parameters and the volume: the larger the volume a particle presents, the less important are its inertia parameters, and its mass distribution tends to be spherical. The third axe distinguishes the particles presenting planar and needle type mass distribution.

We measure the set of previously retained parameters on the simulated particles. Then we project in the real particles PCA space, the simulated particles caracterized by their measurements (Figure 27). We observe that the cloud of simulated particles in red in the space of the PCA is sligthly less dispersed than for the real particles. The variability of observed shapes seems less important. Indeed on the Figure 27 (a), it appears less elongated and voluminous particles in the simulated material. Nevertheless, all particle shapes are retrieved in the simulated material, and all the variability of shape of the studied intermetallic particles can be simulated.

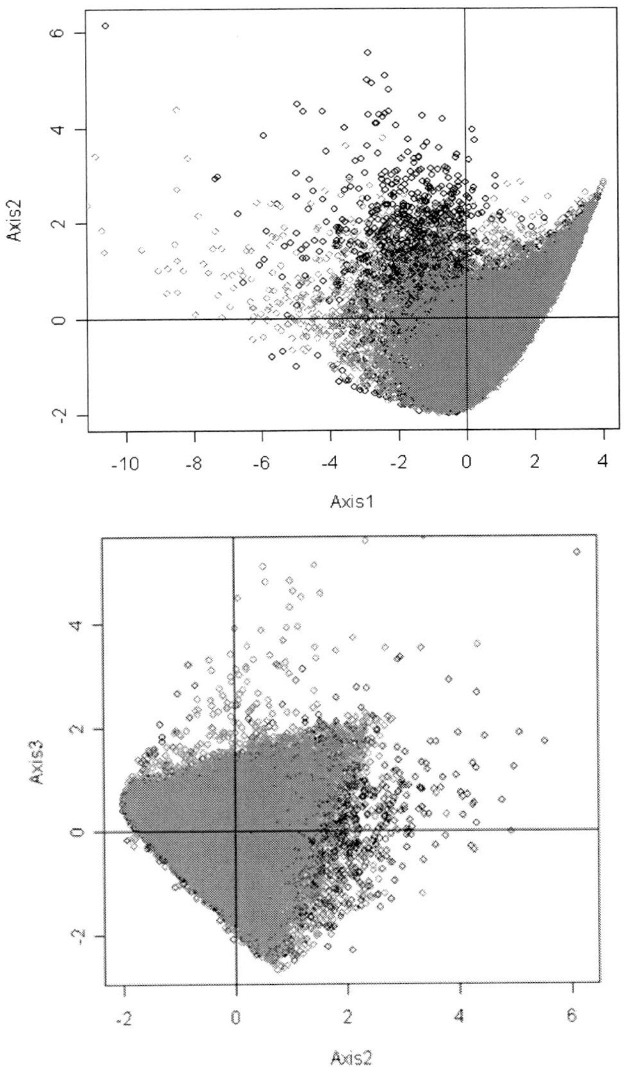

Figure 28. First Planes of the PCA of the intermetallic particles contained in AA 5xxx deformed at 10 %, in black; in grey, projected particles contained in the AA deformed at 82%.

Follow Up of Intermetallic Particles with Regards to the Strain Applied during Hot Rolling Process

In mechanical studies, it is fundamental to understand the break-up of intermetallic particles in aluminium alloys. We propose here to use our morphological shape analysis method (except for the curvature) to compare the two particle populations of an aluminium sample at two deformation stages of a hot rolling process (Parra-Denis et al, 2008). The curvatures are suitable for the study of the particles shape at the beginning of the process, when their shapes are very complex. With the stress applied during hot rolling, particles are flattened and then the classification into 5 clusters of curvature is not any more possible. The first population corresponds to the one previously studied, namely to the beginning of the process (10% deformation). The second one corresponds to the more advanced stage of deformation (82%).

The comparison between two particle populations is made possible by projecting one of them in the PCA representation space of the other one. As the deformation process progresses, the number of particles increases. The Figure 28 plots the 80% data points in the 10% factorial planes.

It is obvious that there are fewer 82% particles in the top left quadrant than 10% particles. This quadrant corresponds to large particles (the different particles trends are shown on Figure 27). As expected, large particles tend to disappear, while needle-like and flat ones tend to appear, which we have checked on a 2D histogram of plane 2–3. Large particles are indeed the most brittle. As the deformation process goes along, they break. Their pieces become new smaller particles with simpler shapes.

CONCLUSION

In this text we show a methodology to carry out the study of complex shapes inclusions observed in an Aluminium material. A prospective approach was proposed to simulate the material. This kind of analysis can be applied to other material microstructure studies.

This analysis was been enriched by mechanical studies (Moulin and al, 2010, Moulin and al, 2009, and Moulin, 2008) of the intermetallic particles. They show a good correlation between the obtained classification of intermetallic according to their shapes and their break-up properties. Thus a good comprehension about the observed phenomenon by image analysis can lead to an efficient study of the material microstructure, but all the measured parameters have to be well chosen with an ongoing effort of comprehension between the community of image analysisis and physicists.

REFERENCES

Alkemper J. and Voorhee P.W. (2001). Three dimensionnal characterization of dendritic microstructures. *Acta materiala,* 49, 897-90.

Besl P.J. and Jain R.C (1986). Invariant surface characteristics for 3D object recognition in range images. Computer Vision Graphics Image Processing, Academic Press Professional, Inc, 33, 33-80.

Benzécri J.P (1973). L'analyse des données. T1 : la taxinomie, vol1,Dunod.

Benzécri J.P (1980). La Pratique de L'analyse Des Données,Bordas, Paris, B. M. *(ed.)*Abrégé Théorique, Études de Cas Modèles, Dunod, II.

Bookstein F.L. (1997). Morphometric tools for landmark data. Cambridge University Press.

Buffiere J-Y., Maire E., Cloetens P., Lormand G. and Fougeres R. (1999). Characterization of internal damage in a MMCp using X-ray synchrotron phase contrast microtomography. *Acta Materialia*, 47, 1613-1625.

Crofton, W.M. (1868). On the Theory of local probability. *Phi.Trans.of Royal Soc.*London, 158, 181-199.

Delarue A. and Jeulin D (2001). Multiscale simulation of spherical aggregates », *Image Anal. Stereol.*, 20, 181-186.

Feuerstein M. (2006). Influence de la microstructure sur les propriétés mécaniques de tôles d'aluminium AA5182. Thèse de doctorat, Ecole des Mines de Saint-Étienne.

Frey P. and George (1999). Maillage : Applications aux éléments finis.P.L. Sciences, H. *(ed.)* Science Publications.

Gardner, R.J.; Jensen, E.B.V. and Volcic A. (2003). Geometric tomography and local stereology. Advances in Applied Mathematics. Academic Press Inc, 30, 397-423.

Holboth A. (2003). The spherical deformation model. Biostatistics, 4, 583-595.

Hoppe H., DeRose T., Duchamp T., McDonald J. and Stuetzle W. (1993). Mesh optimization. ACM SIGGRAPH 1993, 19-26.

Hotelling H. (1933) Analysis of a Complex of Statistical Variables with Principal Components, *J. Educ. Psy., 24*, 498-520.

Jeulin D. (1991). Modèles de structures aléatoires et de changement d'échelle, Thèse de Doctorat d'Etat, Université de Caen.

Lantuejoul, C. and Maisonneuve, F. (1984). Geodesic methods in quantitative image analysis. Pattern Recognition. 17, 177.

Lorensen W.E. and Cline H.E.(1987). Marching cubes: A high resolution 3D surface construction algorithm. SIGGRAPH '87: Proceedings of the 14th annual conference on Computer graphics and interactive techniques, ACM Press, 163-169.

Maire E., Grenier JC, Daniel D., Klöcker H. et BigotA. (2006). Quantitative 3D characterizationof intermetallic particle in Al-Mg industrial alloy by X-ray microtomography, *Scripta materialia*, 55, issue 2, 123-126.

Matheron G. (1967). Eléments pour une théorie des milieux poreux, Masson.

Matheron G. (1972). Ensembles aléatoires, ensembles semi-markovien, et polyèdre poissoniens , *Advances in Applied Probability*, 4, 508-541.

Moreaud M. and Jeulin D. (2005). Multi-scale simulation of random spheres aggregates – application to nanocomposite, 9th European congress on stereology and image analysis, 2, ISBN 83-917834-4-8, 341-348.

Moulin N., Jeulin D., Klöcke H. (2009) Stress concentrations in non-convex elastic particles embedded in a ductile matrix, *International Journal of Engineering Sciences*, Vol. 4 (2), February 2009, 170-191.

Moulin N., Parra-Denis E., Jeulin D., Ducottet C., Bigot A., Boller E., Maire E., Barat C., Klöcker H. (2010). Constituent Particle Break-up during Hot Rolling of AA 5182, *Advanced engineering materials*, 12, n°2, 20-29.

Moulin N. (2008) Modélisation numérique de la fragmentation de particules de formes complexes avec une application au laminage des alliages d'aluminium, PhD thesis of Ecole des Mines de Saint Etienne.

Parra-Denis E., Ducottet, C. and Jeulin, D. (2005) 3D image analysis of non metallic inclusions, 9th European congress on Stereology, 111-122.

Parra-Denis E. and Jeulin D. (2006). Modélisation morphologique 3D des particules intermétalliques dans les alliages d'aluminium, Materiaux 2006, in the congress acts.

Parra-Denis E., Moulin N. and Jeulin D. (2007). Three-dimensional complex shapes analysis from 3D local curvature measurements: application to intermetallic particles in aluminium alloy 5xxx, IAS, 26, 157-164.

Parra-Denis E. (2007). Analyse morphologique 3D de particules de forme complexes: application aux intermétalliques dans les alliages d'aluminium, PhD thesis of Ecole doctorale de Saint Etienne.

Parra-Denis E., Moulin N. and Jeulin D. (2008). 3D complex shape characterization by statistical analysis : Application to aluminium alloy, *Materials characterization*, 59, 338-343.

Pearson K. (1901). On Lines and Planes of Closest Fit to Systems of Points in Space Philosophical magazine, 2, 559-572.

Stokely E.M. and Wu S.Y. (1992). Surface Parametrisation and Curvature Measurement of Arbitrary 3-D Objects: Five Practical Methods, IEEE Trans. Pattern Anal. Mach. Intell., *IEEE Computer Society*, 14, 833-840.

Voorhees P.W. Alkemper J. and Mendoza R. (2002). Morphological Evolution of dendritic microstructures, *Advanced Engineering materilas*, 4, 481-489.

Ward, J.H. (1963). Hierarchical grouping to optimize an objective function. *Journal of the American Statistical Association*, 58, 234-244.

In: Aluminum Alloys
Editor: Erik L. Persson
ISBN: 978-1-61122-311-8

Chapter 2

ISOMERS IN THE CHEMISTRY OF ALUMINIUM COMPLEXES

Milan Melnik, Mária Kohútová and Ferdinand Devínsky
Department of Chemical Theory of Drugs,
Faculty of Pharmacy, Comenius University,
832 32 Bratislava, Slovak Republic

ABSTRACT

The chemistry of aluminium complexes covers quite bulk fields, as shown by a survey covering the crystallographic and structural data of over one thousand and two hundred examples. About nine percent of those complexes exist as isomers and are summarized in this paper. Included are distortion (92.2 %), ligand (4.3 %), *cis – trans* (2.5 %) and polymerisation (1 %) isomerism. These are discussed in terms of coordination about aluminium atom, and correlations are drawqn between donor atoms, bond lengths and interbond angles. Distortion isomers differ only by degree of distortion in Al-L bond distances and L-Al-L bond angles are the most common.

ABBREVIATIONS

acac	acetylacetonate
Bu^n	n-butyl
Bu^t	terc-butyl
c	cubic
$C_{11}H_{22}NPCl$	9 - phosphabarbalone
$C_{22}H_{28}PSi$	benzeneannulate phosphene silicium
$C_{30}H_{62}N_7P$	1,3-bis[bis(diethylamino)methylene]-1,3-diazonia-2$1^3$-phospha-cyclobutane
colH	collidineH(+)
Et	ethyl
$(Et_2N)_2PSe$	di(ethylamine)phosphineselenide

hmpa	hexamethylphosphoramide
hms	hydroxymethanesulphonate
hpr	2-hydroxypropanoate
hx	hexagonal
m	monoclinic
Me	methyl
MePO	methylphosphineoxide (2-)
$Me_2Si(NBu^t)_2P$	dimethylsilyldi(tercbutylamine)phosphine
2,4,6-$Me_3C_6H_2O$	2,4,6-trimethylbenzyloxide
Me_3SiO	trimethylsiloxide
or	orthorhombic
Ph	phenyl
Ph_3PSe	triphenylphosphineselenide
PMe_3	trimethylphosphine
Pr^i	iso-propyl
pyH	pyridinium
rh	rhombohedral
salen	N,N'-ethylenebis(salicylideneaminate)
thf	tetrahydrofurane
tmap	3,4,5,6-tetramethyl-2-azapyrylium
tmen	N,N,N',N'-tetramethylethylenediamine
tr	triclinic
trg	trigonal

I. Introduction

Aluminium plays an important role as the pure metal and its alloys while its compounds have extensive applications as structural and medical ceramics, electronic and optical materials, catalysts, ionic conductors, coagulants for water purification, and reagents [1, 2]. The chemistry of aluminium has long been an active area of study and the relationship between structures; reactivity and catalytic activity have of major industrial importance. Systematic studies in the field of stereoselectivity of complex compounds over the last 50 years have become of increasing interest. Stereoselectivity in complex compounds is very often related to important stereospecifity of biological systems, catalysis, and stereochemical effects in technical processes. It is well known, that isomers are substances that have the same number and kinds of atoms arranged differently. Because their structures are different, isomers have different behaviour. Isomers can be broadly classed into two main categories, structural and stereoisomers. The former can be divided into hydrate, ionisation, coordination number, linkage, bonding and polymerisation sub – categories. The latter can be divided into geometric (*cis – trans*, *fac – mer*), optical, ligand and distortion isomerism.

Over six hundred aluminium coordination compounds have been surveyed [3] with over seventy isomeric examples noted. The organoaluminium compounds as shown by a survey covering the crystallographic and structural data of over three hundred and fifty example [4]. About 6.5% of those compounds exist as isomers. In our review article [5] we analysed and classified almost three hundred heterometallic aluminium compounds and about 5.6% if these compounds exist as isomers.In this review we analyse and classify these isomeric examples with those that have been reported up to the end of 1998. The primary source has been the Cambridge Crystallographic Data Base.

II. Isomers of Aluminium Coordination Compounds

An almost six hundred aluminium coordination compounds has been surveyed by us [3] with over sixty isomeric examples noted. In these chapters we analyse and classify these examples. Included are distortion, ligand and polymerisation isomerism.

1. Distortion Isomerism

The coexistence of two or more species differing only by degree of distortion of M – L bond distances and L – M – L bond angles is typical of the general class of distortion isomers [6]. There are over fifty such examples in the chemistry of aluminium coordination compounds. The aluminium oxidation state in these isomers is found only in the oxidation state + 3.

1.1. Isomeric Forms

There are seven examples [7 – 19] which exists in two isomeric forms and one [20 – 23] in three isomeric forms and their crystallographic and structural data are gathered in Table 1. In Al(MePO) [16, 17] both isomeric forms belong to the homo – trigonal classes. The remaining examples differ from each other not only by degree of distortion but also by crystal class. In three of these one isomer is orthorhombic and the other monoclinic [7, 8, 13 – 15]. In $[Al(H_2O)_6]Cl_3$ one is hexagonal [11] and the other rhombohedral [12]. In $AlF_3{\cdot}H_2O$, one is cubic [18] and the other tetragonal [19]. The AlF_3 which exists in three isomeric forms, one is hexagonal [20, 21] the other rhombohedral [22] and the last one is orthorhombic [23].

In three pairs of the $[AlCl_4]^-$ isomers [7 – 10] each Al(III) atom is surrounded by four chlorine atoms which created a tetrahedral environment with a different degree of distortion (Table 1). Unfortunately, only for one pair of the isomes, namely $[C_{11}H_{22}NPCl][AlCl_4]$ [7] (orthorhombic and monoclinic) all structural data are available. In addition the monoclinic isomer contains two crystallographically independent molecules. Each Al(III) atom has a tetrahedral arrangement created by a four chlorine atoms with a different degree of distortion. The mean Al-Cl bond distances are 2.120 Å (in orthorhombic), 2.109 Å (monoclinic, molecule 1) and 2.114 Å (monoclinic, molecule 2). This indicates that the degree of distortion increases in the order monoclinic (molecule 1) < orthorhombic < monoclinic (molecule 2).

In colourless$[Al(H_2O)_6]Cl_3$ (hexagonal [11] and rhombohedral [12]) six water molecules about each Al(III) atom created an octahedral coordination (AlO_6) with the mean Al-O bond distances 1.88 Å in the former and 1.87 Å in the latter. The latter is somewhat more crowded than the former.

Table 1. Crystallographic and Structural Data for Aluminium Coordination Compounds - Isomeric Forms[a]

Compound (colour)	Cryst. cl. Cryst. gr. Z	a [Å] b [Å] c [Å]	α [°] β [°] γ [°]	Chromo-phore	Al – L [Å]	L – Al – L [°]	Ref.
$(C_{11}H_{22}NPCl)[AlCl_4]$ (colourless)	or $P2_12_12_1$ 4	8.309(4) 15.264(2) 19.936(2)		$AlCl_4$	Cl[b] 2.10(4,7)	Cl,Cl[b] 109.5(2,2.1)	7
$(C_{11}H_{22}NPCl)[AlCl_4]$ (colourless)	m $P2_1/c$ 8	15.921(2) 7.903(12) 32.142(4)	102.49(1)	$AlCl_4$ $AlCl_4$	Cl 2.109(2,5) Cl 2.114(2,11)	Cl,Cl 109.5(1,1.7) Cl,Cl 19.5(1,2.5)	7
$(C_{22}H_{28}PSi)[AlCl_4]$ (colourless)	or Pcab 8	15.437(8) 17.987(9) 20.256(10)		$AlCl_4$	Cl not given		8
$(C_{22}H_{28}PSi)[AlCl_4]$ (yellow)	m $P2_1/c$ 4	13.981(7) 12.632(6) 17.302(9)	113.48(4)	$AlCl_4$	Cl not given		8
$[TeCl_3][AlCl_4]$ (colourless)	m $P2_1/c$ 4	6.600(5) 12.675(8) 13.578(8)	105.72(5)	$AlCl_4$	Cl 2.128(2,41)	Cl,Cl 109.5(3,2.6)	9
$[TeCl_3][AlCl_4]$ (colourless)	tr P-1 4	6.554(1) 16.691(4) 8.391(1)	92.74(1) 97.31(1) 106.58(3)	$AlCl_4$	Cl not given		10
$[Al(H_2O)_6]Cl_3$ (colourless)	hx R3c ?	11.827(6) 11.895(3)		AlO_6	H_2O 1.88(2,0)	O,O 90(1)	11
$[Al(H_2O)_6]Cl_3$ (colourless)	rh R3c?	7.85(3)	97.00(20)	AlO_6	H_2O 1.87		12
$Al(\eta^2\text{-acac})_3$ (colourless)	m $P2_1/c$ 4	14.069(9) 7.568(5) 16.377(10)	99.00(5)	AlO_6	O 1.892(6,19)	O,O 91.8(2,2)[c] 89.4(2,1.8) 179.0(2,9)	13
$Al(\eta^2\text{-acac})_3$ (colourless)	m $P2_1/c$ 4	13.972(3) 7.527(2) 16.307(5)	98.88(2)	AlO_6	O 1.879(2,9)	O,O 91.0(1,1.0)[c] 89.7(1,1.7) 178.7(1,9)	14

Al(η^2-acac)$_3$ (colourless)	or Pna2$_1$ 16	15.699(3) 32.546(7) 13.369(2)		AlO_6	O 1.879(16,0)	O,O 91.3(6)[c]	15
α-Al(MePO) (colourless)	trg P31c ?	13.9949(13) 8.5311(6)		AlO_4 AlO_6	O not given O not given		16
β-Al(MePO) (colourless)	trg R-3c 18	24.650(2) 25.299(5)		AlO_4 AlO_6	O not given O not given		17
$AlF_3{\cdot}H_2O$ (white)	c Pm3m 1	3.610		AlF_6	F 1.805	F,F not given	18
$AlF_3{\cdot}H_2O$ (white)	tg P4/ncc 4	7.734 7.330		AlF_6	F not given	F,F not given	19
AlF_3 (colourless)	hx R-3 6	4.914(3) 12.46(1)		AlF_6	F 1.80(-,9)	F,F not given	20
Compound (colour)	Cryst. cl. Cryst. gr. Z	a [Å] b [Å] c [Å]	α [°] β [°] γ [°]	Chromo- phore	Al – L [Å]	L – Al – L [°]	Ref.
AlF_3 (colourless)	hx R-3 8	4.9254(7) 12.447(5)		AlF_6 AlF_6	F 1.7940(25) F 1.8010(39)	F,F 90.0(-,1) F,F 90.0(-,1)	21
AlF_3 (colourless)	rh R-3 8	7.015		AlF_6	F 1.79	F,F not given	22
AlF_3 (colourless)	or Cmcm 12	6.931(3) 12.002(6) 7.134(2)		AlF_6 AlF_6	F 1.799(1,3) F 1.799(1,2)	F,F 90.0(-,3) F,F 90.0(-,2)	23

Footnotes: a) When more than one chemically equivalent distance or angle is present, the mean value is tabulated. The first number in parenthesis is the e. s. d. , and the second is the maximum deviation from the mean.

b) The chemical identity of the coordinated atom or ligand is specified in these columns.

c) Six - membered metallocyclic ring.

Table 2. Crystallographic and Structural Data for Monomeric Aluminium Coordination Compounds – Independent Molecules (Distortion Isomers)[a]

Compound (colour)	Cryst. cl. Cryst. gr. Z	a [Å] b [Å] c [Å]	α [°] β [°] γ [°]	Chromo- phore	Al – L [Å]	L – Al – L [°]	Ref.
(colH)[AlF_4] (colourless) at 173 K	or Pbcn 24	29.673(6) 16.644(3) 12.439(3)		AlF_4 AlF_4 AlF_6	F[b] 1.649(3,3) F 1.654(3,14) polymer	F,F[b] 109(2) F,F 109(2) see Table	24
(Me_2NH_2)[Al($Me_3SiO)_4$] (colourless)	m $P2_1/c$ 8	20.003(4) 14.910(3) 20.471(4)	116.82(1)	AlO_4 AlO_4	O 1.753(5,27) O 1.758(6,29)	O,O 109.5(1,4.5) O,O 109.5(1,4.2)	25
(tmap)[$AlCl_4$] (light pink) at 163 K	or Pnma 8	17.871(6) 14.243(2) 10.648(2)		$AlCl_4$ $AlCl_4$	Cl 2.133(2,8) Cl 2.133(2,16)	Cl,,Cl 109.5(1,1.0) Cl,,Cl 109.5(1,4)	26
{($Et_2N)_2PSe$}$_2$[$AlCl_4$]$_2$ not given	tr P-1 2	10.635(7) 12.335(8) 15.159(9)	95.94(8) 93.46(7) 110.99(9)	$AlCl_4$ $AlCl_4$	Cl 2.112(6,13) Cl 2.112(5,33)	Cl,,Cl 109.5(3,1.8) Cl,,Cl 109.5(3,1.5)	27
($C_{11}H_{22}NPCl$)[$AlCl_4$] (colourless)	m $P2_1/c$ 8	15.921(2) 7.903(2) 32.142(4)	102.46(1)	$AlCl_4$ $AlCl_4$	Cl 2.109(2,5) Cl 2.116(2,11)	Cl,,Cl 109.5(1,1.7) Cl,,Cl 109.5(1,2.5)	7
{$Me_2Si(NBu^t)_2P$}·[$AlCl_4$] (not given)	or Pnma 8	30.237(9) 10.010(3) 14.146(5)		$AlCl_4$ $AlCl_4$	Cl 2.124(5,16) Cl 2.126(5,3)	Cl,,Cl 109.5(1,5) Cl,,Cl 109.5(1,1.5)	28
[$SeCl_3$][$AlCl_4$] (yellow)	tr P-1 2	9.87 8.27 9.83	139.9 94.8 93.8	$AlCl_4$ $AlCl_4$	Cl 2.125(16,35) Cl 2.140(16,30)	Cl,,Cl 109.5(-,3.6) Cl,,Cl 109.5(-,2.5)	29
(Se_8)[$AlCl_4$] (red brown)	or $Pca2_1$ 4	14.92(2) 10.67(1) 13.22(1)		$AlCl_4$ $AlCl_4$	Cl 2.125(30,55) Cl 2.113(30,47)	Cl,,Cl 109.5(1.0,4.9) Cl,,Cl 109.5(1,0,3.3)	30
{$N(SCl)_2$}[$AlCl_4$] (yellow)	m $P2_1/m$ 4	12.725(3) 13.568(3) 6.200(2)	97.33(3)	$AlCl_4$ $AlCl_4$	Cl 2.129(1,12) Cl 2.136(1,14)	Cl,Cl 109.5(1,2.4) Cl,Cl 109.5 (1,3.9)	31
(S_5N_5)[$AlCl_4$] (pale yellow)	or Pnmc 8	9.412 13.647 20.761		$AlCl_4$ $AlCl_4$	Cl 2.114(6,21) Cl 2.124(5,3)	Cl,Cl 109.5(2,1.1) Cl,Cl 109.5(2,2.2)	32

$[AlCl_4]_2[Al(MeCN)_5Cl]\cdot$ $\cdot MeCN$ (colourless)	or $P2_12_12_1$ 4	10.121(6) 14.562(13) 21.053(13)		$AlCl_4$ $AlCl_4$ AlN_5Cl	Cl 2.118(5,9) Cl 2.121(5,20) N_{eq} 1.986(8,13) N_{ax} 2.021(8) Cl_{ax} 2.196(4)	Cl,Cl 109.5(2,1.6) Cl,Cl 109.5(2,2.9) N,N 87.7(3,3.9) 171.4(4,4) N,Cl 94.3(3,7) 178.3(3)	33
$[AlCl_4][Al(thf)_4Cl_2]$ (colourless)	m $P2_1/n$ 8	9.032(2) 39.134(9) 15.319(5)	91.22(3)	$AlCl_4$ $AlCl_4$ AlO_4Cl_2 AlO_4Cl_2	Cl 2.117(9,23) Cl 2.09(1,3) O_{eq} 1.94(1,2) Cl_{ax} 2.227(6,7) O_{eq}1.94(1,2) Cl_{ax} 2.234(6,7)	Cl,Cl 109.5(4,2.1) Cl,Cl 109.5 (4,1.7) O,O 90.0(5,1.3) 178.9(4,5) O,Cl 89.99(4,5) Cl,Cl 179.9(1) O,O 90.0 (5,4) 179.5(5,2) O,Cl 90.0(3,5) Cl,Cl 179.3(3)	34
$[AlCl_4][Al(thf)_4Cl_2]$ (colourless)	m $P2_1/n$ 8	15.299(5) 9.011(3) 39.101(2)	91.22(2)	$AlCl_4$ $AlCl_4$ AlO_4Cl_2 AlO_4Cl_2	Cl 2.110 Cl 2.084 O_{eq} 1.922(-,42) Cl_{ax} 2.222(-,17) O_{eq} 1.946(-,18) Cl_{ax} 2.226(-,8)	Cl,Cl not given Cl,Cl not given O,O 90.0(-,4) Cl,Cl 179.6 O,O 90.0(-,7) Cl,Cl 179.5	35
$[Hg_2(Me_6C_6]\cdot[AlCl_4]_2\cdot$ $\cdot PhMe$ (colourless)	m $P2_1/c$ 4	21.16(1) 10.951(7) 18.80(1)	104.9(1)	$AlCl_4$ $AlCl_4$	Cl 2.14(1,3) Cl 2.12(1,9)	Cl,Cl 109.5(5,3.9) Cl,Cl 109.5(6,3.8)	36
$[W_2(Me_3PC)_2(PMe_3)_4Cl_4]\cdot$ $\cdot[AlCl_4]$ (not given)	m $P2_1/c$ 2	9.773(4) 24.797(14) 12.633(7)	92.85(4)	$AlCl_4$ $AlCl_4$	Cl 2.10(3,8) Cl 2.12(4,3)	Cl,Cl not given Cl,Cl not given	37
$\{(pyH)_3Br\}[AlBr_4]_2$ (not given)	or Pbca 8	13.655(2) 16.160(2) 27.837(3)		$AlBr_4$ $AlBr_4$	Br 2.286(5,13) Br 2.290(4,3)	Br,Br 109.5(2,1.8) Br,Br 109.5(2,1.6)	38
$Al(BH_4)_3(NH_3)$ (colourless)	m ? ?	11.147(5) 11.417(5) 11.974(5)	101.3(6)	AlB_3N AlB_3N	B 2.240(-,18) N 1.951(-) B 2.218(-,26) N 1.941	B,B 105.4(-,1), 132.2 B,N 95.8(-,8), 124.6 B,B 105.9(-,8), 133.0 B,N 94.8(-,7), 124.8	39

Table 2. (Continued)

Compound (colour)	Cryst. cl. Cryst. gr. Z	a [Å] b [Å] c [Å]	α [˚] β [˚] γ [˚]	Chromo- phore	Al – L [Å]	L – Al – L [˚]	Ref.
$AlCl_3\{(Me_2N)_3SiCl\}$ (colourless) at 238 K	or Pbca 8	14.718(14) 12.379(7) 33.442(21)		$AlCl_3N$ $AlCl_3N$	Cl 2.118(1,3) N 1.990(5) Cl 2.114(3,4) N 1.972(9)	Cl,Cl 109.9(1,2.7) Cl,N 109.0(2,3.6) Cl,Cl 110.6(1,1.5) Cl,N 108.2(2,2.2)	40
$AlCl_3(Ph_3PSe)$ (colourless)	tr P-1 4	8.967(2) 12.626(4) 18.242(4)	84.83(2) 89.02(2) 85.67(2)	$AlCl_3Se$ $AlCl_3Se$	Cl 2.116(3,3) Se 2.452(2) Cl 2.112(3,6) Se 2.421(2)	Cl,Cl 111.9(1,1.1) Cl,Se 106.8(1,6.6) Cl,Cl 111.0(1,1.5) Cl,Se 107.7(1,11.5)	41
Al(η^4-salen)· ·(2,4,6-$Me_3C_6H_2O$) (colourless)	tr P-1 4	11.833(1) 12.170(1) 17.835(2)	75.57(1) 75.29(1) 65.66(1)	AlO_3N_2 AlO_3N_2	η^4O 1.793(3,6) η^4N 1.997(3,3) LO 1.737(3) η^4O 1.794(3,3) η^4N 1.998(3,0) LO 1.741(2)	O,O 102.4(1,13.3) O,N 89.6(1,1)[c] 91.4 – 161.5(1) N,N 78.8(1)[d] O,O 102.9(1,12.5) O,N 89.6(1,3)[c] 91.4 – 164.3(1) N,N 79.0(1)[d]	42
$AlCl_3(Me_2NH)_2$ (colourless)	m $P2_1/b$ 8	6.430(2) 13.968(5) 24.49(1)	95.24(5)	$AlCl_3N_2$ $AlCl_3N_2$	Cl_{eq} 2.187(2,5) N_{ax} 2.062(3,4) Cl_{eq} 2.178(2,6) N_{ax} 2.069(3,9)	Cl,Cl 120.0(1,4.7) Cl,N 90.0(1,1.6) N,N 177.6(2) Cl,Cl 120.0(1,6.3) Cl,N 90.0(1,1.5) N,N 176.8(1)	43
$AlCl_3(Me_2NH)_2$ (colourless	m $P2_1/c$ 8	6.422(1) 24.491(4) 13.963(2)	96.27(1)	$AlCl_3N_2$ $AlCl_3N_2$	Cl_{eq} 2.176(1,3) N_{ax} 2.062(3,11) Cl_{eq} 2.185(1,6) N_{ax} 2.054(3,3)	Cl,Cl 10.0(1,6.3) Cl,N90.0 (1,2.0) N,N 176.5(1) Cl,Cl 10.0(1,4.7) Cl,N 90.0(1,2.3) N,N 176.8(1)	44
$AlCl_3(\eta^2-C_{30}H_{62}N_7P)$ (not given)	m $P2_1/c$ 8	15.547(8) 19.365(5) 26.333(9)	101.03(4)	$AlCl_3N_2$ $AlCl_3N_2$	Cl not given N 1.979(6,34) Cl not given N 1.950(6,13)	N,N 72.2(2)[e] Cl,N 159.7(2) N,N 71.8(2)[e] Cl,N 151.0(2)	45

$AlF_2\{Me_3C)_2SiNCMe_3\}_2Cl$ (not given)	or Pnna 8	16.063(2) 21.896(2) 17.987(2)		AlF_2N_2Cl AlF_2N_2C	F 2.085(2,0) N 1.853(4,0) Cl 2.151(3) F 2.103(3,0) N 1.842(3,0) Cl 2.151(3)	F,F 152.3(2) N,N 141.4(2) F,N 74.7, 96.1(1) N,Cl 109.3(1) F,Cl 103.9(1) F,F 156.0(2) N,N 138.4(2) F,N 74.8, 96.6(1) N,Cl 110.8(1) F,Cl 102.0(1)	46
$\{C(NH_2)_3\}_3\{AlF_6\}$ (colourless)	c $P2_1/a3$ 8	13.953(2)		AlF_6 AlF_6	F 1.818(1,0) F 1.817(1,0)	F,F 90.42(8) 180.00(9) F,F 90.34(8) 180.00(9)	47
$(qnH)[Al(H_2O)_6]SO_4$	tg P31m 3	11.738(2) 8.951(2)		AlO_6 AlO_6	H_2O 1.872(6,1) H_2O 1.877(6,5)	O,O 90.0(2,2.0) O,O 90.0(2,2.3)	48
$[Al(H_2O)_6]_2(SO_4)_3\cdot$ $\cdot 4.4\ H_2O$ (colourless)	tr P-1 2	7.425(2) 26.975(2) 6.0608(5)	90.03(1) 97.66(2) 91.94(1)	AlO_6 AlO_6	H_2O 1.889(11,16) H_2O 1.878(12,27)	O,O 90.0(3,2.0) 178.1(3,5) O,O 90.0(3,2.4) 177.9(3,1.4)	49
$[Al(H_2O)_6]_2(SO_4)_3\cdot$ $\cdot 5\ H_2O$ (colourless)	tr P-1 2	7.420(6) 26.97(2) 6.062(5)	89.57(5) 97.34(5) 91.53(5)	AlO_6 AlO_6	H_2O 1.885(9,12) H_2O 1.890(9,13)	O,O 90.0(4,2.3) O,O 90.0(4,2.5)	50
$[Al(H_2O)_6](NO_3)_3\cdot$ $\cdot 3\ H_2O$ (colourless)	m $P2_1/c$ 4	113.892(2) 9.607(1) 10.907(2)	 95.5(2)	AlO_6 AlO_6	H_2O 1.876(2,6) H_2O 1.878(2,16)	O,O 90.0(2,4) O,O 90.0(1,1.9)	51
$Al(\eta^2\text{-hpr})_3$ (colourless)	m $P2_1$ 4	14.277(2) 9.265(1) 10.828(1)	 100.7(1)	AlO_6 AlO_6	O 1.880(4,22) O 1.882(4,26)	O,O 82.5(1,4)^d 170.0(2,2.3) O,O 82.3(1,2)^d 170.3(2,1.0)	52
$[Al(\eta^4\text{-salen})(\eta^1\text{-hmpa})_2]\cdot$ $\cdot Cl\cdot 2$ thf (colourless)	tr P-1 2	10.821(2) 11.234(2) 32.974(7)	84.99(3) 82.73(3) 81.38(3)	AlO_4N_2 AlO_4N_2	η^4,O_{eq} 1.828(4,3) η^4N_{eq} 2.008(4,4) η^1,O_{ax}1.924(4,9) η^4,O_{eq} 1.818(4,1) η^4N_{eq} 2.003(4,5) η^1,O_{ax}1.930(4,29)	not given not given	53

Table 2. (Continued)

Compound (colour)	Cryst. cl. Cryst. gr. Z	a [Å] b [Å] c [Å]	α [°] β [°] γ [°]	Chromo- phore	Al – L [Å]	L – Al – L [°]	Ref.
$[Al(\eta^4\text{-salen})(\eta^1\text{-hmpa})_2]\cdot I$ (colourless)	m Pn 2	10.552(2) 26.060(5) 14.030(3)	 104.39(3)	AlO_4N_2 AlO_4N_2	η^4,O_{eq} 1.815(6,1) η^4N_{eq} 2.023(7,2) η^1,O_{ax}1.924(6,4) η^4,O_{eq} 1.822(6,3) η^4N_{eq} 2.013(7,5) η^1,O_{ax}1.915(6,3)	not given not given	53
$[Al(H_2O)_6](IO_3)_2\cdot(HI_2O_6)$ (HIO_3) (not given)	hx $P6_3$ 6	16.107(2) 12.378(2)		AlO_6 AlO_6 AlO_6	H_2O 1.890(8,4) H_2O 1.876(8,12) H_2O 1.882(8,7)	O,O not given O,O not given O,O not given	54
$[Al(H_2O)_6](IO_3)_3\cdot(HIO_3)_2$ (not given)	hx $P6_3$ 6	16.126(3) 12.398(2)		AlO_6 AlO_6 AlO_6	H_2O 1.890(8,4) H_2O 1.876(8,12) H_2O 1.882(8,7)	O,O not given O,O not given O,O not given	55
$[Al(H_2O)_6](hms)_3$ (not given)	m P2/n 8	10.821(1) 22.503(4) 14.439(3)	 91.58(2)	AlO_6 AlO_6 AlO_6	H_2O 1.88(2,3) H_2O 1.90(3,3) H_2O 1.892(15,5)	O,O not given O,O not given O,O not given	56

Footnotes: a) When more than one chemically equivalent distance or angle is present, the mean value is tabulated. The first number in parenthesis is e. s. d. , and the second is maximum deviation from the mean.

b) The chemical identity of the coordinated atom or ligand is specified in these columns.

c) The six – membered metallocyclic ring.

d) The five – membered metallocyclic ring.

e) The four – membered metallocyclic ring.

In another colourless Al(η^2-acac)$_3$ (monoclinic [13, 14] and orthorhombic [15]) three homo – bidentate acetylacetonates form about each Al(III) atom an octahedral arrangement with the mean Al-O bond distances of 1.892 and 1.879 Å, respectively.

Two polymeric, α – Al(MePO) [16] and β – Al(MePO) [17] contain non – equivalent aluminium(II) atoms, which have either tetrahedral (AlO_4) and pseudo – octahedral (AlO_6) geometries. Unfortunately, the structural data are not available in the original papers.

In white $AlF_3{\cdot}H_2O$ (cubic [18] and tetragonal [19]) six fluorine atoms created the pseudo – octahedral arrangement about each Al(III) atom in polymeric chains.

In colourless AlF_3, which exists in three isomeric forms, hexagonal [20, 21], rhombohedral [22], and orthorhombic [23], the aluminium(III) atoms are coordinated to six fluorine atoms in polymeric chains. The mean Al-F bond distances are about 1.80 Å (Table 1).

1.2. Independent Molecules

1.2.1. Monomeric Compounds

The crystallographic and structural data for monomeric aluminium(III) complexes which contain two – [7, 26–53], or even three – [54–56] crystallographically independent molecules differ mostly by degree of distortion are gathered in Table 2. There are thirty two examples, which contain two crystallographically independent molecules with chromophores: AlF_4 [24], AlO_4 [25], $AlCl_4$ [7, 26–32, 36, 37], $AlCl_4$ + AlN_5Cl [33], $AlCl_4$ + AlO_4Cl_2 [34, 35], $AlBr_4$ [38], AlB_3N [39], $AlCl_3N$ [40], $AlCl_3Se$ [41], AlO_3N_2 [42], $AlCl_3N_2$ [43 – 45], AlF_2N_2Cl [46], AlF_6 [47], AlO_6 [48–52] and AlO_4N_2 [53]. In addition there are three examples, which contain three such molecules, all with the chromophore AlO_6 [54–56]. These derivatives exist in the following crystal classes: monoclinic (x15) > orthorhombic (9) > triclinic (x7) > hexagonal (x2) > cubic, tetragonal each (x1) (Table 2).

Crystal structure of colourless (colH)[AlF_4] consists [24] of well separated $colH^+$ cation and AlF_4^- anion. The anion contains monomeric plus polymeric units with non – equivalent aluminium atoms within the same structure. The monomeric unit contains two crystallographically independent molecules. Each Al(III) atom is a tetrahedrally coordinated by four fluorine atoms (AlF_4) with the mean Al-F bond distances of 1.649 and 1.654 Å, respectively. In the polymeric unit, each Al(III) atom has a pseudo – octahedral environment, created by six fluorine atoms with the mean Al-F bond distance of 1.749(2) Å which is about 0.100 Å shorter than that of Al-F (bridge) bond distance (mean, 1.849(2) Å). This derivative is an example of mixed – isomerism (distortion and polymerisation).

In $[Al(Me_3SiO)_4]^-$ [25] which contains two crystallographically independent molecules, four unidentate $Me_3SiO_4^-$ ligands created a tetrahedral arrangement around each Al(III) atom (AlO_4) with a different degree of distortion.

The mean Al-O bond distances are 1.753 and 1.758 Å, respectively. The maximum deviation from the ideal tetrahedral angle (109.5°) is 4.5° (molecule 1) and 4.2° (molecule 2). This indicates that the former is somewhat more distorted than the latter.

There are thirteen derivatives [7, 26–37] which contain two independent [$AlCl_4$] anions. In each of them four chlorine atoms form a tetrahedral arrangement around each Al(III) atom with a different degree of distortion. The mean Al-Cl bond distances found in all thirteen derivatives, more crowded vs less crowded, are 2.112 vs 2.126 Å, respectively.The maximum deviation from the ideal tetrahedral angle, (109.5°) are 2.05° (mean value) vs 2.65° (mean value), respectively. These indicate that the former molecules are more crowded and somewhat less distorted, than the latter which are less crowded and somewhat more distorted (Table 2).

There is one derivative with $[AlBr_4]^-$ anion [38], here four unidentate bromine atoms created a tetrahedral arrangement about each Al(III) atom ($AlBr_4$). The mean Al-Br bond distances and maximum deviation from the ideal tetrahedral angle are 2.286 Å and 1.8° (molecule 1) vs 2.290 Å and 1.6° (molecule 2).

In colourless $Al(BH_4)_3(NH_3)$ [39] three unidentate BH_4^- with NH_3 molecule form a tetrahedral arrangement about each Al(III) atom (AlB_3H). The Al-N and Al-B (mean) bond distances, (molecule 1 vs molecule 2) are 1.951 and 2.240 Å vs 1.941 and 2.218 Å, respectively. The six "tetrahedral angles" fall in the wide range 95 – 132.2° (molecule 1) and 94.1 – 133° (molecule 2), which indicates that there is high degree of distortion in both molecules, but anyhow the latter is somewhat more distorted.

There are two derivatives, $AlCl_3\{(Me_2N)_3SiCl\}$ [40] and $AlCl_3(Ph_3PSe)$ [41], in which besides three ahlorine atoms and unidentate N-donor [40] or Se-donor [41] ligand complete a tetrahedral configuration about each Al(III) atom ($AlCl_3N$ [40], $AlCl_3Se$ [41]). The Al-N and AlCl (mean) bond distances in [40] are 1.990(5) and 2.118(1) Å (molecule 1) vs 1.972(5) and 2.114(3) Å (molecule 2). The maximum deviation of the tetrahedral angle is 3.6° in the former and 2.2° in the latter one, indicates that the former is somewhat more distorted than the latter one.

In [41] the Al-Cl (mean) and Al-Se bond distances molecule 1 vs molecule 2 are 2.116(3) and 2.452(2) Å vs 2.112(3) Å and 2.421(2) Å. The "tetrahedral angles" in the former range from 100.2(1) to 113.0(1)° which is more narrow, that than found in the latter molecule (96.2(1) – 119.2 (1)°), indicates higher degree of distortion in the latter molecule, than in the former.

In colourless $Al(\eta^4$-salen)(2,4,6-$Me_3C_6H_2O$) [42] the geometry around the Al(III) atom is intermediate between square based pyramidal and trigonal bipyramidal (AlO_3N_2). Two molecules are differing by degree of distortion (Table 2).

In $AlCl_3(Me_2NH)_2$ [43, 44] three chlorine atoms in a plane with two N atoms of Me_2NH_2 ligands in an axial positions created a trigonal bipyramidal arrangement about each Al(III) atom ($AlCl_3N_2$). The molecules are differ by degree of distortion.

In the two remaining derivatives, $AlCl_3(\eta^2C_{30}H_{62}N_7P)$ [45] and $AlF_2\{(Me_2C)SiNCMe_3\}_2$ Cl [46], which contain two crystallographically independent molecules, the geometry around each Al(III) atom is intermediate between square based pyramidal and trigonal bipyramidal with the chromophores $AlCl_3N_2$ [45] and AlF_2N_2Cl [46] (Table 2).

The colourless cubic $\{C(NH_2)_3\}_3[AlF_6\}$ [47] six fluorine atoms form an octahedral arrangement about each Al(III) atom (AlF_6).The two molecules are almost identical (Table 2).

There are four colourless derivatives, tetragonal [48], two triclinic [49,50] and monoclinic [51], which contain two crystallographically independent $[Al(H_2O)_6]^{3+}$ cations. Each Al(III) atom is an octahedrally coordinated (AlO_6). The mean Al-O bond distances, more crowded molecules (x6) vs less crowded molecules (x6), are 1.878 vs 1.883 Å. The mean values deviation of *cis*-O-Al-O bond angles are 1.77 and 2.20°, respectively. This indicates that the less crowded molecules are somewhat more distorted than the more crowded molecules.

In monoclinic Al(η^2-hpr)$_3$ [53] three homo – bidentate 2-hydroxypropanoate ligands created a distorted octahedral arrangement around Al(III) atom, with the mean Al-O bond distances (molecule 1 vs molecule 2) 2.880(4) vs 1.882(4) Å, respectively. In each molecule are three five – membered metallocyclic rings with the mean values of 82.5(1) and 82.3(1)°, respectively. There is a marked deviation of the *trans* O-Al-O angle from 180°. This angle ranges from 167.7° to 172.3° (average 170.0°) (molecule 1) and from169.3° to 171.3° (average 170.3°) (molecule 2).

Structures of [Al(η^4-salen)(η^1-hmpa)$_2$]Cl·2thf [53] and [Al(η^4(salen)(η^1-hmpa)]I [53] are very similar. Both contain two crystallographically independent molecules. Each Al(III) atom is six coordinated (AlO_4N_2). In the former, the complex cations are connected via the CH hydrogen – bonded chlorine anions to give a dimeric species (CH...Cl 2.559(6) – 2.843(8) Å.). The sum of all six (Al-O (x4) plus Al-N (x2)) bond distances are 11.50 Å (molecule 1) and 11.52 Å (molecule 2).

Structure of the latter (iodine) complex contains well separated complex cations and iodine anions. The sums of all six Al-L bond distances are exactly the same are those found in the chlorine complex (11.52 and 11.50 Å) (Table 2).

Remainders three examples [54–56] (Table 2) contain three crystallographically independent molecules. All contain $[Al(H_2O)_6]^{3+}$ cations with a six coordinated Al(III) atoms (AlO_6). The mean Al-O distances (molecule 1, molecule 2, molecule 3) are 1.890, 1.876 and 1.882 Å [54], 1.883, 1.895 and 1.881 Å [55], 1.880, 1.900 and 1.892 Å [56].

1.2.2. Di- and Oligomeric Compounds

There are eight derivatives, six dimeric [57–62], one tetrameric [63] and one polymeric [64], which contains two crystallographically molecules and one hexameric derivative [65] which contains even four crystallographically independent hexamers. Crystallographic and structural data of the derivatives are gathered in Table 3.

In the dimeric derivatives of the bridging types observed, the edge – shared tetrahedral structure is the most common. The bridging atoms involved can be two O-donor ligands [57–59]. In [Al{μ-OSi(H)$_2$C$_2$H$_3$}]$_2$ [57] two chlorine atoms besides two O-donor bridging ligands, completed a tetrahedral coordination around each Al(III) atom (AlO_2Cl_2). The mean Al-O and Al-Cl bond distances (molecule 1 vs molecule 2) are 1.805 and 2.081 Å vs 1.810 and 2.082 Å. The Al-Al distances are 2.684(3) and 2.686(3) Å, respectively.

Another colourless derivative, [Al{μ-OSi(H)Me$_2$}I$_2$]$_2$ [58] which contains a central Al_2O_2 ring, a pair of iodine atoms around each Al(III) atom completed a tetrahedral arrangement (AlO_2I_2). The Al-Al distances and the sum of all four, (Al-O (x2) and Al-I (x2)), bond distances (molecule 1 vs molecule 2) are 2.691(5) and 8.56 Å vs 2.699(5) and 8.57 Å.

In [Al(μ-OBu^t)(Bu^{t}O)$_2$]$_2$ [59] two OBu^t ligands serve as bridges between two Al(Bu^{t}O)$_2$ moieties, with the Al-Al bond distances of 2.777(2) and 2.779(2) Å, respectively. The mean Al-O$_{(br)}$ and Al-O$_{(t)}$ bond distances (molecule 1 vs molecule 2) are 1.827(2) and 1.690(2) Å vs 2.829(2) and 1.685(2) Å.

Crystal structure of another colourless derivative [60], consists of well separated [ClAl(μ-OH)$_2$AlCl(η^6-18-crown-6]$^{2+}$. The structure of one of the crystallographically independent cations is shown in Figure 1. The cation resides on a crystallographic centre of inversion. Each aluminium atom exhibits octahedral coordination with three crown ether oxygen atoms arranged in a meridional fashion.

Table 3. Crystallographic and Structural Data for Di- and Oligomeric Aluminium Coordination Compounds – Distortion Isomers[a]

Compound (colour)	Cryst. cl. Cryst. gr. Z	a [Å] b [Å] c [Å]	α [°] β [°] γ [°]	Chromo- phore	Al – L [Å]	Al – Al [Å] Al – L – Al [Å] μL – Al – μAl [°]	L –Al – L [°]	Ref.
$[Al\{\mu\text{-}C_2H_3(H)_2SiO\}Cl_2]_2$ (colourless) at 223 K	tr P-1 2	6.287(1) 9.063(1) 14.420(2)	82.25(1) 78.12(1) 80.37(1)	AlO_2Cl_2 AlO_2Cl_2	μO[b] 1.805(4,3) Cl 2.081(3,6) μO[b] 1.810(4,5) Cl 2.082(2,0)	2.684(3) 96.0(3) 84.0(2) 2.686(3) 95.8(3) 84.7(7)	O,O[b] not given not given	57
$[Al(\mu\text{-}Me_2HSiO)I_2]_2$ (colourless) at 223 K	tr P-1 2	8.391(3) 8.393(2) 14.811(5)	78.25(2) 74.41(5) 85.81(2)	AlO_2I_2 AlO_2I_2	μO 1.814(5,3) I 2.469(2,10) μO 1.814(8,14) I 2.473(2,4)	2.691(5) 95.9(4) 84.1(3) 2.699(5) 96.2(5) 83.8(3)	μO,I 112.2(2,3) I,I 116.7(1) μO,I 113.4(2,7) I,I 116.2(1)	58
$[Al(\mu\text{-}Bu^tO)(Bu^tO)_2]_2$ (colourless) at 118 K	tr P-1 2	9.946(3) 9.755(3) 16.322(5)	88.89(2) 73.81(1) 88.84(1)	AlO_4 AlO_4	μO 1.827(2,3) O 1.690(2,8) μO 1.829(2,2) O 1.685(2,4)	2.777(2) 98.89(12) 81.11(12) 2.779(2) 98.85(12) 81.15(12)	μO,O 114.0(1,4.2) O,O 115.1(1,0) μO,O 114.0(1,3.8) O,O 115.0(1,3.8)	59
$[\{ClAl(\mu\text{-}OH)\}_2(\mu\text{-}18\text{-}C\text{-}6)]\cdot[AlCl_4]_2\cdot 2.3\ PhNO_2$ (colourless)	tr P-1 3	13.344(2) 17.742(3) 18.318(3)	68.69(1) 81.98(1) 68.00(1)	AlO_5Cl AlO_5Cl $AlCl_4$	μHO 1.86(1,15) O 1.93(1,6) Cl 2.24(11) μHO 1.78(1,2) O 1.95(2,9) Cl 2.20(1) Cl not given	not given 100.7 79.3(7,3.7) not given 106 74(1)	μO,O 92.5(6,8.4) O,O 81.3(6,1.9)[c] μO,Cl 94.6(5,2.9) O,Cl 93.9(5,7.1) μO,O 95(1,5) O,O 81.7(9,1.4)[c] μO,Cl 100(1) O,Cl 92.4(7,3.7) Cl,Cl not given	60
$[Al(\mu\text{-}acac)(acac)Cl]_2\cdot$ $\cdot[AlCl_4]$ (colourless)	m C2/c 8	18.81(1) 11.30(2) 27.70(1)	 106.53(4)	AlO_4Cl AlO_4Cl $AlCl_4$	μO 1.885(29,49) O 1.777(34,10) Cl 2.104(18) μO 1.821(29,20) O 1.876(33,4) Cl 2.108(21) Cl 2.103(28,55)	not given 107.5(1,1.3) 70.7(11) not given 107.8(12) 73.6(12)	O,O 93.3(15) μO,Cl 104.6(1,2.0) O,Cl 106.6(1,6) O,O 93.6(1,1.2) μO,Cl 105.8(1,1.0) O,Cl 104.4(1,2.9) Cl,Cl not given	61
$[Al_2(\mu\text{-}Cl)Cl_6]_2\cdot$ $\cdot[(Me_6C_6)_3Zr_3Cl_6]$	m C2/c	14.167(3) 27.779(7)	 94.27(4)	$AlCl_4$	μCl 2.263(5) Cl 2.084(7,17)	not given 116.1(4)	μCl,Cl 104.5(2,4.1)	62

(not given)	4	15.721(3)		$AlCl_4$	μCl 2.276(5) Cl 2.083(7,3)	not given 112.1(5)	Cl,Cl 114.0(5,1.1) μCl,Cl 104.9(2,3.4) Cl,Cl 113.6(5,8)	
$[HAl(\mu_3\text{-}Pr^iN]_4$ (colourless)	or Pccn 16	21.967(5) 20.006(9) 19.833(12)		AlN_3H (x4) AlN_3H (x4)	μ_3N 1.914(7,10) H 1.51(8,20) μ_3N 1.912(7,13) H 1.50(8,11)	89.9(2) 89.9(1)	N,N 90.1(2) N,H 125.1(13) N,N 90.0(2,1) N,H 125.1(10)	63
$\{H_3Al(\mu\text{-}\eta^2\text{-tmen})\}_n$ (colourless)	or $P2_12_12_1$ 8	9.554(4) 17.241(9) 11.866(6)		AlH_3N_2 AlH_3N_2	H_{eq} not given N_{ax} 2.215(4,23) H_{eq} not given N_{ax} 2.196(4,4)		N,N 178.0 N,N 176.5	64
$[HAl(\mu_3\text{-}Pr^iN)]_6$ (colourless)	tr P-1 4	17.18}2| 10.68(2) 20.09(2)	123.5(4) 90.0(5) 93.5(5)	AlN_3H (x6) AlN_3H (x6) AlN_3H (x6) AlN_3H (x6)	μ_3N 1.898(1) 1.955(3) H 1.50(2) μ_3N 1.898(3) 1.953(3) H 1.49(3) μ_3N 1.897(5) 1.956(5) H 1.48(3) μ_3N 1.898(4) 1.948(6) H 1.50(1)	88.5(2) 88.5(3) 88.5(2) 88.4(2)	N,N 91.4(2) 116.6(3) N,N 91.4(3) 116.5(3) N,N 91.4(2) 116.3(9) N,N 91.4(2) 116.3	65

Footnotes: a) When more than one chemically equivalent distance or angle is present, the mean value is tabulated. The first number in parenthesis is e. s. d. , and the second is maximum deviation from the mean.

b) The chemical identity of the coordinated atom or ligand is specified in these columns.

c) The five – membered metallocyclic ring.

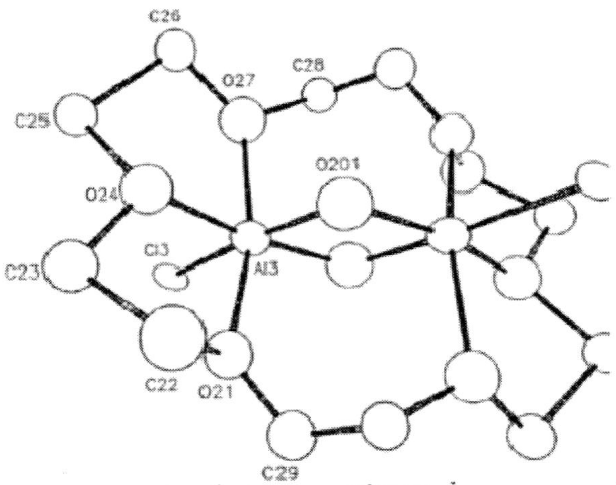

Figure 1. Structure of $[ClAl(\mu\text{-}OH)_2AlCl(\mu\text{-}\eta^6\text{-}18\text{-crown-}6)]^{2+}$ [60].

The aluminium atom shows significant angular deviations from those of exact octahedral symmetry, as expected with a terdentate ligand and bridging groups. The cation may be viewed as comprising $[ClAl(\mu\text{-}OH)_2AlCl]^{2+}$ threaded through the crown ether. The two cations are differing from each other by degree of distortion (Table 3).

There is colourless derivative [61] which contains well separated $[Al(\mu\text{-acac})(acac)Cl]_2^+$ dimeric cations and $[AlCl_4]^-$ anions. In dimeric unit, two Al(acac)Cl moieties are connected by two O-donor acetylacetonate ligands. Each Al(III) atom is five-coordinated (square pyramidal, AlO_4Cl). The mean values of the sum of all (Al-$O_{(x4)}$ plus Al-Cl) bond distances (dimer 1 vs dimer 2) are 9.43 vs 9.50 Å. This indicates that the former dimer is somewhat more crowded than the latter one.

Structure of monoclinic derivative [62] consists of well separated $[Al_2(\mu\text{-}Cl)Cl_6]^-$ anions and $[(Me_6C_6)_3Zn_3Cl_6]^+$ cations. In dimeric complex anion two $AlCl_3$ moieties are bridged by a single chlorine atom and completed a tetrahedral arrangement around each Al(III) atom ($AlCl_4$).

The mean Al-$Cl_{(br)}$, Al-Cl bond distances and Al-Cl-Al bridge angles (dimer 1 vs dimer 2) are 2.263 Å, 2.084 Å, 116.1(4) ° vs 2.276 Å, 2.083 Å and 112.1(5)°.

The remaining five tetrahedral angles range from 100.1(2) – 115.1° vs 101.5 – 114.4°. This indicates that the former dimer is somewhat more distorted thah the latter one.

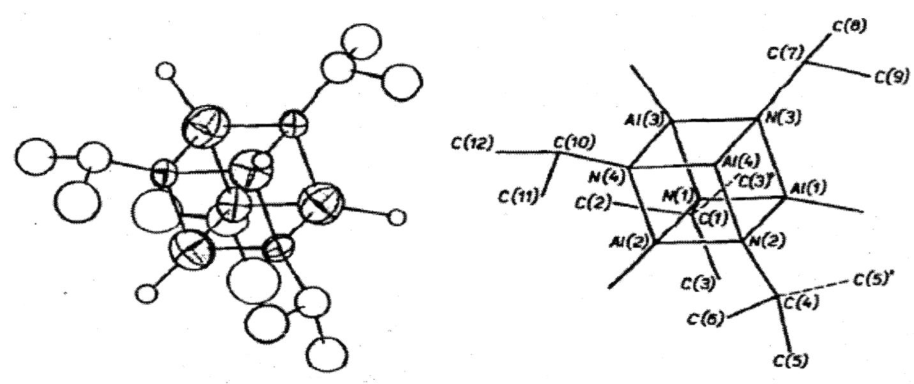

Figure 2. Structure of $[HAl(Pr^iN)]_4$ [63].

A cubane AlN_4 skeleton is found in the structure of $[HAl(Pr^iN)]_4$ [63] which contains two crystallographically independent tetramers (Table 3). Structure of the cubane is shown in Figure 2. Each Al(III) atom is tetrahedrally coordinated (AlN_3H), with the mean Al-N and Al-H bond distances (tetramer 1 vs tetramer 2) of 1.914 Å and 1.51 Å vs 1.912 Å and 1.50 Å.

Two polymeric units have been found in $[H_3Al(\eta^3\text{-tmen})]_n$ [64]. In each of them all Al(III) atoms are five coordinated (trigonal bipyramidal) with various degrees of distortion. Trigonal planar AlH_3 units are bridged by bidentate (non-chelating) N-donors tetramethylethylene ligands, to form infinite chains. The mean Al-N_{axial} bond distances are 2.196 Å (polymer 1) and 2.215 Å (polymer 2) (Table 3).

Remaining example (Table 3) is hexamer, $[HAl(\mu_3\text{-}Pr^iN)]_6$ [65] which contains even four crystallographically independent hexamers. The molecule consists of a hexagonal cage formed by two almost planar six member rings (AlN_3) joined together by six transverse Al-N bonds (Figure 3). Each Al(III) atom is in a distorted tetrahedral environment (AlN_3H).The Al-μ_3N distance range from 1.897(5) to 1.956(5) Å and Al-H bond distance from 1.48(2) to 1.50(2) Å.

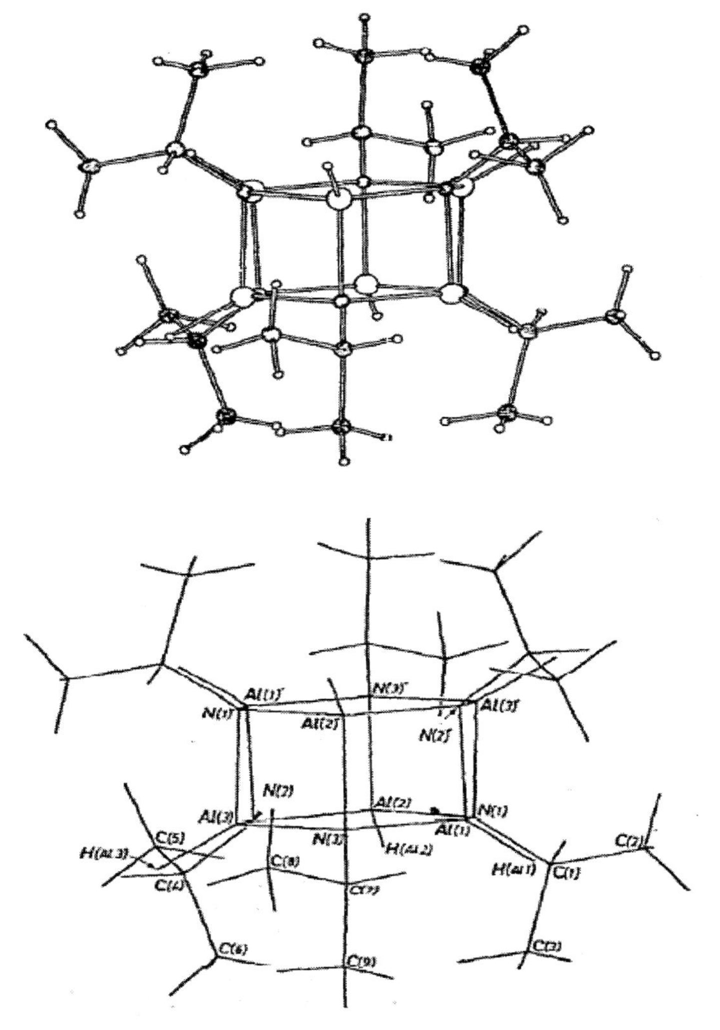

Figure 3. Structure of $[HAl(\mu_3\text{-}Pr^iN)]_6$ [65].

Inspection of the data in Tables 1 – 3 reveals that the sum of Al-L (x n) bond distances elongate with the sum of the covalent radii of donor atoms (xn, in parenthesis), as well as with the increases number of the coordination number (less vs more distorted molecules) in the sequences: 6.615 vs 6.59 Å (F_4, 2.88 Å) < 7.03 vs 7.01 Å (O_4, 2.92 Å) < 7.785 vs 7.77 Å (O_2Cl_2, 3.44 Å) < 8.345 vs 8.31 Å (Cl_3N, 3.72 Å) < 8.50 vs 8.45 Å (Cl_4, 3.96 Å) < 8.575 vs 8.565 Å (O_2I_2, 4.12 Å) < 8.80vs 8.75 Å (Cl_3Se, 4.14 Å) < 9.16 vs 9.14 Å (Br_4,4.56 Å); 9.325 vs 9.315 Å (O_3N_2, 3.69 Å) < 9.50 vs 9.43 Å (O_4Cl, 3.91 Å) < 10.04 vs 10.025 Å (F_2N_2Cl, 3.93 Å) < 10.675 vs 10.66 Å (Cl_3N_2, 4.47 Å); and 10.91 vs 10.90 Å (F_6, 4.32 Å) < 11.295 vs 11.27 Å (O_6, 4.38 Å) < 11.52 vs 11.50 Å (O_4N_2, 4.42 Å) < 11.75 vs 11.61 Å (O_5Cl, 4.64 Å) < 12.23 vs 12.17 Å(O_4Cl_2, 4.90 Å).

2. Ligand Isomerism

There are two Al(III) complexes which exhibit this type of isomerism, colourless $AlCl_3(n\text{-}MeC_6H_4CoCl)$ (n = 2, 3, or 4) [66] and another colourless $[HAl(\mu_3\text{-}NR)]_6$ (R = isopropyl [65] or propyl [67]) (Table 4A).

In the $AlCl_3(n\text{-}MeC_6H_4COCl)$ [66] each Al(III) atom has a tetrahedral geometry with differing degrees of distortion ($AlCl_3O$). The Al-O bond distance reflects the position of methyl group and elongated in the order: 1.824 Å (n = 2) < 1.828 Å (n = 4) < 1.835 Å (n = 3). However, the mean Al-Cl bond distance do not follow this order with the values: 2.093 Å (n = 3) < 2.095 Å (n = 2) < 2.104 Å (n = 4). The sum of all four (Al-O plus Al-Cl(x3)) bond distance increases in the sequence: 8.10 Å (n = 2) < 8.11 Å (n = 3) < 8.14 Å (n = 4).

There are two colourless hexamers, triclinic $[HAl(\mu_3\text{-}NPr^i)]_6$ [65] and trigonal $[HAal(\mu_3\text{-}NPr)]_6$ [67], which are isostructural and are example of the ligand isomerism. In addition, the triclinic hexamer contains four crystallographically independent molecules. Each molecule [65, 67] consists of a hexagonal cage formed by two almost planar six member rings (AlN_3) joined together by six transverse Al-N bonds. Each Al(III) atom is a tetrahedrally coordinate (AlN_3H) with a differing degree of distortion (Table 4A).

3. Polymerisation Isomerism

Structure of (colH)[AlF_4] [24] consists of well separated $(colH)^+$ cations and $[AlF_4]^-$ anions. The anion contains monomeric and polymeric units. In addition the monomeric unit contains two crystallographically independent molecules. In monomeric four fluorine atoms created a tetrahedral geometry around each Al(III) atom (AlF_4). The mean values of Al-F bond distances (molecule 1 vs molecule 2) are 1.649 vs 1.654 Å (Table 4B). In the polymeric units, AlF_2 moieties are connected by fluorine atoms and completed an octahedral coordination around each Al(III) atom (AlF_6). The mean Al-F and Al-$F_{(bridge)}$ bond distances are 1.797 and 1.826 Å.

This derivative is an example of mixed isomerism (polymerisation and distortion).

Table 4. Crystallographic and Structural Data for Aluminium Coordination Compounds – Ligand and Polymerisation Isomers[a]

Compound (colour)	Cryst. cl. Cryst. gr. Z	a [Å] b [Å] c [Å]	α [°] β [°] γ [°]	Chromo- phore	Al – L [Å]	L –Al – L [°]	Ref.
A: Ligand Isomers							
$AlCl_3(2\text{-}MeC_6H_4COCl)$ (colourless)	m $P2_1/c$ 4	9.451(7) 10.756(9) 13.483(12)	115.00(15)	$AlCl_3O$	Cl 2.095(2,9) O 1.824(4)	Cl,Cl[b] 114.3(1,2.9) Cl,O 104.0(1,4.6)	66
$AlCl_3(3\text{-}MeC_6H_4COCl)$ (colourless)	m $P2_1/m$ 4	10.123(12) 7.398(6) 8.731(9)	109.98(8)	$AlCl_3O$	Cl 2.093(2,5) O 1.835(2)	Cl,Cl 112.93(1,7) Cl,O 105.7(1,2.9)	66
$AlCl_3(4\text{-}MeC_6H_4COCl)$ (colourless)	tr P-1 2	7.511(3) 9.825(6) 9.706(6)	107.05(6) 113.86(6) 90.47(6)	$AlCl_3O$	Cl 2.104(1,9) O 1.828(2)	Cl,Cl 113.8(1,9) Cl,O 104.9(1,3.3)	66
$[HAl(\mu_3\text{-}NPr^i)]_6$[c] (colourless)	tr P-1 4	17.18(2) 10.68(2) 20.09(2)	123.5(4) 90.0(5) 93.5(5)	AlN_3H	μ_3N 1.898(4) 1.953(4) H 1.48(3)	not given	65
$[HAl(\mu_3\text{-}NPr)]_6$ (colourless)	trg P-3 3	16.801(2) 6.047(2)		AlN_3H	μ_3N 1.890(4,6) 1.959(8,13) H not given	N,N 91.2(2) 115.2(2)	67
B: Polymerisation isomers							
$(colH)[AlF_4]^-$ (colourless) at 173 K	or Pbcn 24	29.673(6) 16.644(3) 12.439(3)		AlF_4 AlF_4 (monomer) AlF_6 (polymer)	F 1.649(3,3) F 1.654(3,14) μF 1.826(-,6) F 1.797(-,1)	F,F 109(2) F,F 109(2) not given	24

Footnotes: a) When more than one chemically equivalent distance or angle is present, the mean value is tabulated. The first number in parenthesis is e. s. d. , and the second is maximum deviation from the mean. b) The chemical identity of the coordinated atom or ligand is specified in these columns. c) There are four crystallopraphically independent hexamers, the mean values are tabulated.

Table 5. Crystallographic and Structural Data for Aluminium Organometallic Compounds – Distortion Isomers[a]

Compound (colour)	Cryst. cl. Cryst. gr. Z	a [Å] b [Å] c [Å]	α [°] β [°] γ [°]	Chromophore	Al – L [Å]	L – Al – L [°]	Ref.
$[Al(Et)_2Cl_2]\cdot[V_2(PMe_3)_6Cl_3]$ (not given)	or Pnma 4	12.705(2) 12.522(4) 28.554(9)		AlC_2Cl_2	C[b] not given Cl not given	C,C[b] not given	68
$[Al(Et)_2Cl_2]\cdot[V_2(PMe_3)_6Cl_3]$ (not given)	tg $P4_2/nmc$ 8	23.220(5) 18.640(3)		AlC_2Cl_2			68
$Al(Bu^t)_2\{Ph_3Si)_2N\}$ (colourless)	tr P-1 4	12.623(4) 16.880(3) 18.600(2)	93.45(1) 100.25(2) 99.46(2)	AlC_2N AlC_2N	Bu^t,C 2.021(6,5) N 1.880(4) Bu^t,C 2.017(5,2) N 1.878(4)	C,C 117.3(2) C,N not given C,C 117.3(2) C,N not given	69
$Al(Bu^t)_2\{Ph_3)(1\text{-ad})\cdot SiN\}$ (colourless)	tr P-1 4	10.372(3) 17.957(8) 19.184(8)	113.87(3) 99.23(3) 96.71(3)	AlC_2N AlC_2N	Bu^t,C 2.011(3,11) N 1.853(2) Bu^t,C 2.009(3,1) N 1.845(24)	C,C 117.0(1) C,N not given C,C 117.6(1) C,N not given	69
$Na[Al(Et)_4]$ (colourless)	m $P2_1/c$ 8	13.900(2) 13.207(2) 14.443(1)	 117.43(1)	AlC_4 AlC_4	Et,C 2.016(2,7) Et,C 2.013(2,4)	C,C 109.5(1,1.5) C,C 109.5(1,3.1)	70
$Na[Al(\eta^2\text{-}2,7\text{-}C_2B_8H_8)_2]$ (colourless) at 128 K	m $P2_1$ 4	10.035(2) 12.433(3) 11.690(3)	 111.019(7)	AlC_4 AlC_4	C 2.018(5,2) C 2.021(5,5)	C,C 82.1(1,3)[c] 124.3(2,12.0) C,C 82.2(2,1)[c] 123.7(2,12.9)	71
$Al(\eta^1\text{-bht})(\eta^2\text{-}C_{12}H_{23}O_2)(Et)$ (yellow)	m $P2_1$ 2	9.3428(6) 14.3079(6) 11.3322(6)	 91.971(4)	AlO_3C	η^1 O 1.732(3) η^2 O 1.720(5) 1.911(5) Et,C 1.962(9)	O,O 95.8(2)[d] 107.8(3,4.9) O,C 113.9(4,6.1)	72
$Al(Me)_3(\eta^1\text{-tmor})$ (colourless)	m $P2_1/c$ 8	8.452(4) 21.075(4) 12.267(2)	 90.37(2)	AlC_3N AlC_3N	Me,C 1.975(6,4) η^1,N 2.030(4) Me,C 1.968(6,3) η^1,N 2.032(4)	C,C 115.0(3,2.3) C,N 103.1(2,1.7) C,C 114.8(3,2.6) C,N 103.3(2,1.4)	73
$AlCl_3(\eta^1\text{-}C_{15}H_{27}P_3)$ (yellow)	or $P2_12_12_1$ 8	14.619(3) 17.029(7) 17.682(13)		$AlCl_3C$ $AlCl_3C$	not given not given	not given not given	74
$Tl[Al(\eta^5\text{-}C_2B_9H_{11})_2]\cdot(tol)_2$	tr	11.347(2)	92.429(6)	AlB_6C_4	B 2.20(2,6)	not given	75

(colourless)	P-1 ?	11.748(2) 12.708(2)	90874(6) 93.343(5)	AlB_6C_4 AlB_6C_4	C 2.30(2,1) B 2.18(2,4) C 2.27(2,1) B 2,20(2,9) C 2.26(2,3)		
$Al(Bu^t)_2(tbp)$ (colourless) at 130 K	m Pn 12	11.386(10) 29.11(2) 24.969(16)	 102.05(6)	AlC_2O AlC_2O AlC_2O AlC_2O AlC_2O AlC_2O	Bu^t,C 1.992(11,65) tpb,O 1.701(5) Bu^t,C 1.988(12,21) tpb,O 1.687(7) Bu^t,C 1.953(11,11) tpb,O 1.725(7) Bu^t,C 2.005(11,19) tpb,O 1.717(7) Bu^t,C 1.997(14,66) tpb,O 1.728(7) Bu^t,C 2.014(10,30) tpb,O 1.695(8)	C,C 125.9(4) C,O 116.2(4,10.9) C,C 127.7(5) C,O 115.9(5,5.2) C,C 118.9(5) C,O 119.9(4,9.2) C,C 124.1(5) C,O 117.9(5,5.9) C,C 125.3(5) C,O 117.2(5,2.1) C,C 128.7(4) C,O 115.2(4,5.2)	76

Footnotes: a) When more than one chemically equivalent distance or angle is present, the mean value is tabulated. The first number in parenthesis is e. s. d. , and the second is maximum deviation from the mean. b) The chemical identity of the coordinated atom or ligand is specified in these columns. c) Four – membered metallocyclic ring. d) Six – membered metallocyclic ring.The mean Al-C bond distances and the mean C-Al-C (four – membered rings) bond angles (molecule 1 vs molecule 2) are 2.018 Å and 82.1° vs 2.021 Å and 82.3°. The remainders angles range from 112.3 to 136.3° and 110.8 to 136.6°, respectively. The former molecule is somewhat more crowded with higher degree of distortion than the latter molecule.

III. Isomers of Organoaluminium Compounds

The organoaluminium compounds, as show by a survey covering the crystallographic and structural data of over three hundred and fifty examples [4]. About 6.5 percent of those compounds exist as isomers and are summarised in this chapter. There are two types of isomers, distortion and *cis – trans*.

1. Distortion Isomerism

1.1. Monomeric Compounds

There are eleven examples, and their crystallographic and structural data are summarized in Table 5. One example $[Al(Et)_2Cl_2][V_2(PMe_3)_6Cl_3]$ [68] exist in two isomeric forms orthorhombic and tetragonal. In complex $[Al(Et)_2Cl_2]^-$ anion the pair of the respective ligands creates around each Al(III) atom a tetrahedral arrangement. Unfortunately, only the unit cell dimensions are available.

There are six derivatives, which contain two crystallographically independent molecules with inner coordination spheres (chromophores): AlC_2N [69], AlC_4 [70, 71], AlO_3C [72], AlC_3N [73], and $AlCl_3C$ [74]. In colourless $Al(Bu^t)_2\{(Ph_3Si)_2N\}$ and $Al(Bu^t)_2\{(Ph_3)(1\text{-}ad)SiN\}$ (both are triclinic) [69]. In all each Al(III) atom has a trigonal planar environment (AlC_2N), with differing degrees of distortion. In the former, the sum of all three (Al-C (x2) plus Al-N) bond distances are 5.92 Å (molecule1) and 5.91 Å (molecule 2). These values are somewhat higher than those found in the latter derivative, with the values of 5.875 and 5.86 Å, respectively.

In another two colourless, monoclinic $Na[Al(Et)_4]$ [70] and monoclinic $Na[Al(\eta^2\text{-}2,7\text{-}C_2B_8H_8)_2]$ [71], each Al(III) atom has a AlC_4 chromophore. In [70] four ethyl groups created a tetrahedral environment with the mean Al-C bond distances of 2.013 Å (molecule 1) and 2.016 Å (molecule 2). The “tetrahedral angles” deviated from the ideal value of 109.5° by 1.5° (molecule 1) and 3.1° (molecule 2), which indicates that the latter molecule is somewhat more distorted than the former one.

In another derivative [71] a pair of homobidentate $2,7\text{-}C_2B_8H_8$ ligands (C, C' – donors) with two four – membered rings create a tetrahedral arrangement around each Al(III) atom (AlC_4) with a differing degrees of distortion.

In yellow $Al(\eta^1\text{-bht})(\eta^2\text{-}C_{12}H_{23}O_2)(Et)$ [72], a tetrahedral arrangement around each Al(III) is created by two monodentate (bht-O donor) and ethyl group (C – donor) and homo – bidentate $C_{12}H_{23}O_2$ (O, O' – donor) ligand. Unfortunately, only the mean values of the both molecules are available from the original paper (Table 5).

In monoclinic $Al(Me)_3(\eta^1\text{-tmor})$ [73] unidentate ligands form a tetrahedral arrangement around each Al(III) atom (AlC_3N). The sum of all four (Al-C (x3) plus Al-N) bond distances and the range of the six “tetrahedral angles” (molecule 1 vs molecule 2) are: 7.96 Å and 101.4 – 117.3° vs 7.93 Å and 101.9 – 117.4°.

In yellow orthorhombic $AlCl_3(\eta^1\text{-}C_{15}H_{27}P_3)$ [74], also four unidentate ligands create a tetrahedral arrangement around each Al(III) atom ($AlCl_3C$), but only unit cell dimensions were given in the original paper.

The structure of the $[Al(\eta^3-C_2B_9H_{11})_2]^{2-}$ sandwich anion [75] is shown in Figure 4. There are three crystallographically independent molecules, differing mostly by degree of distortion. The sum of all ten (Al-B (x6) plus Al-C (x4)) bond distances elongated in the order: 22.16 Å (molecule 1) < 22.24 Å (molecule 2) < 22.40 Å (molecule 3) (Table 5).

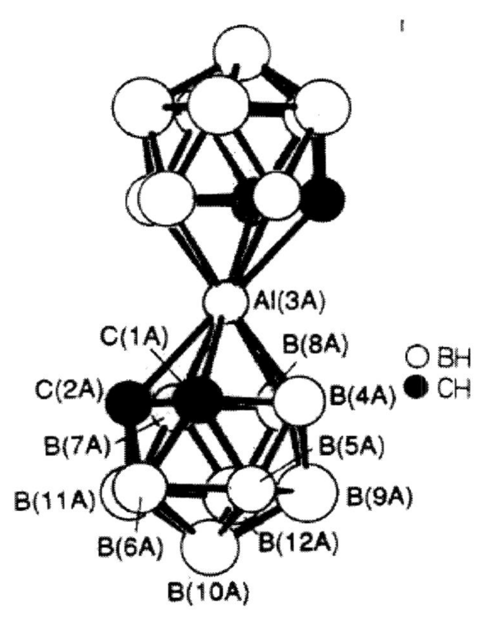

Figure 4. Structure of $[Al(\eta^3-C_2B_9H_{11})_2]^{2-}$ [75].

Monoclinic colourless $Al(Bu^t)_2(\eta^1$-tpb) [76] contains six crystallographically independent molecules. Each Al(III) atom has a trigonal planar geometry, differing mostly by degree of distortion (AlC_2O). The sum of all three (Al-C (x2) plus Al-O) bond distances elongated in the order: 5.631 Å (molecule 1) < 5.663 Å (molecule 2) < 5.685 Å (molecule 3) < 5.722 Å (molecule 4) < 5.723 Å (molecule 5) < 5.727 Å (molecule 6). The L-Al-L bond angles ranges: 110.7 – 129.1° (molecule 1) < 110.4 – 127.7° (molecule 2) < 105.3 – 127.1° (molecule 3) < 115.1 – 125.3° (molecule 4) < 110.0 – 128.7° (molecule 5) < 112.0 – 124.1° (molecule 6).

From the data in Table 5 reveals that the mean values of sum of Al-L (x n) bond distances elongate with the sum of the covalent radii of donor atoms (x n) (in parenthesis) as well as with the increases number of the coordination number, as expected (more vs less crowded units) in the order: 5.66 vs 5.725 Å (C_2O, 2.27 Å) < 5.87 vs 5.92 Å (Cl_2N, 2.29 Å) < 7.935 vs 7.955 Å (C_3N, 3.06 Å) < 8.06 vs 8.08 Å (C_4, 3.08 Å) < 22.20 vs 22.40 Å (B_6C_4, 8.00 Å).

1.2. Di- and Oligomeric Compounds

There are thirteen derivatives (dimeric (x12) and hexameric (x1)), which contain two crystallographically independent molecules and their crystallographic and structural data are gathered in Table 6.

Structure of dark green derivative [77] consists of well separated $[Al_2(\eta^1\text{-trip})_4]^{2+}$ cations, $[Li(tmeda)_2]^-$ anions and ether molecules. In the complex cation two AlC_2 units are held together by direct Al-Al bond (2.470(2) Å). Unfortunately, only mean values of the both dimers are given in the original paper.

Table 6. Crystallographic and Structural Data for Di- and Oligomeric Organoaluminium Compounds – Distortion Isomers[a]

Compound (colour)	Cryst. cl. Cryst. gr. Z	a [Å] b [Å] c [Å]	α [°] β [°] γ [°]	Chromo- phore	Al – L [Å]	Al – Al [Å] Al – L – Al [°] μL – Al – μAl [°]	L –Al – L [°]	Ref.
$[Al_2(\eta^1$-trip$)_4]\cdot[Li(tmeda)_2]\cdot Et_2O$ (dark green) at 130 K	tr P-1 4	15.362(14) 19.401(2) 28.011(2)	71.873(7) 83.823(7) 79.824(7)	AlC_2	C^b 2.021(1,0)	2.470(2)	C,C^b 107.3(9) 116.8(7)	77
$[Al(\mu$-mto$)(Bu^i)_2]_2$ (colourless)	tr P-1 2	10.838(4) 12.792(6) 15.787(7)	81.56(4) 83.04(4) 73.01(3)	AlO_2C_2 AlO_2C_2	μO 1.842(8,10) C 1.95(1,6) μO 1.84(1,1) C 1.95(2,2)	2.815(3) 99.8(4,2) 80.3(4,1) 2.843(6) 101.2(4,0) 78.8(4,6)	C,C 122.8(6,1.1) C,O 111.5(5,4.8) C,C 121.5(6,8) C,O 112.2(5,5.8)	78
$[Al(\mu$-mtp$)(Me)_2]_2$ (colourless)	tr P-1 3	9.828(20 13.547(2) 14.164(2)	62.15(2) 87.26(2) 75.05(2)	AlO_2C_2 AlO_2C_2	μO 1.865(6) 1.946(6) C 1.952(8,18) μO 1.861(6,5) 1.972(6,9) C 1.949(9,24)	2.977(3) 102.7(3) 77.3(3) 2.985(3) 102.3(3,5) 77.6(3,1)	C,C 127.5(4) C,O not given C,C 131.2(4,4) C,O not given	79
$[Al_2(\mu$-H$)(\mu$-$CH_2)\cdot$ $\cdot\{(Me_3Si)CH\}_4\cdot$ $\cdot$ $[Li(tmeda)_2]$ (colourless) at 173 K	tr P-1 4	14.706(4) 21.683(7) 22.005(6)	80.09(2) 84.39(2) 77.37(2)	AlC_3H AlC_3H	μH 1.87(2,1) μH_2C 1.971(4,5) C 2.022(4,10) μH 1.85(2,4) μH_2C 1.968(4,5) C 2.025(4,11)	2.744(1) 92(1,4) 86.6(7,1) 2.748(2) 92.2(8,3.6) 2.025(4,11)	C,C 119.0(2,5.4) C,H 106.2(7,5.4) C,C 117.4(2,2.7) C,H 106.7(6,3.8)	80
$[Al(\mu$-bbp$)(Me)]_2$ (colourless)	tr P-1 2	9.309(2) 9.860(1) 22.132(4)	81.03(2) 81.43(2) 73.84.1)	AlN_2C_2 AlN_2C_2	μN 1.965(6,8) C 1.933(8) Me,C 1.966(11) μN 1.965(7,9) C 1.945(8) Me,C 1.944(9)	not given 91.9(2) 88.1(3) not given 92.2(3) 87.8(3)	C,C 115.5(4) C,N 99.8(3)[c] 116.6(3,4.3) C,C 116.2(3) C,N 98.6(3)[c] 116.9(3,4.4)1	81
$[Al(\mu$-PhS$)(mes)_2]_2$ (colourless)	tr P-1 4	11.068(5) 12.470(3) 17.654(5)	90.97(2) 107.77(3) 112.23(3)	AlC_2S_2 AlC_2S_2	C 1.970(6,12) μS 2.394(2,23) C 1.967(2,3) μS 2.397(2,2)	not given 94.4(1) 85.6(1) not given 92.8(1) 87.2	C,C 118.7(2) C,C 121.7(3)	82

$Cs[Al_2(\mu\text{-}N_3)(Me)_6]\cdot$ $\cdot$p-xylene (colourless)	m C2/m 4	19.143(6) 16.227(6) 10.392(5)	114.06(2)	AlC_3N AlC_3N	μN 1.98(1) C 2.00(1,2) μN 1.98(1) C 2.01(1,2)	not given 131.4(9) not given 132.5(9)	C,C 112.4(6,11.7) C,N 106.2(6,2.4) C,C 115.6(6,9.0) C,N 102.2(6,5.8)	83
$[Al_2\{\mu\text{-}(Me_2NCH_2)_2\}\cdot$ $\cdot\{MeC(Me_3Si)CH\}_4Cl_2]$ (colourless)	tr P-1 2	9.811(3) 15.252(8) 15.166(9)	108.39(3) 91.14(3) 95.95(3)	AlC_2NCl AlC_2NCl	C 1.962(5,1) N 2.045(3) Cl 2.163(2) C 1.959(5,1) N 2.049(4) Cl 2.167(2)		C,C 97.2(2)[c] C,N 109.4(2,3.4) C,Cl 121.5(2,2.1) N,Cl 97.7(1) C,C 96.8(2)[c] C,N110.5(2,2.0) C,Cl120.4(2,1.4) N,Cl 98.6(2)	84
$Al_2(\mu\text{-dc-[18]-C-6})(Me)_6$ (colourless)	m $P2_1/a$ 4	16.423(7) 9.812(5) 20.935(8	107.41(5)	AlC_3O AlC_3O	C 2.00-2.01 O 1.960(8) C 2.00-2.01 O 1.936(8)		not given	85
$Al_2(\mu\text{-}\eta^3\text{-}CS_2)\cdot$ $\cdot\{(Me_3Si)_2CH\}_4$ (colourless) at 173 K	tr P-1 4	15.291(7) 17.130(8) 20.18(1)	99.47(4) 97.10(4) 106.54(4)	AlC_3 AlC_2S_2 AlC_3 AlC_2S_2	C 2.05(1) C 1.945(10,3) C 1.965(10,15) S 2.390(5) C 2.05(1) C 1.95(1,1) C 1.995(10,5) S 2.393(5,1)		C,C 120.0(5,7.2) C,C 118.9(5) S,S 73.7(2)[d] C,S 114.0(4,1.4) C,C 120.0(5,9.0) C,C 121.6(5) S,S 74.2(2)[d] C,S 113.0(4,4.0)	86
$K[Al_2(\mu\text{-}\eta^2\text{-}NO_3)(Me)_6]$ (colourless)	m $P2_1/m$ 8	7.975(8) 24.512(12) 14.601(10)	100.55(8)	AlC_3O AlC_3O	C 1.99(8,13) O 1.98(5,3) C 1.97(8,6) O 2.006(6,2)		C,C 102(3.9) C,O 116(3,10) C,C 101(3,5) C,O 117(3,8)	87
$[Al(\mu_3\text{-}O)(Bu^t)_6$ (colourless)	tr P-1 2	11.847(3) 12.208(2) 13.711(4)	103.83(2) 109.02(2) 92.37(2)	AlO_3C AlO_3C	μ_3O 1.783(6,23) 1.891(4,14) Bu^t,C 1.931(10,29) μ_3O 1.795(6,14) 1.888(4,13) Bu^t,C 1.949(9,12)	 93.0(3,1.1) 127.4(3,3) 92.2(2,1.9) 127.6(2,1.1)	O,O 86.5(2,1.3) 112.2(3,1) O,C 120.7(3,3.7) O,O 87.6(2,1.0) 112.4(2,8) 120.3(3,1.9)	88

Footnotes: a) When more than one chemically equivalent distance or angle is present, the mean value is tabulated. The first number in parenthesis is e. s. d., and the second is maximum deviation from the mean. b) The chemical identity of the coordinated atom or ligand is specified in these columns. c) Six – membered metallocyclic ring d) Four – membered metallocyclic ring.

The distorted edge – shared tetrahedral arrangement is the most common form of bridging observed in this series. The bridging links found are: two OL ligands [78, 79], H plus C donor ligand [80], two NL ligands [81], two SL ligands [82]. In $[Al(\mu\text{-mto})(Bu^{i})_2]_2$ [78] and $[Al(\mu\text{-mtp})(Me)_2]_2$ [79], a pair of μ- O (mto) ligands bring two $Al(Bu^{i})_2$ units within the Al-Al bond distances of 2.815(3) Å (dimer 1) and 2.843(6) Å (dimer 2). The mean Al-O-Al and μO-Al-μO bond angles are 99.8 and 80.3° in the former and 101.2 and 78.8° in the latter one. Each Al(III) atom is tetrahedrally coordinated (AlO_2C_2) with a different degree of distortion.

In $[Al(\mu\text{-mtp})(Me)_2]_2$ [79], two $Al(Me)_2$ units are connected by a pair of μ-O mtp ligands. The Al-Al bond distances, mean Al-O-Al and μO-Al-μO bond angles (dimer 1 vs dimer 2) are 2.977(3), 102.7 and 77.3° vs 2.985(3), 102.3 and 97.6°.

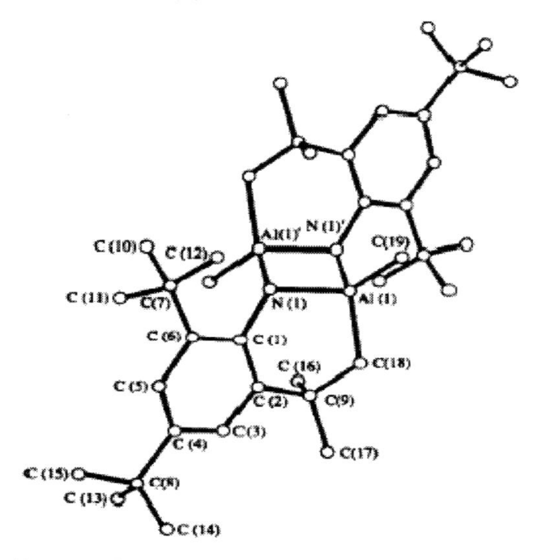

Figure 5. Structure of $[Al(\mu\text{-bbp})(Me)_2]_2$ [81].

In another dimer [80], two $Al\{(Me_3Si)CH\}_2$ units are held together by μ-H and μ-CH_2 and bring two units within Al-Al bond distances of 2.744(1) Å (dimer 1) and 2.748(2) Å (dimer 2) Å. Each Al(III) atom is tetrahedrally coordinated (AlC_3H). The sum of four (Al-C(x3) plus Al-H) bond distances are 7.885 and 7.868 Å. The mean Al-X-Al and μX-Al-μX bond angles (dimer 1 vs dimer 2) are 92.2 and 86.6° vs 92.2 and 87.8°. The remaining "four tetrahedral" angles range from 100.8 – 124.4° vs 102.9 – 120.1°. This indicates that the former dimer is somewhat more distorted than the dimer 2.

In $[Al(\mu\text{-bbp})(Me)_2]_2$ [81] a pair of N-donor bbp ligands serve as a bridges between two $AlMe_2$ units. Each bbp ligand in addition bonding Al(III) atom by C – atom (Figure 5). The mean sum of four (Al-N (x2) plus Al-C (x2)) bond distances and the range of six "tetrahedral angles" (dimer 1 vs dimer 2) are: 7.83 Å, and 88.1 – 120.9° vs 7.82 Å and 87.8 – 121.3°. This indicates that the dimer 2 is somewhat more distorted than the dimer 1.

In colourless $[Al(\mu\text{-PhS})(mes)_2]_2$ [82] two $Al(mes)_2$ units are double bridged by a pair of PhS ligands and completed a tetrahedral arrangement around each Al(III) atom (AlC_2S_2). Noticeable, while the mean sum of four (Al-C (x2) plus Al-S (x2)) bond distances are equal (8.728 Å), the μS-Al-μS, Al-S-Al and C-Al-C (C-Al-S are not available) bond angles range from 85.6(1) to 118.7(2)° (dimer 1) and 87.2(1) – 121.7(2)° (dimer 2).

In $Cs[Al_2(\mu\text{-}N_3)(Me)_6]$·p-xylene [83] two $Al(Me)_3$ units are bridged by a single N-donor atom of N_3 ligand. Each Al(III) atom is tetrahedrally coordinated (AlC_3N). The "tetrahedral angles" range from 103.8(6) to 131.4(9)° (dimer 1) and 96.4 – 132.5(9)° (dimer 2). The mean sum of four (Al – C (x3) plus Al-N (x1)) bond distances are 7.98 and 8.01 Å, respectively. The dimer 2 is somewhat more distorted, than the dimer 1.

In $[Al(\mu\text{-}Me_2NCH_2)_2\{MeC(Me_3Si)CH\}_2Cl]_2$ [84] bidentate N-donor $Me_2NCH_2CH_2NMe_2$ ligand with four atoms serve as bridge in the manner (Al-N-C-C-N-Al) between two $Al\{\eta^2$-$MeC(Me_3Si)CHCH(SiMe_3)CMe\}$-(Cl) units. Each Al(III) atom has a tetrahedral geometry (AlC_2NCl). The mean Al-C, Al-N and Al-Cl bond distances (dimer 1 vs dimer 2) are: 1.962, 2.045, and 2.163 Å vs 1.959, 2.049, and 2.167 Å. The six "tetrahedral angles" range from 97.2(2) to 124.6(2)° and from 96.8(2) to 121.8(2)°, respectively.

In $Al_2(\mu\text{-dc-[18]-C-6})(Me)_6$ [85] the dibenzo-18-crown-6 ligand serve as bridge between two $Al(Me)_3$ units. Each Al(III) atom is tetrahedrally coordinated (AlC_3O). The sum of four (Al-C (x3) plus Al-O (x1)) bond distances are 7.950 Å (dimer 1) and 7.975 Å (dimer 2). This indicates that the dimer 1 is somewhat more crowded than the dimer 2.

Unique bridge has been found in $[Al_2(\mu\text{-}\eta^3\text{-}CS_2)\{(Me_3Si)_2CH\}_4]$ [86]. Two $Al\{(Me_3Si)_2CH\}_2$ units are held together by CS_2 ligand, which serve as bridge, by C-donor atom to one $Al\{(Me_3Si)_2CH\}_2$ unit and by two S-donor atoms to another one unit. Two Al(III) atoms are differ from each other, one has a trigonal planar coordination (AlC_3) and the other one a tetrahedral (AlC_2S_2). Two crystallographically independent dimers are differing by degree of distortion.

The sum of three (Al-C (x3)) and four (Al-C (x2) plus (Al-S (x2)) bond distances (dimer 1 vs dimer 2) are 5.94 and 8.71 Å vs 5.95 and 8.78 Å. The deviation of the L-Al-L bond angles from the ideal value of trigonal planar arrangement (120° AlC_3,) is 7.2° (dimer 1) and 9.0° (dimer 2). Five of the six "tetrahedral angles" (AlC_2S_2) fall in the narrow range 112.6(4) – 118.9(5)° (dimer 1) and 109.0(4) – 121.6(5)° (dimer 2), while the angle which is part of the four – membered metallocycle (-SCS-) is significantly smaller (73.7(2) and 74.2(2)°), respectively. This indicates that the dimer 2 is somewhat more distorted than the dimer 1.

In $K[Al_2(\mu\text{-}\eta^2\text{-}NO_3)(Me)_6]$ [87] a pair of $Al(Me)_3$ units is connected by homo – non – chelating bidentate NO_3^- group. Each Al(III) atom is tetrahedrally coordinated (AlC_3O). The sum of four (Al-C (x3) plus Al-O (x1) bond distances are 7.981 Å (dimer 1) and 7.95 Å (dimer 2). The six "tetrahedral angles" fall in the narrow range 96 - 125° and 93 – 126°, respectively.

Structure of $[Al(\mu_3\text{-}O)(Bu^t)]_6$ [88] contains Al_6O_6 core which can be described as a hexagonal prism with altering Al and O atoms (Figure 6). The Al-O bonds in the six – membered rings (mean 1.783 Å (hexamer 1) and 1.795 Å (hexamer 2), are significant shorter than the transverse bonds joining the rings (1.891 Å and 1.888 Å, respectively. Each Al(III) atom has a tetrahedral arrangement (AlO_3C) (Table 6).

Inspection of the data in Tables 5 and 6 reveals that the distortion isomers are found in the following crystal clases: tetragonal (x1) < orthorhombic (x2) < monoclinic (x8) < triclinic (x12). The mean values of sum of Al-L (x n) bond distances elongate with the sum of the covalent radii of donor atoms (x n) as well as with the increases number of the coordination number (less vs more distorted units) in the order: 4.042 Å (only mean value was published) (AlC_2, 1.54 Å) < 5.94 vs 5.95 Å (AlC_3, 2.31 Å) < 7.496 vs 7.52 Å (AlO_3C, 2.96 Å) < 7.65 vs 7.655 Å (AlO_2C_2, 3.00 Å) < 7.82 vs 7.83 Å (AlN_2C_2, 3.04 Å) < 7.93 vs 7.96 (AlC_3O, 3.04 Å) < 7.98 vs 8.01 Å (AlC_3N, 3.06 Å) < 8.72 vs 8.75 Å(AlC_2S_2, 3.58 Å).

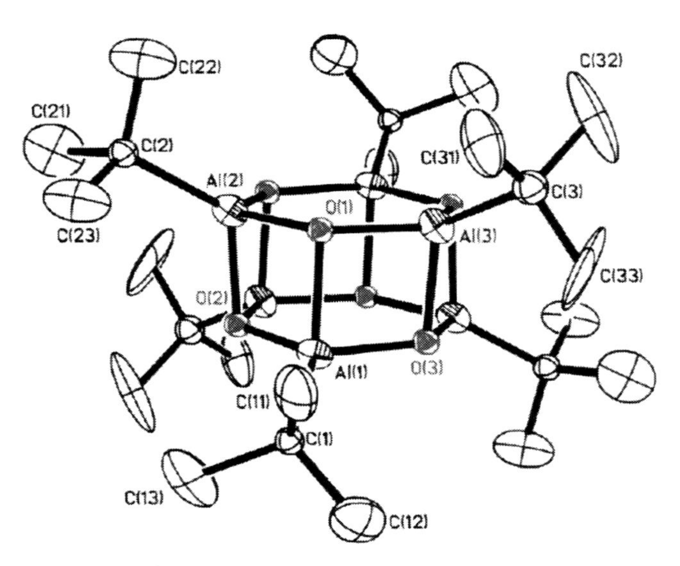

Figure 6. Structure of $[Al(\mu_3\text{-}O)(Bu^t)]_6$ [88].

There are four dimeric examples, in which two Al(III) atoms are direct bond to each other with the following Al-Al bond distances (dimer 1 vs dimer 2): 2.470 Å (only mean value was published) [77], 2.815 vs 2.843 Å [78], 2.977 vs 2.985 Å [79] and 2.744 vs 2.748 Å [80].

3. Cis - Trans Isomerism

Crystallographic and structural data for these isomers are gathered in Table 7. Both *cis* and *trans* isomes are found within the same crystal of $[Al(\mu\text{-}Pr^iNH)(Me)_2]_2$ [89] The two $Al(Me)_2$ units are connected by two Pr^iNH ligands, which serve as bridges. Each Al(III) atom is tetrahedrally coordinated (AlN_2C_2).

The Al-Al bond distances are 2.789 Å (in *trans*) and 2.800 Å (in *cis*). The sum of four (Al-N (x2) plus the Al-Cl (x2)) bond distances are 7.85 Å (in *trans*) and 7.81 Å (in *cis*).

The molecule $[Al(\mu\text{-}MeNH)(Me)_2]_3$ [90], exists in two isomeric forms, *cis* and *trans*.

The *cis* form even contains two crystallographically independent molecules, differing mostly by degree of distortion, within the same crystal.

The rhombohedrical *cis* – isomer crystallizes in space group R3. The molecules have crystallographic symmetry C_3 with six – membered $(AlN)_3$ rings of alternate aluminium and nitrogen atoms, all methyl substituents on nitrogen atom are equatorial. Each Al(III) atom is tetrahedrally coordinated (AlN_2C_2).

The sum of four (Al-N (x2) plus Al-C (x2)) bond distances in the both independent molecules is equal (7.826 Å). The Al . . . Al separations of 3.403 and 3.394 Å, ruled out a direct metal – metal bond. The six "tetrahedral angles" range in the narrow range from 101.9(5) to 117.6(6)° and 102.4(4) to 116.2(6)°, respectively.

The monoclinic *trans* isomer [90] crystalliyes in space group C2/c. The molecules have similar molecular dimensions as those of the rhombohedral *cis* – isomer, but the configuration at one of the nitrogen atom is reversed, and the conformation of the $(AlN)_3$ rings is of the skew – boat types.

Table 7. Crystallographic and Structural Data for Di- and Oligomeric Organoaluminium Compounds – cis – trans Isomers[a]

Compound (colour)	Cryst. cl. Cryst. gr. Z	a [Å] b [Å] c [Å]	α [°] β [°] γ [°]	Chromo-phore	Al – L [Å]	Al – Al [Å] Al – L – Al [°] μL – Al – μAl [°]	L – Al – L [°]	Ref.
$[Al(\mu\text{-}Pr^iNH)(Me)_2]_2$ (colourless)	tr P-1 3	8.586(3) 11.366(5) 13.440(5)	107.48(3) 90.52(2) 99.21(3)	AlN_2C_2 (trans)	μN[b] 1.959(85,7) C 1.964(9,2)	2.789 90.0(2,80 90.0	C,C[b] 120.6(3) C,N 110.7(3,2.6)	89
				AlN_2C_2 (cis)	μN 1.949(5,13) C 1.956(8,16)	2.800 91.9(2,4) 87.2(2,3)	C,C not given C,N 110.8(2,1.6)	
cis-$[Al(\mu\text{-}NHMe)\cdot(Me)_2]_3$ (colourless)	rh R3 2	9.983(4)	104.39(3)	AlN_2C_2	μN 1.941(12,10) C 1.972(12,3)	3.403(10) 122.4(5) 101.9(5)	C,C 117.6(6) C,N 108.3 (5,1.8)	90
				AlN_2C_2	μN 1.939(9,9) C 1.974(14,11)	3.394(10) 122.1(4) 102.4(4)	C,C 116.2(6) C,N 109.6(5,1.1)	
trans-$[Al(\mu\text{-}NHMe)\cdot(Me)_2]_3$ (colourless)	m C2/c 4	11.897(8) 15.904(12) 9.778(7)	107.50(3)	AlN_2C_2	μN 1.901(12,68) C 1.982(20,23)	3.331(8,18) 124.9(9,5.0) 100.9(6,8)	C,C 117.5(9,2) C,N 108.9(7,1.8)	90

Footnotes: a) When more than one chemically equivalent distance or angle is present, the mean value is tabulated. The first number in parenthesis is the e. s. d. , and the second is the maximum deviation from the mean. b) The chemical identity of the coordinated atom or ligand is specified in these columns.

Table 8. Crystallographic and structural data for di- and polymeric heterometallic aluminium compounds – distortion isomers[a]

Compound (colour)	Cryst. cl. Cryst. gr. Z	a [Å] b [Å] c [Å]	α [°] β [°] γ [°]	Chromo-phore	M – L [Å]	Al – M [Å] Al – L – M [°] L – M – L [°]	L – M – L [°]	Ref.
A: Dimers								
$Cp_2Ti(\mu\text{-}H)_2{\cdot}Al(Et_2O)Cl_2$ (not given)	or Pbca 16	12.322(2) 15.090(3) 37.941(7)		AlH_2Cl_2O	μH[b] 1.85(7,15) Cl 2.161(5,12) O 1.956(7)	2.747(4) 99.4(2) 78(3)	Cl,Cl[b] 110.1(2) H,Cl 103(2,8) 136(2) H,O 80(2), 158(2) Cl,O 95.3(3,9)	91
				$TiC_{10}H_2$	μH 1.78(8,6) cp,C 2.34(1,3)		H,H 80(3)	
				AlH_2Cl_2O	μH 1.64(5,18) Cl 2.152(5,9) O 1.979(8)	2.755(4) 105(3) 78(3)	Cl,Cl 109.1(2) H,Cl 98(2,2) 126(2,3) H,O 80(2), 158(2) Cl,O 94.7(3,1.0)	
				$TiC_{10}H_2$	μH 1.81 cpC 2.33(1,5)		H,H 90(2)	
$(\eta^2\text{-acac})Cu(\mu\text{-}Me_3SiO)_2{\cdot}Al(Me_3SiO)_2$ (blue violet)	tr P-1 2	11.107(6) 14.956(5) 19.739(6)	100.77(5) 94.41(5) 107.0(3)	AlO_4	μO 1.815(9,4) O 1.711(9,3)	2.778(4) 94.3(4,1) 90.2(4)	μO,O 112.5(4,4) O,O 114.3(4)	92
				CuO_4	μO 1.973(8,0) O 1.894(9,3)		μO,μO 81.3(3) μO,O 91.4(4,2) O,O 95.8(4)	
				AlO_4	μO1.802(9,3) O 1.683(9,5)	2.784(4) 94.2(4,2) 91.1(4)	μO,O 111.9(5,1.2) O,O 115.7(5)	
				CuO_4	μO 1.994(8,4) O 1.891(8,2)		μO,μO 80.4(3) μO,O 92.6(4,7) O,O 94.5(4)	
$cp_2Zr(\mu\text{-}H)(\mu\text{-}C_6H_4){\cdot}Al(Bu^i)_2$ (colourless)	m $P2_1/n$ 8	16.749(3) 13.833(1) 20.178(2)	 90.74(1)	AlC_3H	μH 1.85 μC 2.09(1) Bu^i,C 2.01(2,7)	not given 101.3,79.1	μH,μC 95.6 μC,C 108.0(5,8) C,C 117.9(7)	93
				$ZrC_{12}H$	μH 1.90 μC 2.436(9) C 2.166(9) cp,C 2.47-2.54		 μH,μC 83.7 μH,C 117.8 μC,C 33.9(4)[c]	
				AlC_3H	μH 1.80	not given	μH,μC 92.8	

				$ZrC_{12}H$	μC 2.08 Bu^i,C 1.99(1,6) μH 1.81 μC 2.425(9) C 2.1690(9) cp,C 2.47-2.54	105.9,79.1	μC,C 109.5(5,1.2) C,C 116.1(5) μH,μC 82.1 μH,C 116.1 μC,C 34.1[c]	
cpFe{μ-$C_5H_4CH_2$-(Me_2)N}$AlMe_3$ (yellow)	m $P2_1/a$ 8	12.969(2) 13.103(2) 20.411(3)	91.45(2)	AlC_3N FeC_{10} AlC_3N FeC_{10}	N 2.062(5) MeC 1.987(7,4) not given N 2.053(5) MeC1.960(7,13) C not given		C, C 113.5(3,7) C,N 105.0(2,8) not given C,C 114.2(3,2.2) C,N 104.1(2,4) not given	94
B: Trimers								
[(thf)$_2$Mg{(μ-OPh)$_2$·Al(OPh)$_2$}$_2$] (colourless)	m $P2_1$ 4	9.334(8) 32.96(2) 16.853(10)	95.57(6)	AlO_4 MgO_6 AlO_4 MgO_6	μO 1.75(2,3) O 1.70(2,4) μO 2.115(20,35) O 2.11(2,2) μO 1.79(2,2) O 1.70(2,2) μO 2.10(2,5) O 2.07(2,2)	2.95(2,2) 99.0(7,2.8) 89.3(8,1.2) 3.04(2,5) 100.4(7,1.6) 86.6(8,1.4)	μO,O 112.2(9,5.8) O,O 115.8(9,1.1) μO,μO 71.2(6,4) 99.3(6,4.3) μO,O 92.7(8,6.1) O,O 87.2(7) μO,O 113.1(8,4.4) O,O 114.8(10,7) μO,μO 71.4(6,0) 101.1(6,1) μO,O 92.5(7,3.6) O,O 89.2(7)	95
W{μ-NBu^t)$_2$·$AlCl_2$}$_2$ (purple blue)	or $P2_12_12_1$ 4	10.660(1) 15.387(3) 16.990(6)		AlN_2Cl_2 WN_4 AlN_2Cl_2 WN_4	μN 1.915(14,22) Cl 2.098(9,4) μN 1.860(13,3) μN 1.943(13,22) Cl 2.100(9,14) μN 1.852(12,5)	not given 90.2(6,6) not given 89.6(5,4)	N,N 88.016 N,N 91.0(6) 119.4(5) N,N 87.6(5) N,N 93.5(2) 123.2(5)	96

Table 8. (Continued)

Compound (colour)	Cryst. cl. Cryst. gr. Z	a [Å] b [Å] c [Å]	α [°] β [°] γ [°]	Chromo- phore	M – L [Å]	Al – M [Å] Al – L – M [°] L – M – L [°]	L –M – L [°]	Ref.
$(C_6H_6)Ti\{(\mu\text{-}I)_2{\cdot}AlI_2\}_2$ (black)	m $P2_1/b$ 4	13.964(3) 14.010(3) 11.926(2)	 95.47(2)	AlI_4 TiC_6I_4 AlI_4 TiC_6I_4	μI 2.567(8,1) I 2.489(8,2) μI 2.972(4,27) C 2.51 μI 2.569(7,8) I 2.494(7,2) μI 2.998(4,12) C 2.51	not given 87.1(2,6) not given 87.5(2,4)	μI,μI 98.2(2) I,I 115.8(3) μI,μI 81.5(2) 130.3(1) μI,μI 96.8(2) I,I 113.8(3) μI,μI 79.7(2) 134.1(1)	97
C: Tetramers								
$(\eta^6\text{-m-}Me_2C_6H_4)Sm{\cdot}\{(\mu\text{-}Cl)_2AlCl_2\}_3$ (yellow)	m $P2_1/n$ 8	18.36(5) 16.37(4) 20.31(6)	 114.11(2)	$AlCl_4$ SmC_6Cl_6 $AlCl_4$ SmC_6Cl_6	μCl 2.196(8,30) Cl not given μCl 2.826(8,36) η^6,C 2.90(3,4) μCl 2.175(1,15) Cl not given μCl 2.836(4,4.2) η^6,C 2.88(2,15)	 95.5(3,2.7) 95.8(1,1.9)	μCl,μCl 97.4(3,2.8) Cl,Cl 117.5(1) μCl,μCl 74.9(2,6.9) 138.1(2,2.1) μCl,μCl 97.5(1,1.9) Cl,Cl 117.6(1) μCl,μCl 74.9(2,7.1) 138.0(2,5.2)	98
$[(\eta^2\text{-dmpe})_2Mo(\mu\text{-}H)\text{-}AlH_4]_2$ (yellow)	tr P-1 2	12.783(8) 20.283(11) 9.572(5)	98.34(5) 10.358(5) 10.805(5)	AlH_5 MoP_4H_2 AlH_5 MoP_4H_2	μH 1.75(27,1) H not given μH not given η^2,P 2.409(7,19) μH 1.65(26,1) H not given μH not given η^2,P 2.410(7,12)	2.565(9) 2.644(14)[d] 98.3(14) 2.578(9) 2.661(15)[d] 107.6(13)	μH,μH 81.7(14) P,P 80.0(2,3)[e] 96.0(2,9),147.0(2) 170.0(2) μH,μH 75.2(12) P,P 79.7(3,4)[e] 96.2(3,1.2),148.3(2) 168.7(2)	99

[$Me_2Al(\mu$- $OBu^t)_2\cdot Mg(\mu$-$OBu^t)_2]_2$ (colourless)	m $P2_1/c$ 4	15.447(5) 14.578(4) 18.822(6)	110.04(3)	AlO_2C_2 MgO_4 AlO_2C_2 MgO_4	μO 1.855(8,0) MeC 1.96(2,0) μO 1.954(7,1) 2.043(8,0) μO 1.870(9,0) MeC 1.95(2,0) μO 1.966(8,14) 2.035(8,0)	2.888(5) 2.886(7) 2.895(6) 2.901(8)	μO,μO 89.2(3) C,C 116.0(7) μO,C 111.7(5) μO,μO 81.9(3,1.9) 125.1(3,1.9) μO,μO 88.1(4) C,C 114.9(8) μO,C 113.0(6) μO,μO 82.0(3,2.9) 124.5(4,2.0)	100
{$Li_2[(Me_3SiCH_2)_3Al]_2\cdot(\mu_4$- $CH_2)$} (colourless)	tg P-4 4	15.6666(8) 17.029(2)		AlC_4 LiC_3 AlC_4 LiC_3	μ_4C 2.060(5) μC 2.051(8,2) C 1.985(8) μ_4C 2.13(2) μC 2.18(1,1) μ_4C 2.071(5) μC 2.061(8,11) C 1.995(8) μ_4C 2.10(1) μC 2.20(1,6)	2.65(1,1) 137.5(6) 77.3(4,1) 3.51(2) 110.7(8) 2.62(1,3) 135.1(6) 75.6(7,1) 3.50(2) 112.7(7)	μ_4C,μC 105.3(3) μ_4C, C 112.1(4) μC,μC 108.1(3) μC,C 112.8(3,1.8) μ_4C,μC 98.6(5,2) μC,μC 140.5(9) μ_4C,μC 105.7(3,3) μ_4C,C 113.8(4) μC,μC 107.6(3) μC,C 111.8(3,1.0) μ_4C,μC 99.7(5,1.7(7) μC,μC 135.7(7)	101
{$[cp_2Sm(\mu_3$-$H)]_2\cdot[(\mu$- $H)_2AlH(NEt_3)]_2$} (colourless)	tr P-1 4	11.012(3) 11.911(3) 14.314(4)	88.76(2) 78.65(2) 96.17(2)	AlH_4N $SmC_{10}H_4$ AlH_4N $SmC_{10}H_4$	μ_3H 1.86 μH 1.69(x2) H 1.59 Et_3N 2.11(1) μ_3H 2.41(x2) μH 1.76(x2) cp C 2.464(-,6) μ_3H not given Et_3N 2.11(1) μ_3H not given cp,C 2.465(-,3)	3.276(5,8) 5.113[d] 4.096(4)[f] 3.280(5,3) 5.114[d] 4.068(5)[f]	H,H 118.6(-,4.4) H,N 85.9 101.8(-,6) H,H 143.9 cp,cp 123.6 H,N 102.4 cp,cp 124.4	102
D: Polymers								
α-$BaAlF_5$ (colourless)	or $P2_12_12_1$ 4	13.710 5.604 4.930		AlF_6 BaF_x	F 1.785(6,24) 1.585(6,10) 2.624-3.815(6)		F,F 90.0(2,3.1)	103

Table 8. (Continued)

Compound (colour)	Cryst. cl. Cryst. gr. Z	a [Å] b [Å] c [Å]	α [°] β [°] γ [°]	Chromo- phore	M – L [Å]	Al – M [Å] Al – L – M [°] L – M – L [°]	L –M – L [°]	Ref.
β-$BaAlF_5$ (colourless)	m $P2_1/m$ 8	5.148 19.556 7.553	 92.4(3)	AlF_6				103
γ-$BaAlF_5$ (colourless)	m $P2_1/m$ 4	5.528 9.724 7.366	 90.8(9)	AlF_6				103
$KAlF_4$ (colourless) by neutron diffraction	tg P4/mbm 2	5.043(2) 6.164(6)		AlF_6 KF_x	F 1.752(1) 1.817(1) F 2.848(1)		F,F not given	104
$KAlF_4$ (colourless) at 4 K	m $P2_1/m$ 4	7.340 7.237 6.407	 106.8	AlF_6 AlF_6 KF_6	F 1.76 1.87(-,3) F 1.76 1.82(2) F not given		F,F 90.0(-,2.0) F,F 90.0(-,1.7)	105

Footnotes: a) When more than one chemically equivalent distance or angle is present, the mean value is tabulated. The first number in parenthesis is the e. s. d. , and the second is the maximum deviation from the mean. b) The chemical identity of the coordinated atom or ligand is specified in these columns. c) Four – membered metallocyclic ring. d) Al – Al distance e) Five – membered metallocyclic ring f) Sm – Sm distance.

Each Al(III) atom is tetrahedrally coordinated (AlN_2C_2). The sum of four (Al-N(x2) plus Al-C (x2)) bond distance of 7.766 Å is about 0.060 Å smaller than that found in the *cis* – isomer. The Al . . . Al separation is also somewhat shorter with the value of 3.331 Å. The six "tetrahedral angles" range from 100.1(6) to 117.7(9)°.

IV. Isomers of Heterometallic Aluminium Compounds

In our review article [5] we analysed and classified almost three hundred heterometallic aluminium compounds and about 5.6 % of these compounds exist as isomers and are summarized in this chapter.

1. Distortion Isomerism

In the heterometallic aluminium chemistry, only distortion isomerism has been found. The crystallographic and structural data for these isomers are gathered in Table 8. There are sixteen derivatives (heterobi- (x4), heterotri- (x3), heterotetra- (x5) and heteropolymeric (x5)) derivatives.

1.1. Heterodimeric Compounds

Orthorhombic $cp_2Ti(\mu\text{-}H)_2Al(Et_2O)Cl_2$ [91] which contains two crystallographically independent dimers, the $Al(Et_2O)Cl_2$ and cp_2Ti units are connected by two hydrogen atoms, with Al-Ti bond distances of 2.747(4) Å (dimer 1) and 2.755(4) Å (dimer 2). The mean Al-H-Ti bond angles are 99.4(2) and 105(3)°, respectively. Each Al(III) atom is penta – coordinated (AlH_2Cl_2O). The mean Al-μH, Al-Cl and Al-O bond distances (dimer 1 vs dimer 2) are 1.85, 2.161 and 1.956 Å vs 1.64, 2.152 and 1.979 Å. Each titanium atom is sandwiched with additional two hydrogen atoms which are deposited in the respective sandwich ($TiC_{10}H_2$).

In blue violet triclinic derivative [92] the $Al(Me_3SiO)_2$ and Cu(η^2-acac) moieties are bridged by two O-donor Me_3SiO ligands. There are two crystallographically independent dimers. While each Al(III) atom is tetrahedrally coordinated (AlO_4), each Cu(II) atom is square – planar coordinated (CuO_4). The Al-Cu bond distances are 2.778(4) and 2.784(4) Å, respectively. The sum of four Al-O and Cu-O bond distances (dimer 1 vs dimer 2) are 7.052 and 7.734 Å vs 6.970 and 7.770 Å.The six "tetrahedral (O-Al-O) angles)" range from 90.2(4) to 114.3(4)° in dimer 1 and 91.1(4) – 115.7(5)° in dimer 2. The dimer 1 is somewhat more crowded as well as more distorted than the dimer 1.

Colourless monoclinic AlZr dimer [93] also contains two crystallographically independent molecules. Two non-equivalent units, $Al(Bu^i)_2$ and $Zrcp_2$, are connected by hydrogen atom plus a C-donor atom of C_6H_4 ligand. The Al-H-Zr and Al-C-Zr bond angles are 101.3 and 79.1° (dimer 1) and 105.9° and 79.1° (dimer 2). Each Al(III) atom is tetreahedrally coordinated (AlC_3H) with the values of the sum of four (Al-C (x3) plus Al-H) bond distances of 7.96 and 7.86 Å, respectively. The tetrahedral angles are in the range from 95.6 to 117.9° (dimer 1) and 92.8 – 116.1° (dimer 2). The dimer 2 is somewhat more distorted as well as more crowded, than the dimer 1.

In yellow monoclinic AlFe dimer [94] the AlMe and Fecp units are connected by $C_5H_4CH_2NMe_2$ ligand. This ligand is coordinated to Al(III) atom through its N-donor atom and five C-donor atoms are bond to the iron atom (Figure 7). Each Al(III) atom is tetrahedrally coordinated (AlC_3N) and each iron atom is sandwiched (FeC_{10}). The sum of four (Al-C (x3) plus Al-N) bond distances are 8.023 Å (dimer 1) and 7.933 Å (dimer 2). The six "tetrahedral angles" are in the range 104.2(2) – 114.2(3)° and 103.7(2) – 116.4(3)°, respectively. The tetrahedral in dimer 2 is somewhat more distorted as well as more crowded, than dimer 1.

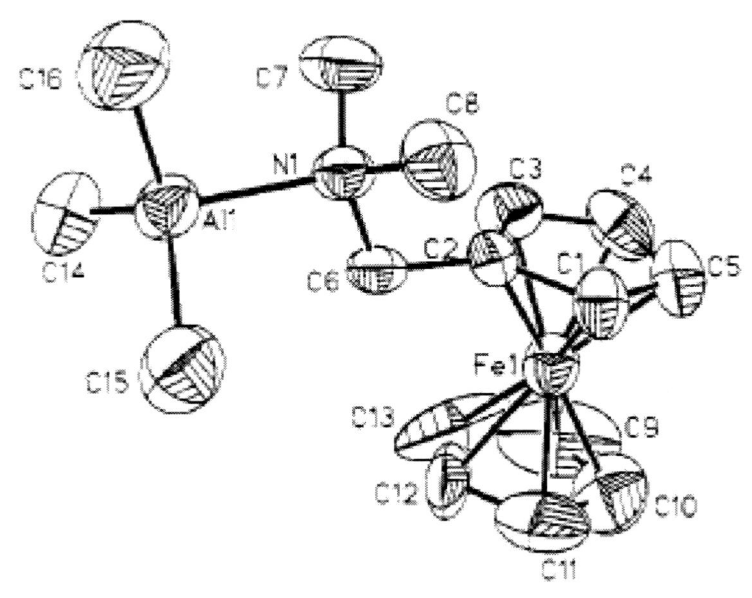

Figure 7. Structure of [cpFe(μ-$C_5H_4CH_2NMe_2$)$AlMe_3$] [94].

1.2. Heterotrimeric Compounds

In colourless monoclinic Al_2Mg trimer [95] which contains two crystallographically independent molecules a "central" metal atom is magnesium and edge shares with a tetrahedral aluminium unit on each side (Figure 8). The bridging atoms are two O-donor OPh ligands. The mean Al-Mg distances, Al-O-Mg and μO-Al-μO bond angles (trimer 1 vs trimer 2) are: 2.95 Å, 99.0 and 89.3° vs 3.04 Å, 100.4 and 86.6°. While each Al(III) atom is tetrahedrally coordinated (AlO_4), The Mg(II) atom is an octahedrally coordinated (MgO_6) (Table 8C).

In purple blue orthorhombic Al_2W trimer [96], which also contains two crystallographically independent molecules, a "central" W(VI) atom edge shares with a tetrahedral aluminium unit on each side.The bridged atoms are two N-donor atoms of NBu^t ligands. Each W(VI) as well as Al(III) atom are tetrahedrally coordinated (WN_4 and AlN_2Cl_2) (Table 8). The mean values of sum (Al-N (x2) plus Al-Cl (x2), and W-N (x4)) bond distances (trimer 1 vs trimer 2) are 8.03 and 7.44 Å vs 8.09 and 7.41 Å.

A black monoclinic $(C_6H_6)Ti\{(\mu\text{-}I)_2AlI_2\}_2$ [97] has a central "open sandwich" $Ti(III)(C_6H_6)$ unit edge – shared with two tetrahedral aluminium atoms in a similar manner to that outlined above. The mean Al-I bond distances terminal vs bridging are 2.489 and 2.567 Å (trimer 1) and 2.797 and 2.569 Å (trimer 2). Each Ti(III) atom has a TiC_6I_4 chromophore.

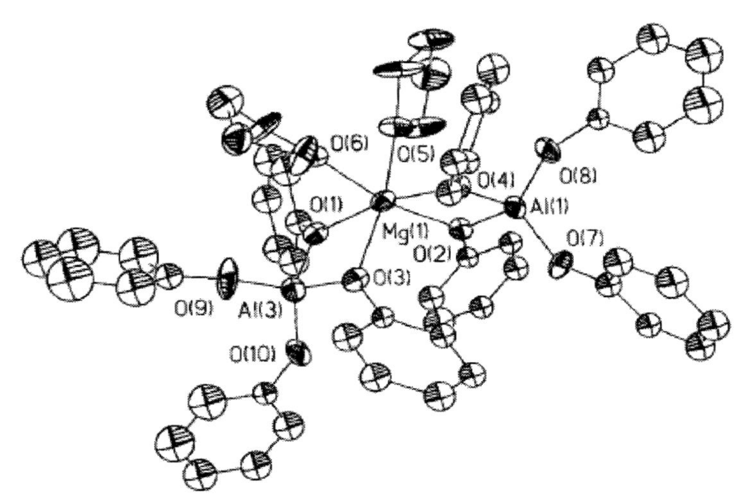

Figure 8. Structure of $[(thf)_2Mg\{(\mu\text{-}OPh)_2Al(OPh)_2\}_2]$ [95].

1.3. Heterotetrameric Compounds

There are five tetrameric derivatives [98–102] which contain two crystallographically independent molecules (Table 8C). Yellow monoclinic $(\eta^6\text{-m-}Me_2C_6H_4)Sm\{(\mu\text{-}Cl)_2AlCl_2\}_3$ [98], which contains two crystallographically independent molecules, has three tetrahedrally coordinated aluminium atoms ($AlCl_4$) edge – shared with a central Sm(III) atom (SmC_6H_6). The mean Al-Cl (bridge) bond distances are 2.196 Å (molecule 1) and 2.175 Å (molecule 2). Unfortunately, the Al-Cl (terminal) bond distances were not given in the original paper. The six "tetrahedral angles" are in the range 94.6(3) – 117.5(1)° and 95.6(1) – 117,6(1)°, respectively.

There is yellow triclinic $[(\eta^3\text{-dmpa})_2Mo(\mu\text{-}H)AlH_3]_2$ complex [99] in which the linkage between the Mo and Al atoms is accomplished via a double hydrogen bridge $Mo(H)_2Al$, with the Al centres linked by $Al(H)_2Al$ bridges. The coordination environment of each aluminium atom is a distorted trigonal bipyramid (AlH_5).

The mean Al-Mo and Al-Al bond distances (molecule 1 vs molecule 2) are 2.565 and 2.644 Å vs 2.578 and 2.661 Å. The mean Al-H (bridge) bond distances are 1.75 and 1.65 Å, respectively.

The colourless Al_2Mg_2 cluster [100] contains three orthogonal metal-(μ-O)-metal planes. The Al-Mg-Mg-Al backbone is almost linear with an Al-Mg-Mg angle of 178.44°. The Al and Mg atoms exist in distorted tetrahedral environments (AlO_2C_2 and MgO_4). The Al-Mg (mean) and Mg-Mg bond distances (molecule 1 vs molecule 2) are 2.888 and 2.886 Å vs 2.896 and 2.901 Å. The six "tetrahedral angles" (L-Al-L) are in the range 89.2(3) – 116.0(7)° and 88.1(4) – 114.9(8)°, respectively. The O-Mg-O bond angles are in much wide range: 80.0(3) – 126.1(3)° and 79.1(3) – 126.5(4)°, respectively.

Structure of $\{[cp_2Sm(\mu_3\text{-}H)]_2[(\mu\text{-}H)_2AlH(NEt_3)]_2\}$ [102] is shown in Figure 9. The cluster contains a heterocyclic eight member ring $(AlHSmH)_2$. There are also two μ_3-H atoms bridging three metal atoms (2 Sm and Al). The mean Al-Sm, Sm-Sm and Al-Al separations (molecule 1 vs molecule 2) are: 3.276, 4.096 and 5.113 Å vs 3.280, 4.069 and 5.114 Å. Each Al(III) atom has a square – pyramidal arrangement (AlH_4N) with N donor atom in an apical position. The bent sandwich cp_2Sm has a staggered conformation.

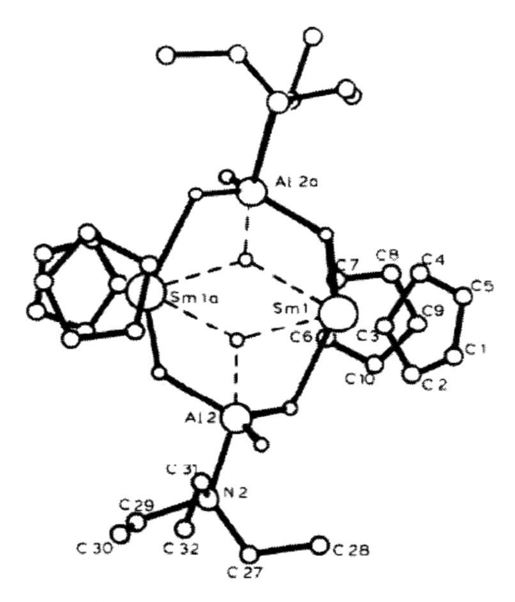

Figure 9. Structure of $\{[cp_2Sm(\mu_3\text{-}H)]_2[(\mu\text{-}H)_2AlH(NEt_3)]_2\}$ [102].

1.4. Heteropolymeric Compounds

Colourless $BaAlF_5$ exists in three isomeric forms, one orthorhombic and two monoclinic [103]. In the polymers aluminium and fluorine atoms are grouped in two different kinds of chains, both with the formula $(AlF_5)^{2-}$, linear and bent. Unfortunately, only for the α-isomer (orthorhombic) structural data are available (Table 8). Complex of the composition $KAlF_4$ exists in two isomeric forms, tetragonal [104] and monoclinic [105]. Their structures may be described as a sequence of $[AlF_{4/2}F_2]^-_\infty$ layers of AlF_6 octahedra sharing four corners in the (001) plane with K^+ ions between the layers. The mean Al-F bond distances in the tetragonal form range from 1.752(1) to 1.817(1) Å and 1.76 to 1.90 Å in the monoclinic form. Inspection of the data in the Table 8 reveals the variation in Al(III)-L mean bond length, the following trends are apparent

4 – coordinate: 1.698 Å (OL) < 1.972 Å (CL) < 2.057 Å (NL) < 2.100 Å< (Cl) < 2.497 Å (I);

1.767 Å (μ-H) < 1.815 (μ-OL) < 1.944 Å (μ-NL) < 2.088 (CL) < 2.567 Å (μ-I).

5 – coordinate: 1.59 Å (H) < 1.73 Å (μ-H) < 1.86 Å (μ_3-H) < 1.97 Å (OL) < 2.11 Å (NL) <

2.16 Å (Cl).

6 – coordinate: 1.76 Å (F) < 1.84 Å (μ-F).

The Al-M distances (molecule 1 vs molecule 2) elongated in the order: 2.565 vs 2.578 Å (Al-Mo) [99]< 2.620 vs 2.650 Å (Al-Li) [101]< 2.747 vs 2.755 Å (Al-Ti) [91] < 2.778 vs 2.784 Å (Al-Cu) [92] < 2.888 vs 2.895 Å (Al-Mg) [100] < 2.95 vs 3.04 Å (Al-Mg) [95] < 3.276 vs 3.280 Å (Al-Sm) [102].

The shortest Al-Al bond distances are 2.644 Å (molecule 1) and 2.661 Å (molecule 2) [99]. The mean values of sum of Al-L (xn) bond distances elongate with the sum of the covalent radii of donor atoms (xn) as well as with the increases number of the coordination number (more vs less crowded molecules) in the order: 6.935 vs 7.016 Å (O_4, 2.92 Å) < 7.630 vs 7.640 Å (O_2C_2, 3.00 Å) < 7.933 vs 8.023 Å (C_3N, 3.06 Å) < 8.147 vs 8.188 Å (C_4, 3.08 Å) < 9.563 vs 9.770 Å (H_2Cl_2O, 3.45 Å) < 10.112 vs 10.126 Å (I_4, 5.32 Å). There are two types of the chromophores AlO_4 and AlN_2Cl_2, which follow the order with the values: 9.935 vs 7.016 Å (O_4, 2.92 Å) and 7.630 vs 7.640 Å (O_2Cl_2, 3.00 Å).

CONCLUSION

An analysis of over one thousand and two hundred aluminium complexes shows that some of 8.75 % of them exists in isomeric forms. Aluminium complexes are for the most part colourless, but there are some coloured examples, some of which are due to ligand absorption and others due to charge – transfer bands. The lowest and highest coordination number found in the aluminium chemistry of isomers is two and six, respectively. The frequency of occurrence increases in the order: two – (AlC_2) < three – (AlC_2O, AlC_2N) < six – (AlF_6, AlO_6, AlO_5Cl, AlO_4Cl_2) < five – (AlH_4N,AlO_4Cl, AlH_3N_2, AlO_3N_2, $AlCl_3N_2$, AlH_2Cl_2), AlF_2N_2Cl) < four – (AlH_4, AlF_4, AlO_4, AlC_4, $AlCl_4$, $AlBr_4$, AlI_4, AlO_3C, AlN_3H, AlC_3H, AlC_3O, AlC_3N, AlB_3N, $AlCl_3C$, $AlCl_3Se$, AlO_2C_2, AlO_2Cl_2, AlO_2I_2, AlN_2C_2, AlN_2Cl_2, AlC_2S_2, AlC_2NCl) coordinate. The most common are four coordinate examples with variously distorted tetrahedral arrangement. In the series of five coordinater examples by far prevail a trigonal – bipyramidal with a different degree of distortion. Consideration of the nuclearity of the isomers shows a range of possibilities: mono – (x60) > di – (x27) > poly (x15) > tetra – , hexa – (each x4) > trimer (x3).

There are four types of isomerism – distortion (x102), ligand (x5), *cis – trans* – (x4) and polymerisation (x1). There are two examples which occur in three isomeric forms, AlF_3 (hexagonal [20, 21], rhombohedral [22] and orthorhombic [23]), and $BaAlF_5$ (orthorhombic and two are monoclinic [103]). Nine derivatives, ($C_{11}H_{22}NPCl$)[$AlCl_4$] (orthorhombic and monoclinic [7]); ($C_{22}H_{28}PSi$)[$AlCl_4$] (orthorhombic and monoclinic [8]); ($TeCl_3$)[$AlCl_4$] (monoclinic [9], triclinic [10]); [$Al(H_2O)_6$]Cl_3 (hexagonal [11], rhombohedral [12]); Al(η^2-acac)$_3$ (monoclinic [13,14], orthorhombic [15]); Al(MePO) (both trigonal [16,17]); $AlF_3 \cdot H_2O$ (cubic [18], tetragonal [19]); [$Al(Et)_2Cl_2$][$V_2(PMe_3)_6Cl_3$] (orthorhombic, tetragonal [68]) ; and $KAlF_4$ (tetragonal [104], monoclinic [105]); exist in two isomeric forms.

In Al(Bu^t)$_2$(tbp) (monoclinic) [76] six crystallographically independent molecules has been found. In another two derivatives, [HAl(μ_3-NPr^i)]$_6$ (triclinic) [65] and (thf)$_2$Mg{(μ-OPh)$_2$Al(OPh)$_2$}$_2$ (monoclinic) [95] four such molecules are present. The three such molecules have been found in the following four derivatives, [$Al(H_2O)_6$](IO_3)$_2$(HI_2O_6)(HIO_3) (hexagonal) [54], [$Al(H_2O)_6$](IO_3)$_2$(HIO_3) (hexagonal) [55], [$Al(H_2O)_6$](hm)$_3$ (monoclinic) [56], and Tl[Al(η^5-$C_2B_9H_{11}$)$_2$(tol)$_2$] (triclinic) [75]. Remainders (Tables 2, 3, 5, 6, 8) contain two crystallographically independent molecules. All these isomers as well as independent molecules differ by degree of distortion involving both Al-L distances and L-Al-L angles and in the oligomers also Al-Al (M) distances. The existence of two or more species, differing by degree of distortion, is typical of the general class of distortion isomerism [6].

There are five aluminium complexes, which belong to the ligand type of isomerism; $AlCl_3(n\text{-}MeC_6H_4COCl)$, (n = 2, monoclinic; n = 3, monoclinic and n = 4, triclinic [66], $[HAl(\mu_3\text{-}NPr^i)]_6$ (triclinic) [65] and $[HAl(\mu_3\text{-}NPr)]_6$ (trigonal) [67].

Colourless (colH)$[AlF_4]$ (orthorhombic) [24] which contains monomer (AlF_4) and polymer (AlF_6) units, is an example of polymerisation isomerism.

Colourless $[Al(\mu\text{-}NPr^i)(Me)_2]_2$ [89] which contains in one crystal *cis* – and *trans* – isomers (Table 7) is extraordinary example. Another colourless, $[Al(\mu\text{-}NHMe)(Me)_2]_3$ exists in two isomeric forms, *cis* – (rhombohedral) and *trans* (monoclinic) [90] (Table 7).

There is a wide variety of metallocyclic rings, and the effects of both steric and electronic factores can be seen from the mean values of the L-Al-L bond angles which open in the order: 71.6° (-NCN-) < 74.0° (-SCS-) < 79.0° ($-NC_2N-$) < 82.0° ($-OC_2O-$) < 82.3° ($-CC_2C-$) < 89.5° ($-OC_3N-$) < 92.5° ($-OC_3O-$) < 98.0° ($-CC_3N-$). There are at least two contributing factors to the size of the L-Al-L chelate bond angles, bond ligand based. One is steric constraint imposed on the ligand and the other is the need accomodate the imposed ring size.

Many factors affecting chemical equilibrium (total solution concentration, solvent, pH, molar ratio of the components, temperature, pressure, etc.) play an important role in the preparation of isomers. The above factors are important because they exhibit a great influence on the composition and structure of the coordination sphere in the initial reactants. The other external factors stabilizing a distorted configuration (one of many equivalent configuration) can be any weak low symmetry perturbation, e.g. the influence of the next (subsequent) coordination sphere, intramolecular (Van der Waals) interactions, crystal forces, hydrogen bonds, etc. The effect of distortion of the neighbouring moieties on a given center also represents at low symmetry perturbation. Low symmetry outside influences lead to the deformation of the complex, which in the case of weak contribution is very small.

Special consideration is needed for distortion isomers, whose formation is conditioned by the stabilization of not one, but two or several distorted configurations. The isomer cases discussed above seem to be due to such a compromise of the "soft" properties of the aluminium coordination sphere and the stabilizing properties of the crystal lattice, that the various configurations so formed have nearly equal energies and therefore the isomers exist simultaneously at the same (room) temperature. It is necessary to emphasize that the requirement of very close energy values for the isomers, which is essential for the possibility of their observation at the same (or approximately the same) temperature, narrows very much the number of possible observations of distortion isomers which themselves are in thermodynamic minima.

It is hoped that such a review will help to focus attention on areas of aluminium chemistry, that could be enhanced by further study, and assist in allowing comparative behaviour of the aluminium atom in the situations which can arise from the widespread use of aluminium.

These data have been retrieved in large part by using the Cambridge Crystal Database as a pointer to the original literature. However, some relevant material does get buried in the literature and is not visible from automated retrieval searches, and some is passed on in incorrect form from one paper to another, and even into the database.

ACKNOWLEDGMENTS

The authors thank the Ministry of Education of the Slovak Republic, Grant Agency of the Slovakia (VEGA grant No. 1/3416/06 and 1/0353/08)

REFERENCES

[1] H. W. Roesky, *Inorg. Chem.*, 43 (2004) 7284.
[2] H. Cowley, *J. Organomet. Chem.*, 689 (2004) 3866.
[3] E. Holloway and M. Melník, *Main Group Met. Chem.*, 19 (1996) 619.
[4] E. Holloway and M. Melník, *J. Organomet. Chem.*, 543 (1997) 1.
[5] E. Holloway and M. Melník, *Main Group Met. Chem.*, 19 (1996) 411.
[6] M. Melník, *Coord. Chem. Rev.*, 47 (1982) 239.
[7] H. Cowley, R. A. Kemp, J. G. Lasch, N. C. Norman, C. A. Stewart, B. R. Whitlesey, and T.C. Wright, *Inorg. Chem.*, 25 (1986) 740.
[8] U. Heim, H. Pritzlow, U. Fleischer, and H. Grützmacher, *Angew. Chem.* (Int. Ed. Engl.), 32 (1993) 1359.
[9] B. Krebs, B. Buss, and D. Altena, *Z. Anorg. Allg. Chem.*, 386 (1971) 257.
[10] B. H. Christian, M. G. Collins, R. J. Gillespie and J. F. Sawyer, *Inorg. Chem.*, 25 (1986) 777.
[11] D. R. Buchanan and P.M. Harris, *Acta Crystallog.*, *Sect. B*, 24 (1968) 953.
[12] K. R. Andress and C. Carpenter, *Z. Kristallogr.*, 87 (1934) 446.
[13] P. K. Hon and C.E. Pfluger, *J. Coord. Chem.*, 3 (1973) 67.
[14] Rahman, S. N. Ahmed, M. A. Khair, E. Zagrando, and L. Randaccio, *J. Bangladesh Acad. Sc.*, 14 (1990) 161.
[15] W. McClelland, *Acta Crystallogr., Sect-B*, 31 (1975) 2496.
[16] K. Maeda and J. Akimoto, *Angew. Chem.* (Int. Ed. Engl.) 43 (1995) 1199.
[17] K. Maeda, J. Akimoto, Y. Kiyozumi, and F. Misu ami, *J. Chem. Soc., Chem. Commun.*, 1995, 1033.
[18] R. Chandross, *Acta Crystallogr.*, 179 (964) 1477.
[19] R. D. Freeman, *J. Phys. Chem.*, 60 (1956) 1152.
[20] J. A. A. Ketcloar, *Nature*, 22 (1931) 303; *J. Kristallogr.*, 85 (1933) 119.
[21] E. Hoppe and D. Kissel, *J. Fluorine Chem.*, 24 (1984) 327.
[22] F. Hanic, K. Matiašovský, and D. Štempelová, *Acta Chim. Hung.*, 32 (1962) 309.
[23] Le Bail, C. Jacoboni, M. Leblanc, R. De Pape, H. Huroy, and J. L. Fourquet, *J. Solid State Chem.*, 77 (1988) 96.
[24] N. Herron, D. L. Thorn R. L. Harlow, and F. Davidson, *J. Amer. Chem. Soc.*, 115 (1993) 3028.
[25] M. H. Chisholm, J. C. Huffman, and J. L. Weseman, *Polyhedron*, 10 (1991) 1367.
[26] P. B. Broxterman, H. Hogeveen, R. F. Kingma, and F. van Bolhuis, *J. Amer. Chem. Soc.*, 107 (1985) 5722.
[27] M. Burford, R. E. v. H. Spence, and R. D. Rogers, *J. Chem. Soc., Dalton Trans.*, 3611 (1990).
[28] M. Veith, B. Bertsch, and V. Huch, *Z. Anorg. Allg. Chem.*, 559 (1988) 73.

[29] B. A. Stork – Blaisse and C. Romers, *Acta Crystallogr., Sect. B,* 27, (1971) 386.
[30] R. K. Mc Mullan , D. J. Prince, and J. D. Corbett, *Inorg. Chem.*, 10 (1971) 1749.
[31] O. Glemser, E. Kindler, B. Krebs, R. Meros, F. M. Schnepel, and J. Wegener, *Z. Naturforsch.*, 35b (1980) 657.
[32] J. Banister, P. J. Dainty, A. C. Hazell, R. G. Hazell, and J. G. Lomborg, *J. Chem. Soc., Chem. Commun.*, 1187 (1969); A. G. Hazell and R. G. Hazell, *Acta Chem. Scand.*, 26 (1972) 1987.
[33] R. Beattie, P. J. Jones, J. A. K. Howard, L. E. Smart, C. J. Glimore, and J. W. Akitt, *J. Chem. Soc., Dalton Trans.*, 528 (1979).
[34] N. C. Means, C. M. Means, S. G. Bott, and J. L. Atwood, *Inorg. Chem.*, 26 (1987) 1466.
[35] N. R. Streľcová, V. K. Beľskii, L. V. Ivakina, P. A. Storozhenko, and B. M. Bulychev, *Koord. Khim.*, 13 (1987) 1101.
[36] W. Frank and B. Dincher, *Z. Naturforsch.*, 42b (1987) 828.
[37] S. J. Holmes, R. R. Schrock, M. R. Churchill, and H. J. Wasserman, *Organometalics*, 3 (1984) 476.
[38] H. Burger, K. Hensen, and P. Pickel, *Z. Naturforsch.*, 43b (1988) 963.
[39] E. B. Lobkovskii, V. B. Polyakova, S. P. Shilkin, and K. N. Semenenko, *Zh. Strukt. Khim.*, 16 (1975) 77; Engl. Ed. P. 66.
[40] M. Coroley, M. C. Cushner, and P. E. Riley, *J. Amer. Chem. Soc.*, 102 (1980) 624.
[41] N. Burford, B. W. Royan, R. E. v. H. Spence, and R. D. Rogers, *J. Chem. Soc., Dalton Trans.*, 1990, 2111.
[42] P. L. Gurian, L. K. Cheatham, J. W. Ziller, and A. R. Barron, *J. Chem. Soc., Dalton Trans.*, 1991, 1449.
[43] E. B. Lobkovskii, I. I. Korobov, and K. N. Semenenko, *Zh. Strukt. Khim.*, 19 (1978) 1063.
[44] Ahmed, W. Schwarz, J. Weidlein, and H. Hess, *Z. Anorg. Allg. Chem.*, 434 (1977) 207.
[45] V. D. Romanenko, T. V. Sarina, M. Sanchez, A. N. Chernega, A. B. Rozhenko, M. R. Mazienes, and M. I. Povolotski, *J. Chem. Soc., Chem. Commun.*, 1993, 963.
[46] W. Clegg, U. Klingebiel, J. Neemann, and G. M. Sheldrick, *J. Organomet. Chem.*, 249 (1983) 47.
[47] M. Grottel, A. Kozak, H. Maluszynska, and Z. Pajak, *J. Phys. Condens. Matter.*, 4 (1992) 1837.
[48] E. C. Lingafelter, P. L. Orioli, B. J. B. Schein, and J. M. Stewart, *Acta Crystallogr.*, 20 (1966) 451; B. J. B. Schein, E. C. Lingafelter, and J. M. Stewart, *J. Chem. Phys.*, 47 (1967) 5183.
[49] S. Menchetti and C. Sabelli, *Tschermarks Min. Petr. Mitt.*, 21 (1974) 164.
[50] Jen Ho Fang and P. D. Robinson, *Amer. Mineral.*, 61 (1976) 311.
[51] D. Lazar, B. Ribar, and B. Prelesnik, *Acta Crystallogr., Sect. C*, 47 (1991) 2282.
[52] G. G. Bombi, K. Corain, A. A. Sheikh – Osman, and G. C. Valle, *Inorg. Chim. Acta*71 (1990) 79.
[53] M. G. Davidson , C. Lambert, I. Lojoz – Solera, P. R. Raithly, and R. Snaith, *Inorg. Chem.*, 14 (1995) 3765.
[54] P. D. Cradwick and A. S. de Enedredy, *J. Chem. Soc., Dalton Trans.*, 1977, 146.
[55] H. Kùppers, W. Schäfer, and G. Will, *Z. Kristallogr.*, 159 (1982) 231.

[56] T. S. Cameron, W. J. Chute, G. Owen, J. Aherne, and A. Linden, *Acta Crystallogr., Sect. C*, 46 (1990) 231.
[57] P. Bissinger, M. Paul, J. Riede, and H. Schmidbaur, *Chem. Ber.*, 126 (1993) 2579.
[58] P. Bissinger, P. Mikulcik, J. Riede, A. Schrier, and H. Schmidbaur, *J. Organomet. Chem.*, 446 (1993) 37.
[59] R. H. Cayton, M. H. Chisholm, E. R. Davidson, V. F. Di Stasi, P. Du, and J. C. Huffman, *Inorg. Chem.*, 30 (1991) 1020.
[60] L. Atwood, S. G. Bott, and M. T. May, *J. Coord. Chem.*, 23 (1991) 313.
[61] Levinski, S. Pasynkiewicz, and J. Lipkowski, *Inorg. Chim. Acta*, 179 (1990) 113.
[62] F. Stollmaier and U.Thewalt, *J. Organomet. Chem.*, 208 (1981) 327.
[63] G. Del Piero, M. Cesari, G. Dozzi, and A. Mazzei, *J. Organomet. Chem.*, 129 (1977) 281.
[64] G. J. Palenik, *Acta Crystallogr.*, 17 (1964) 1573.
[65] Cesari, G. Perego, G. Del Piero, S. Cucinella, and E. Cernia, *J. Organomet. Chem.*, 78 (1974) 203.
[66] B. Chevrier, J. M. Le Canpertier, and R. Weiss, *Acta Crystallogr., Sect. B*, 28 (1972) 2659.
[67] G. Del Piero, M. Cesari, G. Piero, S. Cucinella, and E. Cernia, *J. Organomet. Chem.*, 129 (1977) 289.
[68] F. A. Cotton, S. A. Duraj, W. J. Manzer, and W. J. Roth, *J. Amer.Chem. Soc.*, 107 (1985) 3850.
[69] A. Petrie, K. Ruhlandt – Senge, and P. P. Power, *Inorg. Chem.*, 32 (1993) 1135.
[70] J. H. Medley, F. R. Fronczek, N. Ahmad, M. C. Day, R. D. Rogers, C. R. Kerr, and J. J. Atwood, *J. Crystallogr. Spectrosc. Res.*, 15 (1985) 99.
[71] D. M. Schubert, C. B. Knobler, and M. F. Hawthorne, *Organometalics*, 6 (1987) 1353; D. M. Schubert, M. A. Bandman, W. S. Rees, Jr., C. B. Knobler, P. Lu, W. Nam, and M. F. Hawthorne, *Organometalics*, 9 (1990) 2046.
[72] B. Power, A. W. Apblett, S. G. Bott, J. L. Atwood, and A. R. Barron, *Organometalics*, 9 (1990) 2529.
[73] M. Taghiof, D. G. Hendershot, M. Barber, and J. P. Oliver, *J. Organomet. Chem.*, 431 (1992) 271.
[74] B. Briet, U. Bergsträsser, G. Maas, and M. Regitz, *Angew. Chem. Int. Ed. Engl.*, 31 (1992) 1055.
[75] M. A. Bandman, C. B. Knobler, and M. F. Hawthorne, *Inorg. Chem.*, 27 (1988) 2399.
[76] M. A. Petrie, M. M. Olmstead, and P. P. Power, *J. Amer. Chem. Soc.*, 113 (1991) 8704.
[77] R. J. Wehmschultz, K. Ruhlandt – Senge, M. M. Olmstead, H. Hope, B. E. Sturgeon, and P. P. Power, *Inorg. Chem.*, 32 (1993) 2983.
[78] M. L. Sierra, R. Kumar, V. Srini, J. de Mel, and J. P. Oliver, *Organometalics*, 1 (1992) 206.
[79] D. G. Hendershot, M. Barber, R. Kumar, and J. P. Oliver, *Organometalics*, 10 (1991) 3302.
[80] W. Uhl and M. Layb, *Z. Anorg. Allg. Chem.*, 620 (1994) 856.
[81] B. Hitchcock, H. A. Jasim, M. F. Lappert, and H. D. Williams, *Polyhedron*, 9 (1990) 245.
[82] M. Taghiof, M. J. Heeg, M. Bailey, D. G. Dick, R. Kumar, D. G. Hendershot, H. Rahbarnooki, and J. P. Oliver, *Organometalics*, 14 (1995) 2903.

[83] J. L. Atwood, W. E. Hunter, R. D. Rogers, and J. A. Weeks, *J. Inclusion Phenom.*, 3 (1985) 113.
[84] M. G. Gardiner and C. L. Raston, *Organometalics*, 12 (1993) 81.
[85] G. H. Robinson, W. E. Hunter, S. G. Bott, and J. L. Atwood, *J. Organomet. Chem.*, 326 (1987) 9.
[86] W. Uhl, A. Wester, and W. Hiller, *J. Organomet. Chem.*, 443 (1993) 9.
[87] J. L. Atwood, K. D. Crissinger, and R. D. Rogers, *J. Organomet. Chem.*, 155 (1978) 1.
[88] M. R. Mason, J. M. Smith, S. G. Bott, and A. R. Barron, *J. Amer. Chem. Soc.*, 115 (1993) 4971.
[89] S. Amirkhalili, P. B. Bitchcock, A. D. Jenkins, J. Z. Negathi, and J. D. Smith, *J. Chem. Soc., Dalton Trans.*, 1981, 377.
[90] G. M. Mc Laughlin, G. A. Sim, and J. D. Smith, *J. Chem.Soc., Dalton Trans.*, 1972, 2197.
[91] E. B. Lobkovskii, G. L. Soloveichik, B. M. Bulychev, R. G. Gerr, and Y. T. Struchkov, *J. Organomet. Chem.*, 270 (1984) 45.
[92] C. Sirio, O. Poncelet, I. G. Hubert – Pfalzgraf, J. C. Daran, and J. Vaissermann, *Polyhedron*, 11 (1992) 177.
[93] G. Erker, M. Albrecht, C. Krüger, S. Werner, P. Bringer, and F. Langhauser, *Organometalics*, 11 (1992) 3517.
[94] R. Kumar, H. Rahbarnoochi, M. J. Heegm D. G. Dick, and J. P. Oliver, *Inorg. Chem.*, 33 (1994) 1103.
[95] J. A. Meese – Marktscheffel, R. E. Cramer, and J. W. Gilje, *Polyhedron*, 13 (1994) 1045.
[96] A. Danopoulos, G. Wilkinson, B. Hussain – Bates, and M. B. Hursthouse, *J. Chem. Soc. Dalton Trans.*, 1990, 2753.
[97] S. I. Trojanov and V. B. Rybakov, *Metalloorg. Khim.*, 1 (1988) 1282; Engl. Ed., p. 700.
[98] Fen Bao – Chen, Shen Qi, and Lin Yong – Hua, *J. Organomet. Chem.*, 376 (1989) 61.
[99] V. Saboonchian, G. Wilkinson, B. Hussain – Bates, and M. B. Hursthouse, *Polyhedron*, 10 (1991) 595.
[100] Chung – Cheng Chang, Wen – Hwa Lee, Tzeng – Yih Hu, Gene – Hsiang Lee, Shie – Ming Peng, and Yu Wang, *J. Chem. Soc., Dalton Trans.*, 1994, 315.
[101] W. Uhl, M. Layh, and W. Massa, *Chem. Ber.*, 124 (1991) 1511.
[102] V. K. Belskii, Y. K. Gunko, B. M. Bulychev, and G. L. Soloveichik, *J. Organomet. Chem.*, 419 (1991) 299.
[103] R. Domesle and R. Hoppe, *Z. Anorg. Allg. Chem.*, 495 (1982) 16.
[104] J. Nouet, J. Pannetier, and J. L. Fourguet, *Acta Crystallogr., Sect B*, 37 (1981) 32; A. Gibaud, A. Le Bail, and A. Bulou, *J. Phys. C; Solid State Phys.*, 19 (1986) 4623.
[105] J. M. Launay, A, Bulou, A. W. Hewat, A. Gibaud, J. Y. Laval, and J. Nouet, *J. Phys. (Paris)*, 49 (1985) 771.

In: Aluminum Alloys
Editor: Erik L. Persson
ISBN: 978-1-61122-311-8

Chapter 3

ROLLING OF ALUMINUM: THE INFLUENCE OF THE CASTING ROUTE ON MICROSTRUCTURAL EVOLUTION DURING DOWNSTREAM PROCESSING

Stavros Katsas
Novelis Foil Innovation Centre, 41 rue du Brill, L-4422,
Belvaux, Luxembourg

ABSTRACT

Aluminum sheet is widely preferred for its light weight, strength and corrosion resistance. It is used in numerous applications such as automotive, transport, packaging, construction and the printing industry to name but a few. The two most popular processes for its production are direct chill (DC) casting of ingots and twin roll casting (TRC) of coil slabs.

DC casting involves modest solidification rates resulting in a rather coarse cast grain size. Further downstream processing consists of homogenization of the ingot followed by a series of hot and cold rolling steps. The main advantage of this route is that the elevated temperature processing steps allow the elimination of casting segregation phenomena, consequently resulting in homogeneous properties at final gauge. When high performance service properties are required (e.g. formability), DC cast material is preferred.

In TRC, aluminum exits the caster rolls in the form of a coil slab and is then directly cold rolled to final gauge. The process offers low capital investments, savings in energy and low operational cost together with some metallurgical advantages (e.g. higher solidification rates/solute leves and finer constituent particles). However, if an intermediate homogenization treatment is not introduced, metal produced via this route can be rather unstable at final gauge.

With continuous pressure to reduce the industry's carbon footprint, TRC is gaining popularity and more work is now focused on the optimization of this processing route to produce material with performance similar to DC.

In this review, an overview of the two casting methods along with the associated downstream processing for the production of rolled aluminum sheet and their effect on the microstructural evolution is presented.

1. Introduction

Aluminum sheet is widely preferred for its light weight, strength and corrosion resistance. It is used in numerous applications such as automotive, transport, packaging, construction and the printing industry to name but a few. The two major routes for the production of aluminum sheet are (a) Conventional Direct Chill (DC) casting, and (b) Continuous casting of thin slab or sheet (Figure 1 and Table 1). Continuous casting is mainly divided in Twin Roll Casting (TRC) and Flexcasting, with the latter being used mainly in North America and Japan. Here, focus will be on the comparison of aluminum sheet produced via DC and TRC casting only. Flexcasting is mentioned in Figure 1 and Table 1 only for reference.

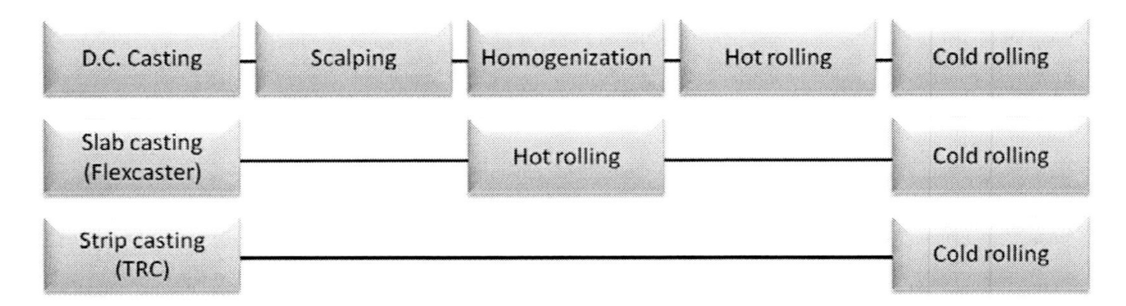

Figure 1. Summary of the three most common processes for manufacturing aluminum rolled sheet; Direct Chill (DC) casting, Flex casting and Twin Roll Casting (TRC).

Table 1. Comparison of the three most common processes for manufacturing of aluminum rolled sheet; Direct Chill (DC) casting, Flex casting and Twin Roll Casting (TRC)

	DC	TRC	Flexcaster
As cast thickness	400-800mm	4-8mm	10-25mm
As cast length	Up to 10m	Endless	Endless
As cast speed	50-150mm/min	1-1.5m/min	7-9m/min
Solidification rate	Low	Very high	High
Scalping	Yes	No	No
Homogenization	Yes	No	No
Hot rolling	Yes	No	In line with the process down to 1-2.5mm

2. Casting

A description of the two main casting routes (Direct Chill Casting and Twin Roll Casting) for the production of rolled aluminum sheet is presented here. Purpose of this chapter is to introduce the microstructural features present after casting for each route, in order to explain its response to further processing.

2.1. Direct Chill (DC) Casting

The DC casting process is considered a batch process and is the most common technique for casting of rolled aluminum products. The typical size of the produced ingots is 500mm thickness, 1000mm width and 5000-10000mm length. Dimensions vary depending on the casting technology used in each cast house and the requirements of downstream processing.

During casting, the molten metal is poured into a water-cooled mould, which is closed at the bottom with a metallic block (Figure 2). The moment solidification occurs with the mould sides and the bottom block, this is lowered with constant speed (i.e. casting speed) with means of a ram, while keeping metal level at a constant height inside the mould by controlling the metal flow. As the solidified metal is lowered under the mould, water is sprayed at the sides to enhance cooling and solidification. The casting process finishes when the ingot reaches the bottom of the casting pit.

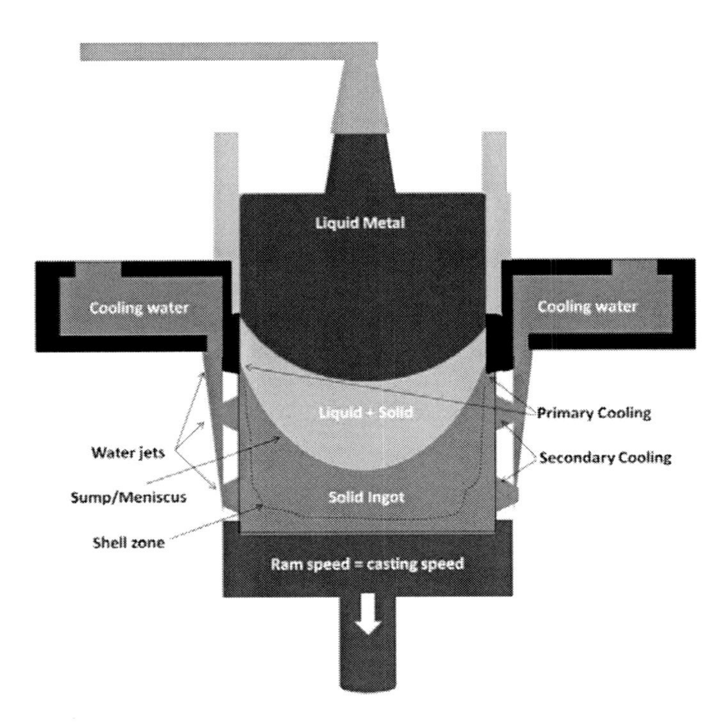

Figure 2. Schematic of the Direct-Chill (DC) casting process (dimensions not to scale).

Upon first solidification in contact with the mould sides, the metal will contract slightly creating a small air gap between the mould and the ingot (shrinkage gap), thus reducing the cooling rate. At the central part of the ingot, metal will remain in a semi-solid/liquid state for a longer period of time due to the slower heat extraction rate. Based on the differences experienced by the sides and the central part of the ingot during solidification, the solidification process and the resultant microstructure are categorized in primary and secondary. Primary solidification describes the process of the formation of the shell zone at the surface, whereas secondary solidification refers to heat extraction from the bulk. These two zones have a significant effect on the developing microstructure and downstream processing as the primary solidification rate is different than the secondary and hence the casting structure at the centre is expected to be coarser. For example, in a DC cast ingot of AA3004 the difference between secondary dendrite arm spacing and eutectics size between the surface and the centre was found to be threefold [1]. Therefore, it is obvious that if no corrective actions are taken after casting to compensate for this, not only variation in properties across width will exist but issues with workability may also occur. An example is given in [2] for an aluminum alloy AA8079 used in conversion and food packaging applications at thicknesses <50μm where edge cracking was observed during hot rolling as a result of casting variations.

In the primary solidification zone the shell solidifies relatively fast in contact with the mould and the water jets. However, in the case of the bulk, heat extraction takes place much slower through the shell zone. Depending on the casting temperature and the latent heat in the bulk, this can very often result in phenomena such as remelting of the shell zone, precipitation or dissolution of intermetallic phases and segregation. This is because, the type of intermetallics forming during solidification, do not depend solely on the alloy composition, but also on the local solidification rate. It is obvious that such events can lead to surface defects, which in turn can have an effect on the optical properties of the final product. One of the most common example is the so called fir tree formation on the surface of the ingot caused by non uniform shell zones [3]. To avoid such defects, especially for surface aspect critical products (e.g. 1xxx alloys for lithographic or 5xxx automotive sheet), part of the surface is scalped away prior further processing.

The secondary cooling zone describes mainly the development of the sump and is governed by the efficiency of the water spray. The lower part of the sump at the central portion of the ingot is formed, in most cases, in the range of secondary and downstream cooling where the heat transfer is the highest. There, the heat is extracted from this part of the billet mostly through the solid phase that has higher heat conductivity than the liquid. In the central–bottom part of the liquid/solid zone, the colder melt is driven upwards, additionally cooling the melt in this section of the billet [4-5]. The cooling rates encountered in a commercial size DC cast ingot through thickness and across width vary between 0.2 to 5 °C/s [6]. An example of how the varying cooling rates in sequential positions from the surface of the ingot affect the grain and intermetallics size in a DC AA8006 is given in Figures 3-4.

The main process parameters that influence the ingot properties are:

- *Casting temperature*: This needs to be adjusted for each alloy to accommodate for the liquidus. The casting temperature has a strong effect on the volume fraction and distribution of non-equilibrium eutectic. For example, during casting from high melt temperatures, more eutectic is concentrated at the centre of the billet. In this case, the coarseness of the structure, in combination with more active melt flows, may facilitate a deeper penetration of the solute-rich melt into the mushy zone in the centre of the billet, effectively increasing the amount of eutectic [4-5].
- *Metal level*: The level of the metal in the mould affects the pressure exerted on the sump, as well as the temperature distribution and the cooling zones.
- *Casting speed*: This refers to the speed of the ram and can have an effect on the surface quality. In addition, the casting speed directly influences convection in a proportional way.

A higher casting speed deepens the sump and favors higher temperature gradients and hence internal stresses [5].

- *Cooling water*: Parameters such as the flow rate, the temperature, the pH of the cooling water will also have an effect on the heat extraction in the secondary solidification zone.

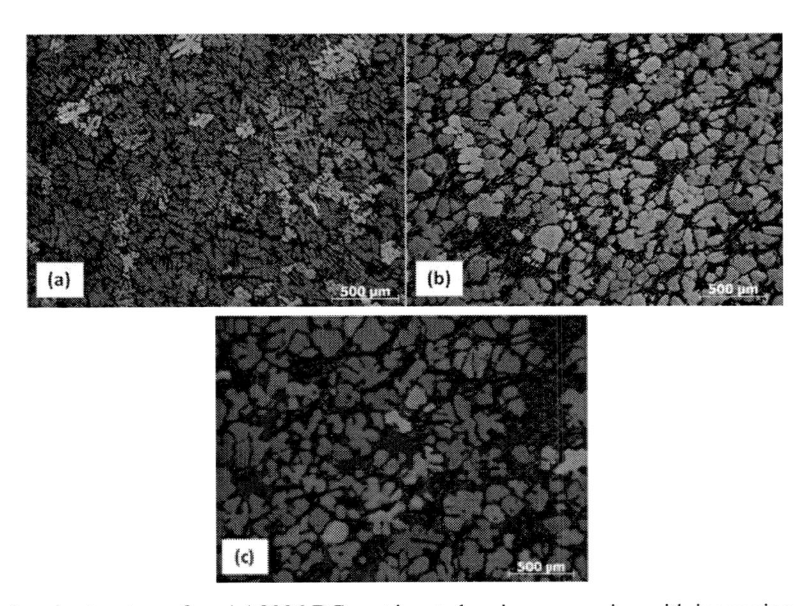

Figure 3. Typical grain structure of an AA8006 DC cast ingot showing coarsening with increasing distance from the surface. Distance from surface (a) 100mm, (b) 1000mm, (c) 2000mm.

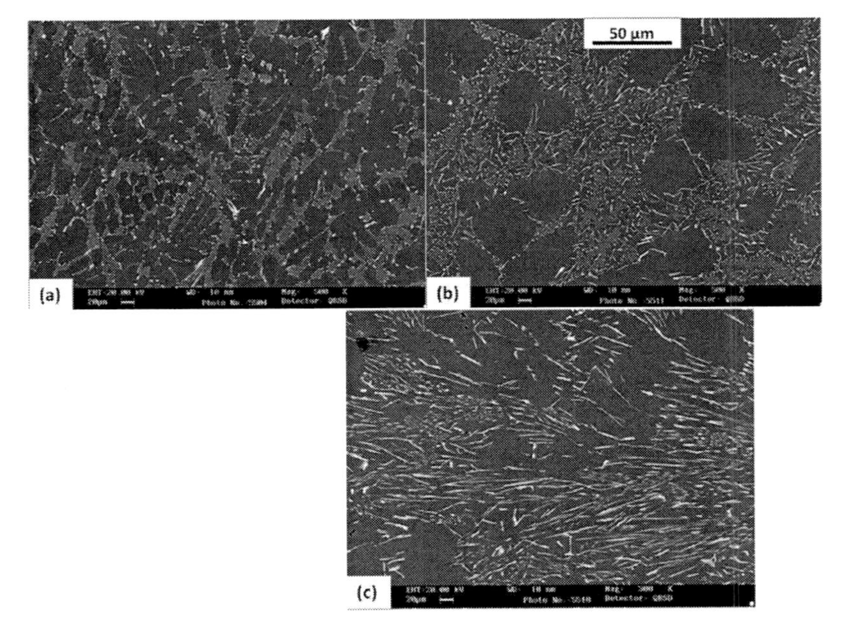

Figure 4. Typical size and morphology of intermetallics of an AA8006 DC cast ingot showing coarsening with increasing distance from the surface. Distance from surface (a) 100mm, (b) 1000mm, (c) 2000mm.

2.2. Twin Roll Casting

The production of coilable strips with twin roll casting (TRC) is now a standard practice in the aluminum industry due to the fact that it presents both economic and metallurgical benefits [7-9]. TRC offers great flexibility, lower investment and operation costs with respect to the traditional DC and hot mill method and is thus very attractive in the production of a variety of sheet and foil products. While much effort has been devoted to the optimization of casting parameters to improve the cast strip quality [10-12], still TRC aluminum strips show marked differences with respect to conventional DC strips and may require different downstream cycles for the manufacture of highly formable grades in particular [13-14]. Today a range of alloys can be cast commercially with this route including 1xxx, 3xxx, 4xxx, 5xxx, 6xxx and 8xxx [15-19].

The high solidification rates available in the process enable a refined microstructure, comprising fine dendritic cells and fine intermetallic particles, together with increased solubility of alloying elements. Downstream processing is rather simple and usually involves only cold rolling to final gauge. Depending on the alloy and the application, a homogenization treatment after casting or annealing at an intermediate gauge can be employed. Both options are exploited either for precipitation of elements from solid solution (e.g. Fe, Mn) with the aim of controlling the microstructure or for softening to allow for rolling to very thin foil gauges (e.g 6μm foil for aseptic applications).

TRC is a continuous process where the metal exits the caster in the form of sheet/slab to be subsequently coiled (Figures. 5-6). Prior casting, the metal is hold in a tundish to ensure constant temperature and flow across the width of the tip. In this case, the rolls also serve the purpose of the mould, giving to the metal its cast slab shape. The metal arrives in liquid form at the tip and solidifies in contact with the chilled rolls, with the cooling rate estimated to be in the order of 500°C/s [20]. It should be mentioned that during solidification, the metal also experiences some degree of hot deformation as its thickness at the entrance of the casting rolls is higher than at the exit.

The most common dimensions of TRC are widths up to 2000mm and thicknesses in the range of 4-8mm. The parameters influencing the microstructure and the overall quality of the sheet are:

- *Casting temperature*: As with DC casting, this needs to be adjusted for different chemical compositions to accommodate for the liquidus of each alloy. The casting temperature will have an effect on the cooling rate and hence on the degree of solute levels and size of intermetallics in the cast metal.
- *Metal level in tundish*: In an analogous way to DC casting, the metal level in the tundish will affect pressure on the meniscus as well as the flow across the tip.
- *Casting speed, tip distance, roll force*: All these parameters will have an affect on the strip profile.
- *Roll material*: The most common material used for the roll shell is steel. However, there are casthouses which use copper shells instead. Due to its higher conductivity and reaction with aluminum, solidification rate and surface appearance of the strip, respectively, will be different. Copper rolls are also employed to increase productivity.

- *Cooling (water flow and lubricant)*: Temperature and flow of water inside the roll will affect the heat dissipation and hence the solidification rate and the resultat grain and intermetallics structure.
- *Line speed*: The line (coiling) speed is adjusted to keep a constant difference with the casting speed. This difference together with the roll force define the hot rolling deformation taking place within the roll bite after solidification.

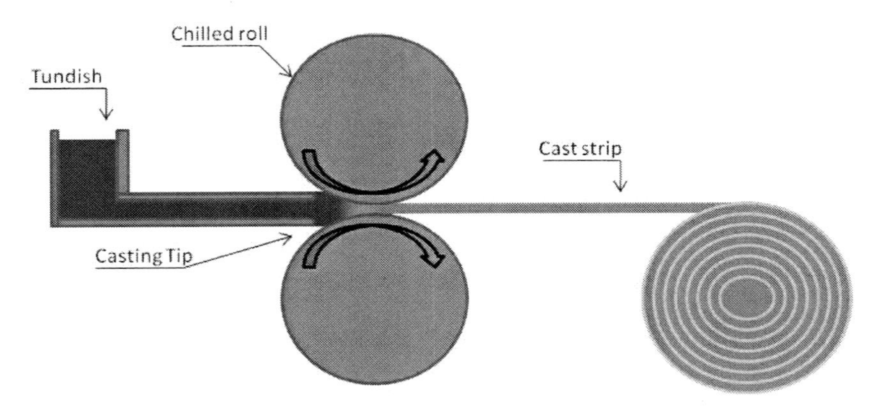

Figure 5. Schematic of Twin Roll Casting (TRC) route.

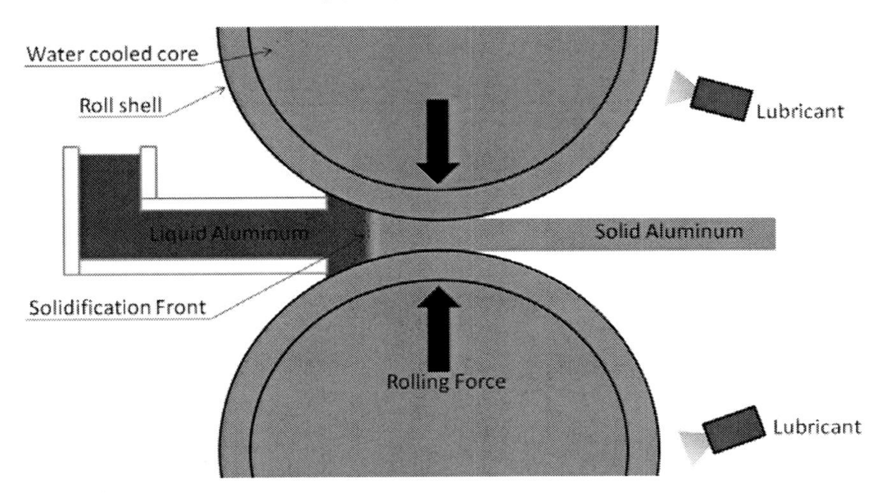

Figure 6. Details of the TRC route (dimensions not to scale).

When compared to the DC casting process, TRC has significantly higher overall solidification rates and smaller temperature gradients through thickness (Figure 7) and hence completely different casting structure. The size of intermetallics averages in the area of 1μm, with the majority of the samples being ≤5μm [11, 13].

The structure is characterized by sheared/columnar grains with angles 5-20° with respect to the surface and coarser, more equiaxed cells, at the centre (Figure 8). Oriented grains are the result of coupling between (i) directional solidification and (ii) the subsequent hot deformation during twin roll casting.

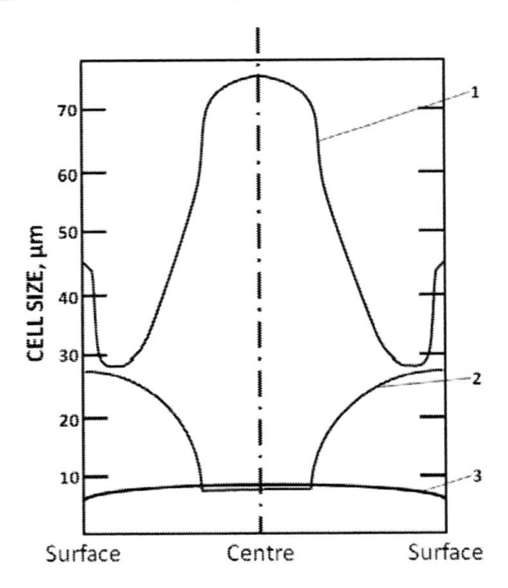

Figure 7. Variation in cell size across cast section for (1) DC block (500mm thick), (2) DC strip (25mm thick) and (3) twin roll cast strip (7mm) (adopted from [23]).

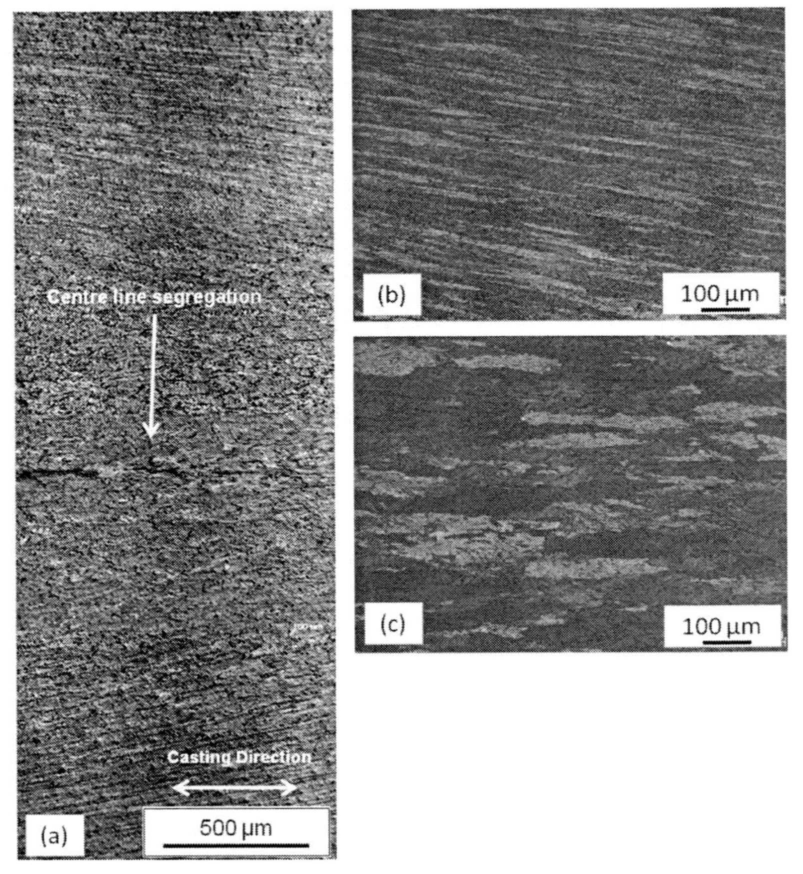

Figure 8. (a) Lower and higher magnification of (b) surface and (c) centre, of AA8006 cast structure produced with TRC. Cross section parallel to the casting/rolling direction.

During solidification, the metal in contact with the chilled rolls will solidifify first, while at the centre will remain at the liquid state for a bit longer and will solidify slower (Figure 7). This occurs because the heat transfer coefficient between the sheet and the roll becomes dramatically larger when sufficient solid forms near the base of the sump. When solid exists throughout the strip, pressure builds up between the rolls increasing the heat transfer coefficient [7]. As a result of the above, the size of second phase particles and solute levels are also influenced. At the surface, second phase particles are much finer than the centre, contrasting the gradual coarsening of the microstructure. Some examples are given by Y. Birol for commercial Al-Fe-Si (AA8011) and Al-Fe-Mn-Si (AA8006) alloys, where the number of eutectic clusters and coarser intermetallic phases were coarser at the centre of the strip.

The opposite was true for the amount of Fe and Mn in solid solution [21-22]. There, Y. Birol also reports that near the surface a very fine grain structure, almost amorphous, is present that does not react to etching as the rest of the bulk. Dynamic recrystallization provides a plausible account for the very fine, equiaxed grain structure in contrast to the elongated grains in the interior. The multi-zone character of the thin strip is also evidenced by through-thickness hardness profiles, showing that higher hardness values at the surface than at the centre of the strip.

In addition, due to the rejection of solute towards the centre, centre line segregation is often present as almost an intrinsic feature of TRC strips (Figure 8). Depending on the application, this may or may not be considered a problem. For example, when rolling to very thin gauges is required (e.g. 10μm househould foil or for other confectionary foil products), centre line segregation can be the source of pinhole formation.

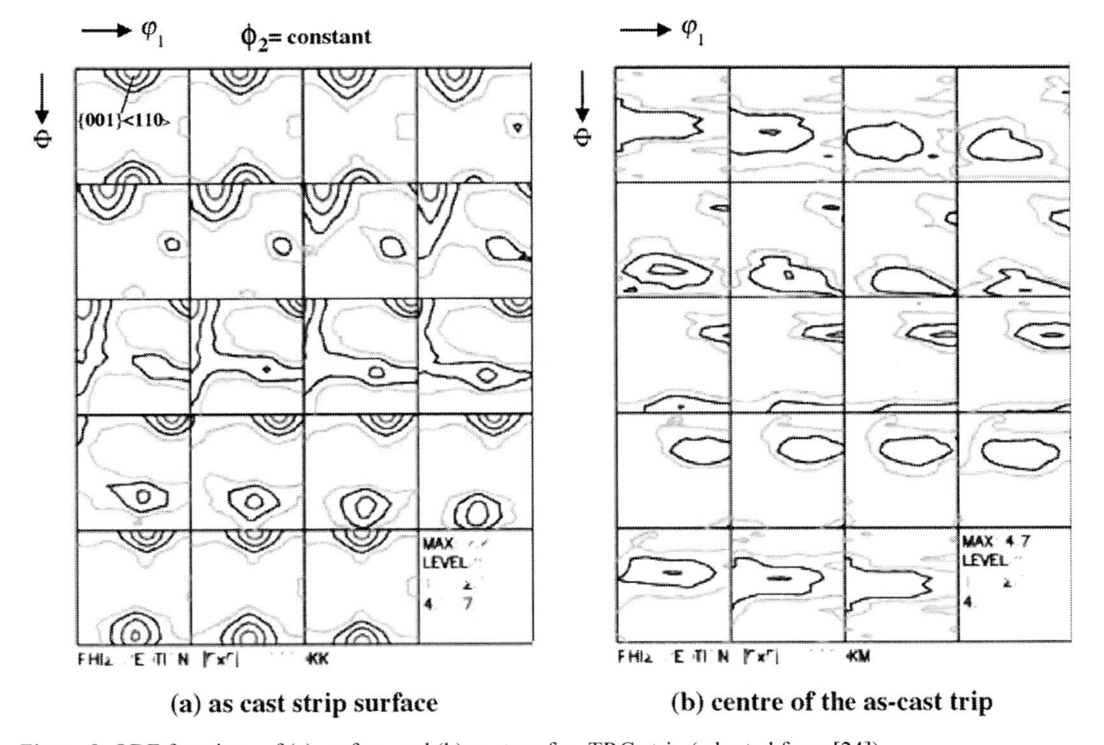

(a) as cast strip surface (b) centre of the as-cast trip

Figure 9. ODF functions of (a) surface and (b) centre of as-TRC strip (adopted from [24]).

It was mentioned earlier that during solidification, the slab is simultaneously subjected to hot rolling as the thickness at the entrance of the rolls is higher than at the exit. As far as crystallographic texture is concerned, several researchers report that the surface region will exhibit a shear component as a result of this low hot rolling reduction, in contrast to the centre through thickness where a weak β-fibre will be present (Figure 9) [24-28].

This difference between the surface and the centre can be further enhanced by several parameters such as (i) differences in travelling velocity between the surface and the centre prior or at the moment of solidification, (ii) lack of lubrication giving rise to high friction, and (iii) at rolling mills with roll diameters (asymetric rolling).

2.3. Summary

In summary, at the finishing of the casting stage for the two main processing routes:

DC casting: An ingot with a size of 500mmX1500mmX10000mm that has solidified relatively slowly and is rather inhomogeneous across width and through thickness with a grain and intermetallics size in the range of 200-300μm and 10-30μm, respectively. In the next processing step, the ingot will be homogenized to be prepared for hot rolling.

TRC: The metal is in the form of a coilable 4-8mm sheet. The microstructure is characterised by high amounts of solute and very fine second phase particles (~1μm), as a result of the high solidification rates. In the next processing step, the coil will be cold rolled to final sheet gauge.

3. Homogenization

The homogenization stage is a necessary part of the DC process route. After casting is complete, the ingots are heated up to temperatures greater than 500°C to improve workability and allow for high rolling reductions. This is achieved by removing the residual stresses, result of the different cooling rates experienced during solidification, thus reducing significantly the occurrence of edge cracking during subsequent hot rolling.

Except the essential improvement of workability, ingot preheating (also termed as homogenization) serves metallurgical purposes at the same time. It was described earlier that there is large variation in terms of solute levels, intermetallics and grain size from the centre to the edge of the ingot. During homogenization, precipitation from solid solution and redistribution of dispersoids via diffusion will take place, reducing microsegregation and improving the response of the ingot in subsequent thermo-mechanical treatments. Homogenization is especially important for Mn containing alloys (e.g. 3xxx and 8xxx alloys) for controlled precipitation of Mn containing particles from solid solution prior hot working, in order to improve thermal stability and avoid grain growth [29-30]. Except precipitation from solid solution, breaking up and spheroidization of the coarse intermetallic particles also occurs [31]. Depending on the alloy and the final application, a simple heat to roll practice or prolonged stay at high temperature is selected.

Although not very common, homogenization practices are often employed in the TRC route. Again, whether a homogenization step is required or not, this will depend on the alloy composition. High Mn containing alloys will require a homogenization step to remove it from solid solution, as it is known to retard recrystallization and often lead to grain growth during hot rolling or final annealing [30].

Hardness measurements are very often used as an indirect measurement of solid solution levels, with higher hardness values indicating faster cooling rates and hence higher amounts of solute in solid solution. A comparison of hardness values before and after homogenization between TRC and DC commercial AA8006 is shown in Figure 10. The graph highlights the degree of supersaturation achieved with each processing route and also the amount of Fe and Mn precititation taking place during homogenization. Figure 11 shows the corresponding microstructure for the hardness values measured in Figure 10.

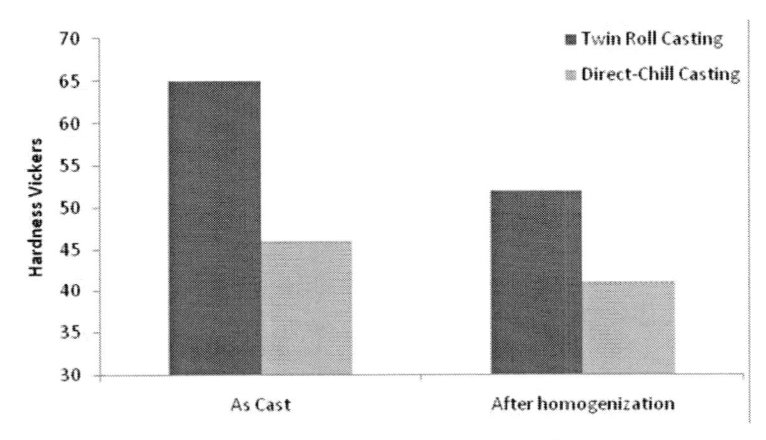

Figure 10. Hardness comparison between DC and TRC cast of commercial aluminum alloy AA8006 before and after homogenization.

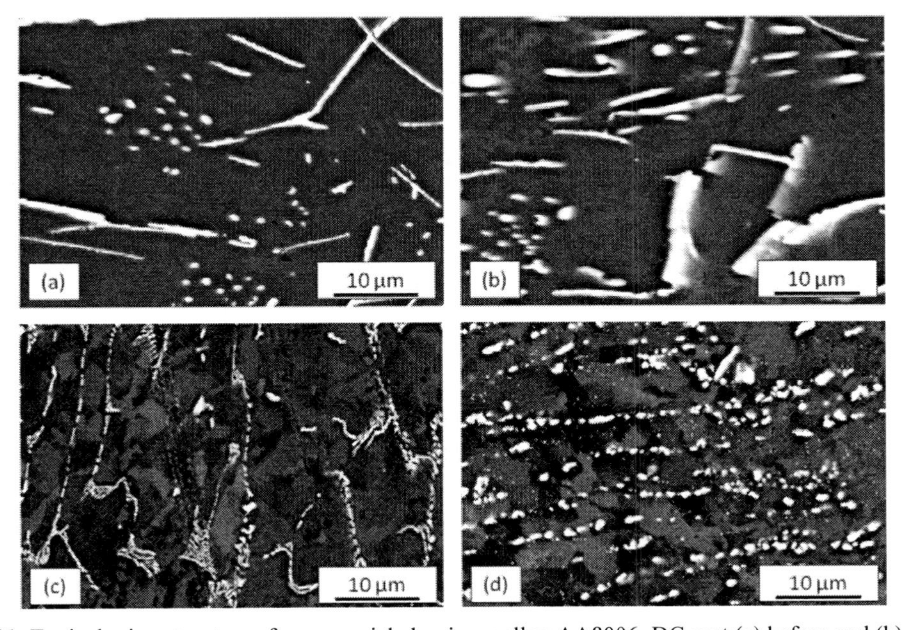

Figure 11. Typical microstructure of commercial aluminum alloy AA8006. DC cast (a) before and (b) after homogenization; TRC cast (c) before and (d) after homogenization. In both cases, micrographs were taken from the centre of the ingot/sheet across width and through thickness.

In the case of DC cast, the structure comprises coarse rod-shaped particles with lengths up to 20μm and a diameter of 1μm (a). During homogenization, coarsening of the rods takes place and the sharp edges of the rods become more rounded. In the TRC variant, the eutectic network is much finer in the as cast condition (c). During homogenization breaking up of the eutectic network and precipitation of submicron particles is much more evident than in the DC cast material due to the higher dislocation density and hence diffusion paths (d).

4. Hot Rolling

A schematic description of the typical hot rolling stages starting from the DC ingot for the production of aluminum sheet are illustrated in Figures 12-13. Before hot rolling, the ingot is scalped to remove 6-20mm from each side and heated either in pusher furnaces or in soaking pits to temperatures between 500°C-620°C for up to 48 hours. The hot ingot is then sourced to a reversing break-down mill to reduce the thickness from the initial 500mm down to 25-45mm. The total number of passes is around 15, reducing each time the thickness by about 50mm. The temperature at the exit of the break-down mill is usually between 450°C - 550°C. Following that, the thickness is further reduced by a four stand tandem mill to 2.5-8mm to be coiled. The temperature at the exit of the hot mill is between 250°C -350°C and is controlled by changing the rolling speed.

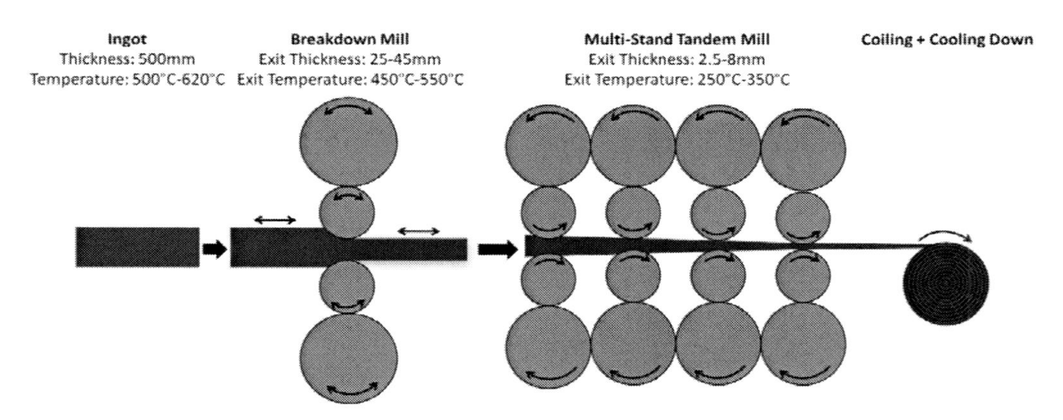

Figure 12. Typical hot rolling stages of aluminum sheet.

The main hot rolling parameters that have an effect on the microstructure are the hot rolling temperature, the rolling reduction after each pass and the time between passes. For example, the higher the rolling temperature and strain rates, the higher the degree of dynamic recovery and recrystallization during each pass. In particular, high hot rolling temperature will reduce deformation stresses and transform the coarse as cast grains to finer ones with cube texture (Figure 14). However, a high degree of cube texture may lead to earing in the final gauge sheet product, especially in canstock alloys [32].

As discussed earlier, TRC sheet also experiences some degree of hot rolling reduction, which depends on the roll force applied during casting and the line speed. Although this can be considered negligible when compared to that experienced in the DC route, it is enough for a shear texture to develop at the surface [24].

Figure 13. Break-down mill at Alunorf Germany (courtesy of Novelis).

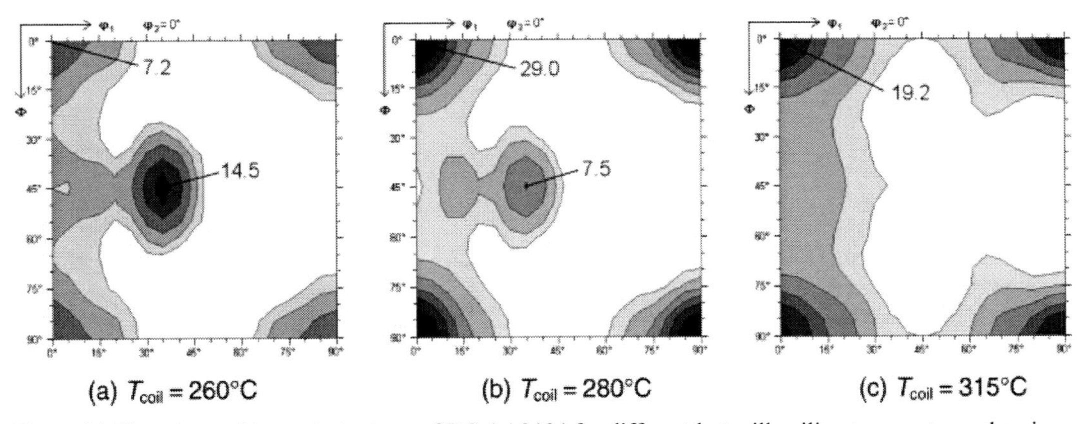

Figure 14. Experimental hot strip textures of DC AA3104 for different hot mill coiling temperatures showing higher Cube components for increasing coiling temperatures (adopted from [32]).

In summary, after the end of hot rolling of the DC ingot, the aluminum sheet can consist of anything between a wrought structure with elongated to fully recrystallized grains, depending on the hot rolling parameters selected. Also the coarse particles present in the DC ingot, are broken to smaller constituents.

5. Cold Rolling

By the general term cold rolling, the rolling process taking place at temperatures below the recrystallization process of each alloy is described. In reality, at the rolling plants, the cold rolling process is performed with the metal starting at room temperature. Still, during the process, the metal can reach temperatures up to 150°C due to friction with the work rolls. In the case of the DC processing route, after the metal exits the hot mill, it is placed in storage racks and is left to cool down to room temperature before being cold rolled to the final gauge. Similarly, the TRC mother coil is also left to cool down prior further processing. The main difference between the two is the degree of cold deformation at the end of the cold rolling schedule. Targetting the same final gauge, the starting cold rolling gauge for DC is usually 2.5-3.5mm, as opposed to 6-8mm for TRC.

As far as the microstructure is concerned, during cold rolling the grains become elongated along the rolling direction and high amounts of stored energy are introduced in the form of dislocation tangles. As with hot rolling, dynamic recovery is also known to occur during cold rolling but to a lesser extent [33]. Thus the energy available in the metal after each rolling pass, will depend on the original grain size, the amount of elements in solid solution, as well as the size and distribution of second phase particles. Consequently, the recrystallization mechanism at final gauge will also be influenced to a great extent by the level of cold rolling reduction [34]. At low to moderate reductions, recrystallization occurs by a discontinuous mechanism at particle stimulated nucleation sites. At high rolling reductions (>90%), which are readily available in foil rolling (<200μm), continuous recrystallization may occur during final annealing. The size distribution of dispersoids and the level of elements in solid solution will also have an effect on the development of the microstructure during cold rolling and annealing. Fine particles will tend to block the dislocation movement and refine the microstructure during recrystallization. Elements in solid solution will inhibit recovery and promote recrystallization during temper annealing [33].

In summary, it is clear that even if the alloy composition is exactly the same, the TRC aluminum sheet will experience a higher degree of cold rolling reduction and therefore its response during temper annealing will be different than that of DC. Also, due to the differeces in size distribution of second phase particles between the two processing routes, recovery and strain hardening during cold rolling will be different. For this reason, it is very common for alloys produced from DC and TRC routes to receive a completely different treatment at final gauge in order to achieve the same temper and mechanical properties.

The effect of the starting cold rolling gauge and the existing microstructure, as a result of the casting route, on the development of the final properties is discussed in the following paragraph.

6. Microstructural Evolution and Comparison of Properties between DC and TRC Aluminum Rolled Sheet

Throughout the previous paragraphs, the differences between the two processing routes for the production of rolled aluminum sheet were highlighted. In this chapter, the effect of these differences and in particular of constituent particle size on final gauge properties such as anisotropy and texture development will be discussed.

The evolution of microstructures and textures during rolling of aluminum sheet has been the subject of numerous investigations the past 40 years [35-49]. In many of the studies a large variety of textures was obtained, covering a wide range from copper to brass orientations. This highlights the fact that microstructure and texture development during rolling of aluminum sheet will depend on several factors among which pre-existing texture, grain size, solute levels and second phase particles [50-51]. For example, the volume fraction and size distribution of second phase particles (number of PSN sites vs dislocation pinning particles) is known to affect the work hardening/recovery rate and hence the development of deformation textures [36]. When considering the role of particles, hot and cold rolling processes should also be taken into account. During hot rolling, only very coarse particles can act as efficient nucleation sites. Therefore, fine particles will not have a significant effect on the development of texture during hot rolling. Consequently, the most important factor for the development of texture during cold rolling is whether the metal was recrystallized at the exit of the hot mill [36]. Grain size has also been shown to have an effect on the development rate of rolling texture [43]. Regarding solute levels, increased levels of Fe in solid solution as those present in TRC material increase the tendency for 0/90 earing [36]. Indirectly, the amount of dissolved Fe also strongly affects the final texture and earing by raising the recrystallization temperature during prior processing.

A few examples will be given here regarding the differece developing between alloys produced with DC and TRC. A first comparison is given for a dilute AA1050 aluminum alloy. In this study Zhou et al [52] examined the mechanical properties of TRC and DC AA1050 after different amounts of cold rolling reductions. The main observations are summarized in that the TRC strip generally strain hardened faster than the DC one and that during temper annealing at final gauge, the former was stronger for the same annealing temperature. This was mainly because the higher degree of Fe supersaturation and finer particle size in the TRC strip retarded recrystallization during temper annealing, pushing the recrystallization temperature of the TRC strip higher when compared to the DC [53]. This means that for the same temperature, the degree of recrystallized grains in the TRC strip when compared to the DC one will be lower, hence the difference in strength. However, when the TRC strip was homogenized at an intermediate gauge prior cold rolling to final gauge, then differences in the mechanical properties after annealing practically dissapeared. This was explained on the basis that during this homogenization treatment, large part of the stored energy differences were removed and also some precipitation of Fe from solid solution took place, coarsening the second phase particles and increasing the available particle stimulated nucleation sites. In this study, it was finally concluded that despite the minor differences in mechanical properties between the two variants and especially the higher strength, the formability of the DC AA1050 was always expected to be better.

In another example, two 5xxx alloys (AA5052 and AA5182) used in more demanding applications such as automotive sheet produced with TRC and DC casting were analyzed at a final gauge of 1mm after annealing to soft temper [54]. The alloys had exactly the same chemical compositions and the only difference was the casting route and the subsequent downstream processing. In accordance to what was described in the previous paragraphs, the density of finer second phase particles was much higher in the TRC AA5052 and AA5182 than in their DC counterparts. After temper annealing to achieve a fully recrystallized microstructure, the grain size was much finer in the TRC alloys than in the DC ones. In terms of texture, in the TRC AA5052 an asymmetric pole figure together with a retained rolling texture were present. In the higher Mg containing TRC AA5182 a strong recrystallization cube texture together with a retained rolling texture developed. In the DC processed variants, the AA5052 had a weak retained rolling and cube texture and the AA5182 a strong cube texture. It can be seen that the DC processed alloys in both cases exhibited stronger recrystallization/cube components, whereas the TRC variants always had a higher volume of deformation/rolling components. There are many explanations why this occurred. One explanation for this is the differences in solid solution levels and especially Fe, which is known not only to favor the strengthening of the rolling texture [39] but also to push the recrystallization temperature higher, as shown in the previous example with AA1050.

Due to the fact that no high temperature processing stages are usually involved in the production of TRC strip, there are no opportunities to remove inhomogeneities like centre line segregation introduced during casting. This is considered one of the main disadvantages of the TRC route and is considered the main reason why DC strips have generally better formability. Gras et al [24] studied the evolution of texture during cold rolling of a TRC AA3105 and focused on the differences between the surface and the centre through thickness. It was observed that before cold rolling, a high volume of shear texture components was present at the surface but not at the centre (Figure 15).

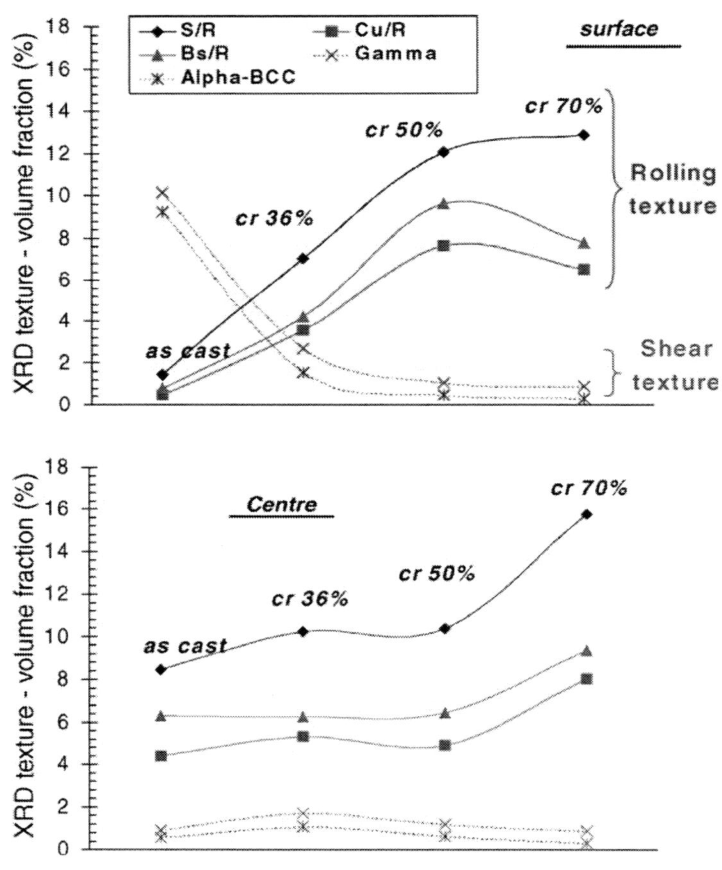

Figure 15. Evolution of the texture volume fraction with the gauge reduction (CR %): (a) surface of the sheet and (b) centre region (adopted from [24]).

With increasing cold rolling reductions, the volume of the rolling texture components (Cu, Bs, S) increased on the expense of the shear ones. In contrast to the surface, the volume of shear components at the centre was close to zero. If this is compared with the texture evolution in a DC3105 [30], due to recrystallization at the exit of the hot mill, the hot mill structure was rather homogeneous through thickness and exhibited a high degree of cube texture. With increasing strains, the cube components weakened and practically disappeared after 90% thickness reduction. The volume fraction of the $Cube_{ND}$ fiber, reflected through {1 0 0} fibers, remained above 10% until 80% cold rolling and then decreased during further cold rolling. Orientations started to flow to the β fiber at Copper, Copper/S, S, Brass/S and Brass positions after 80% cold rolling and then increased during further cold rolling. It is worth noting that the volume fractions of the Copper/S, S, Brass/S and Brass orientations converged to about 12%, while the volume fraction of the Copper orientation, increased by about 15% after 80% cold rolling, and reached 29% after 90% cold rolling. The volume fraction of random orientations dropped rapidly after 80% cold rolling when the β fiber developed. In agreement with the previous example of TRC and DC AA5052 and AA5182 or even other TRC AA8006 [55] and DC AA8006 [56] alloys, the main difference between the two processing routes in terms of major crystallographic components is the higher degree of rolling textures in TRC, partly because of the higher degree of recrystallized structure prior cold rolling in the DC and partly because of the finer particle size and supersaturation that retards recrystallization in the TRC.

Generally, DC aluminum sheet is always expected to have more uniform properties and better formability than TRC and this is why the latter processing route is not widely used in the automotive industry. The main reason behind that, is that the DC processed metal will be homogeneous and have the same response during annealing at final gauge through thickness, due to the numerous high temperature processing steps involved during its production.

On the other hand, in the TRC strip where mostly only cold rolling is available, there are large variation in intermetallics size and stored energy. These differences influence the recrystallization kinetics of the material through thickness [16, 21, 22, 55] and so the metal during annealing is characterized by a dual microstructure with recovered grains at the surface and equiaxed grains at the centre. Prolonged heating of such a microstructure will finally result in grain growth and hence lower formability. A short explanation of this is given below. It was mentioned earlier that the surface substructure of TRC sheets has a much higher density of high angle grain boundaries due to the high shear at the surface during casting [7]. As there is no scalping in TRC processing route, all inhomogeneities introduced during casting will be transferred to the final product. During subsequent cold rolling, the density of high angle grain boundaries at the surface will increase. At sufficiently large strains, the structure at the surface is completely transformed into a substructure only with high angle grain boundaries.

Although such a structure provides an enormous amount of possible recrystallization nuclei, it is rather unstable. In fact, the material will rearrange itself by ordinary grain growth instead of recrystallization [57]. Recrystallization will usually commence around coarse particles, which act as particle stimulated nucleation sites. Close to the surface, the interparticle spacing between these sites is higher than at the centre due to the higher solidification rate during casting and so the distance a newly formed grain will grow before it is hindered by another grain growing from the opposite direction will be greater. So, while generally the surface is much more temperature resistant than the centre, due to the higher volume of fine dislocation pinning particles, from the moment recrystallization will start, grain growth will be the dominating mechanism.

These variations through thickness can be eliminated if an intermediate thermal treatment is introduced. However, this is not always desirable due to the cost increase and the reduction in strength.

CONCLUSION

Direct-chill casting and twin roll casting are the two most common casting routes for the production of aluminum sheet. The main characteristics of each processing route can be summised below:

Direct-chill casting:

- Slow solidification rates resulting in coarse grain and intermetallics cast structure.
- Relatively complex thermomechanical processing route.
- Homogeneous microstructure through thickness at final gauge.
- Used in a variety of applications from simple aluminum foil to demanding automotive and aerospace sheet.

Twin roll casting:

- High solidification rates resulting in refined grain size, second phase particles and high solute levels.
- Simple, more economical and less energy demanding processing route than DC.
- Inhomogeneous microstructure through thickness at final gauge. More temperature resistant during temper annealing.
- Generally less formable than DC cast strip.

REFERENCES

[1] Kamat RG. AA3104 can-body stock ingot: Characterization and homogenization. *Journal of Materials* 48 (1996) 34-38.

[2] Paramatmuni RK, Chang K-M, Kang BS, Liu X. Evaluation of cracking resistance of DC casting high strength aluminum ingots. *Materials Science and Engineering* A379 (2004) 293–301.

[3] Otani T, Arai K, Kato T, Otsuka R. Fir tree structure of 1000- and 5000- series aluminum alloys. Light Metals: Proceedings of Sessions, AIME Annual Meeting (Warrendale, Pennsylvania), 1980, pp. 855-870.

[4] Dua Q, Eskin DG, Katgerman L. The effect of ramping casting speed and casting temperature on temperature distribution and melt flow patterns in the sump of a DC cast billet. *Materials Science and Engineering A* 413–414 (2005) 144–150.

[5] Nadella R, Eskin DG, Du Q, Katgerman L. Macrosegregation in direct-chill casting of aluminum alloys. *Progress in Materials Science* 53 (2008) 421–480.

[6] Chu MG. Nature and Formation of Surface Cracks in DC Cast Ingots. In: Schneider W, editor. Light metals 2002. Warrendale [PA]: TMS; 2002, p. 899–907.

[7] Yun M, Lokyer S, Hunt JD. Twin roll casting of aluminum alloys. *Materials Science and Engineering* A280 (2000) 116-123.
[8] Weschler R. The status of twin roll casting technology. Scandinavian *Journal of Metallurgy* 32 (2003) 58-63.
[9] Cook R, Grocock PG, Thomas PM, Edmonds DV, Hunt JD. Development of twin-roll casting process. *Journal of Materials Processing Technology* 55(1995) 76-84.
[10] Daakand O, Espedal AB, Nedreberg ML, Alvestad I: In R. Huglen, Thin Gauge Twin Roll Casting, Process Capabilities and Product Quality, Light Metals, The Minerals, Metals and Materials Society, Pennsylvania (1997) 745-752.
[11] Birol Y, Kara G, Akkurt S, Romanowski C. The effect of casting parameters on the metallurgical quality of twin-roll casting. InL K. Ekhre, W. Schneider, Editors, Proc. Int. Conf. Continuous Casting DGM-Wiley-VCH (2000) 40-46.
[12] Berg BS, Hansen V, Zagierski PT, Nedreberh ML, Olsen A, Gjonnes J. Gauge reduction in twin-roll casting of an AA5052 alloy-the effects on microstructure. *Journal of Materials Processing Technology* 53 (1995) 65-74.
[13] Birol Y, Karlik M. Microstructure of a thin cast Al-Fe-Mn-Si strip. *Prakt. Metallogr*. 42 (2005) 325-338.
[14] Gras Ch, Meredith M, Hunt JD. Microdefects formation during twin-roll casting of Al-Mg-Mn aluminum alloys. *Journal of Materials Processing Technology* 167 (2005) 66-72.
[15] Haga T, Nishiyama T, Suzuki S. Strip casting of A5182 alloy using melt drag twin roll caster. *Journal of Materials Processing Technology* 133 (2003) 103-107.
[16] Birol Y. Response to annealing treatment of a twin-roll casting thin AlFeMnSi strip. *Journal of Materials Processing Technology* 209 (2009) 506-510.
[17] Sarkar S, Wells MA, Poole WJ. Softening behavior of cold rolled continuous cast and ingot cast aluminum alloy 5754. *Materials Science and Engineering* A421 (2006) 276-285.
[18] Ryu J-H, Lee DN. The effect of precipitation of on the evolution of recrystallization texture in AA8011 aluminum alloy sheet. *Materials Science and Engineerijng* A336 (2002) 225-232.
[19] Haga T, Tkahashi K, Ikawaand M, Watari H. Twin roll casting of aluminum alloy strips. *Journal of Materials Processing Technology* 153-154 (2004) 42-47.
[20] Rolling Aluminum: From the mine through the Mill. The Aluminum Association 3rd edition 2007.
[21] Birol Y, Karlik M. Microstructure of a thin cast Al-Fe-Mn-Si strip. *Prakt. Metallogr*. 42 (2005) 325-338.
[22] Birol Y. Response to annealing treatments of twin-roll cast Al-Fe-Si strips. *Journal of Alloys and Compounds* 458 (2008) 265-270.
[23] Odok AN, Gyongyos I. Continuous Casting Seminar Papers. Aluminum Association. Paper 14. 1975.
[24] Gras Ch, Meredith M, Hunt JD. Microstructure and texture evolution after twin roll casting and subsequent cold rolling of Al-Mg-Mn aluminum alloys. *Journal of Materials Processing Technology* 169 (2005) 156-163.
[25] Hjelen J, Orsund R, Nes E. On the origin of recrystallization textures in aluminium. *Acta Metallurgica et Materialia* 39 (1991) 1377–1404.

[26] Dons AL, Nes E. Nucleation of cube texture in aluminium. *Materials Science Technology* 2 (1986) 8–18.

[27] Nes E, Solberg JK. Growth of cube grains during recrystallization. *Materials Science Technology* 2 (1986) 19–21.

[28] Yun M. A numerical and Experimental Study of the Twin-Roll Casting Process, Department of Materials, Oxford University, Oxford (1992).

[29] Liu WC, Zhai T, Morris JG. Texture evolution of continuous cast and direct chill cast AA3003 aluminum alloys during cold rolling. *Scripta Materialia* 51 (2004) 83-88.

[30] Liu J, Morris JG. Macro-, micro- and mesotexture evolutions of continuous cast and direct chill cast AA3105 aluminum alloys during cold rolling. *Materials Science and Engineering* A357 (2003) 277-296.

[31] Hutchinson WB, Oscarsson A, Karlsson. Control of microstructure and earing behaviour in aluminum alloy AA3004 hot bands. *Materials Science and Technology* 5 (1989) 1118-1127.

[32] Engler O, Lochte L, Hirsch J. Through-process simulation of texture and properties during the thermo-mechanical processing of aluminium sheets. *Acta Materialia* 55 (2007) 5449-5463.

[33] Humphreys FJ, Hatherly M. Recrystallization and related annealing phenomena. 2nd edition. Elsevier 2004.

[34] Ito K. Effects of fine particles in on formation of recrystallization structures and textures in aluminum alloys. In "Homogenization and annealing of aluminum and copper alloys" ed HD Merchant, J Crane, EH Chia (Warrendale PA: TMS 1998) 239-244.

[35] Hirsch J. Virtual fabrication of aluminium products: Microstructural modelling in industrial aluminium fabrication processes. 2006 WILEY-VCH Verlag GmbH.

[36] Oscarsson A, Hutchinson WB, Ekstrom H-E. Influence of initial microstructure on texture and earing in aluminium sheet after cold rolling and annealing. *Materials Science and Technology* 7 (1991) 554-564.

[37] Takeshita T, Kocks UF, Wenk H-R. Strain path dependence of texture development in aluminum. *Acta Metallurgica* 37 (1989) 2595-2611.

[38] Nes E, Embury JD. Influence of a fine particle dispersion on the recrystallization behavior of a two phase aluminium alloy. *Zeitschrift fuer Metallkunde* 66 (1975) 589-593.

[39] Ito K, Musick R, Lücke K. The influence of iron content and annealing temperature on the recrystallization textures of high-purity aluminium-iron alloys. *Acta Metallurgical* 31 (1983) 2137-2149.

[40] Chan HM, Humphreys FJ. The recrystallization of aluminium-silicon alloys containing a bimodal particle distribution. *Acta Metallurgica* 32 (1984) 235-243.

[41] Jensen DJ, Hansen N, Humphreys FJ. Texture development during recrystallization of aluminium containing large particles. *Acta Metallurgica* 33 (1985) 2155-2162.

[42] Hjelen J, Ørsund R, Nes E. On the origin of recrystallization textures in aluminium. *Acta Metallurgica Et Materialia* 39 (1991) 1377-1404.

[43] Engler O, Kong XW, Yang P. Influence of particle stimulated nucleation on the recrystallization textures in cold deformed Al-alloys part I-experimental observation. *Scripta Materialia* 37 (1997) 1665-1674.

[44] Benum S, Nes E. Effect of precipitation on the evolution of cube recrystallisation texture. *Acta Materialia* 45 (1997) 4593-4602.
[45] Engler O, Crumbach M, Li S. Alloy-dependent rolling texture simulation of aluminium alloys with a grain-interaction model. *Acta Materialia* 53 (2005) 2241-2257.
[46] Ryu J-H, Lee DN. The effect of precipitation on the evolution of recrystallization texture in AA8011 aluminium alloy sheet. *Materials Science and Engineering* A336 (2002) 225-232.
[47] Hansen N, Jensen DJ. Deformation and recrystallization textures in commercially pure aluminum. *Metallurgical Transactions* A17 (1986) 253-259.
[48] Engler O, Hirsch J, Lücke K. Texture development in Al 1.8wt% Cu depending on the precipitation state-I. Rolling textures. *Acta Metallurgica* 17 (1989) 2743-2753.
[49] Hirsch J, Nes E, Lücke K. Rolling and recrystallization textures in directionally solidified aluminium. *Acta Metallurgica* 35 (1987) 427-438.
[50] Hutchinson WB, Oscarsson A, Karlsson A. Control of microstructure and earing behaviour in aluminium alloy AA 3004 hot bands. *Materials Science and Technology* 5 (1989) 1118-1127.
[51] Lücke K; Engler O. Effects of particles on development of microstructure and texture during rolling and recrystallisation in fcc alloys. *Materials Science and Technology* 6 (1990) 1113-1130.
[52] Zhou SX, Zhong J, Mao D, Funke P. Experimental study on material properties of hot rolled and twin continuously cast aluminum strips in cold rolling. *Journal of Materials Processing Technology* 134 (2003) 363-373.
[53] Juul Jensen D. In: N. Hansen, D. Juul Jensen, Y.L. Liu and B. Ralph, Editors, Microstructural and crystallographic aspects of recrystallization, Risø National Lab, Roskilde, Denmark (1995), p. 119.
[54] Slamova M, Karlik M, Robaut F, Slama P, Veron M. Differences in microstructure and texture of Al-Mg sheets produced with by twin roll continuous casting and by direct chill casting. *Materials Characterization* 49 (2002) 231-240.
[55] Katsas S, Gras Ch. The effect of recrystallization phenomena during annealing on the final forming properties of TRC AA8006. TMS Light Metals (2009) pp. 1233-1236.
[56] Jazaeri H, Humphreys FJ. The transition from discontinuous to continuous recrystallization in some aluminium alloys II – annealing behavior. *Acta Materialia* 52 (2004) 3251-3262.
[57] Oscarsson et al. Transition from discontinuous to continuous recrystallization in strip-cast aluminium alloys. *Materials Science Forum* 115 (1992) 177-182.

In: Aluminum Alloys
Editor: Erik L. Persson

ISBN: 978-1-61122-311-8

Chapter 4

ALUMINIUM ALLOY ANODES: APPLICATION TOWARDS THE REMOVAL OF BORON FROM DRINKING WATER BY ELECTROCOAGULATION

Subramanyan Vasudevan** and Jothinathan Lakshmi
Electroinorganic Chemicals Division
Central Electrochemical Research Institute
(Council of Scientific Industrial Research)
Karaikudi 630 006, Tamilnadu, India

ABSTRACT

The present study provides an electrocoagulation process for the amputation of boron-contaminated water using aluminium alloy as the anode and stainless steel as cathode. The various parameters like effect of pH, concentration of boron, current density and temperature has been studied. Effect of co-existing anions such as silicate, fluoride, phosphate and carbonate were studied on the removal efficiency of boron. The results showed that the optimum removal efficiency of 90% was achieved at a current density of 0.2 A dm^{-2}, pH of 7.0. The study shows that boron adsorption process was relatively fast and the equilibrium time was 60 min. First and second-order rate equations were applied to study adsorption kinetics. The adsorption process follows second order kinetics model with good correlation. The Langmuir, Freundlich and D-R adsorption models were applied to describe the equilibrium isotherms and the isotherm constants were determined. The experimental adsorption data were fitted to the Langmuir adsorption model. The thermodynamic parameters such as free energy (ΔG^0), enthalpy (ΔH^0) and entropy changes (ΔS^0) for the adsorption of arsenate were computed to predict the nature of adsorption process. Temperature studies showed that adsorption was endothermic and spontaneous in nature.

Keywords: boron removal, aluminium alloy, electrocoagulation, isotherm, kinetics.

---

* To whom correspondence should be addressed. E-mail: vasudevan65@gmail.com

INTRODUCTION

Recently boron has come to the forefront as a possible drinking water contaminant. Boron is widely distributed in the environment, occurring naturally (seawater) or from anthropogenic pollution, mainly in the form of boric acid (H_3BO_3) or borate salts ($NaBO_2$, $Na_2B_4O_7.10\ H_2O$). Borate deposits are rare, being found in dry regions of the world such as the USA, Turkey, Argentine, China, Russia and Chile. Boric acid is a very weak acid with a pK_a of 9.15. In dilute aqueous solution at pH < 7 boron exists predominantly as undissociated boric acid $B(OH)_3$ while at pH > 10 the metaborate anion $B(OH)_4^-$ becomes the main component. Between these two pH values and at high concentration (>0.025 mol/L) highly water soluble polyborate ions such as $B_3O_3(OH)_4^-$, $B_4O_5(OH)_4^-$, and $B_5O_6(OH)_4$ [1-5] are formed. Boron is a major industrial material that is widely used in the production of glass and textile fiberglass, disinfectants, cosmetics, cleaning and bleaching agents, wood preserving materials, fire retardant materials, plastics, ceramics, food preservatives, synthetic rocket fuels, abrasives, absorption units in nuclear reactors, medicines, insecticides, cancer therapy drugs, and agricultural fertilizers [6-10]. Boron in the elemental form is not toxic. The finely divided powder is hard and abrasive, and may cause skin problems indirectly if the skin is rubbed after contact. Trace amounts of boron seem necessary for good growth of plant life. However, in greater amounts boron can be harmful to animal and plant life. Boron accumulated in the body through absorption, ingestion, and inhalation of its compounds affects the central nervous system. Therefore, removal of high levels of boron concentration from water sources is a special concern for living things. In order to consider the toxic effect of boron on humans, the EU and USEPA regulations are suggesting a guideline [11-12] of 1.0 mg L^{-1}. There are a variety of technologies that could be used to remove boron from water includes biological, chemical lime precipitation and sedimentation, ion exchange and reverse osmosis [13-20]. Biological treatment methods cannot be used for boron removal from wastewaters, because inorganic boron compounds are antiseptics. Chemical lime precipitation and sedimentation process will produce a voluminous amount of sludge for disposal and the chemical costs could be prohibitively high. Ion exchange processes by resin offer very good selected ion-exchange efficiency for boron but would require a high capital cost to set up the system for regeneration of effluents. Costly maintenance, replacement of membrane and need of pre/post treatment process contribute significantly to total treatment cost in reverse osmosis. Therefore, the investigation of an alternative technology to remove boron from water is needed. During the last few decades, electrochemical water treatment technologies have undergone swift growth and development. One of these technologies is the electrochemically assisted coagulation that can compete with the conventional chemical coagulation process. Electrochemically generated metallic ions from these electrodes undergo hydrolysis near the anode to produce a series of activated intermediates that are able to destabilize the finely dispersed particles present in the water and waste water to be treated. The advantages of electrocoagulation include high particulate removal efficiency, a compact treatment facility, relatively low cost, and the possibility of complete automation [21-28]. Usually iron and aluminum are used as electrodes in the electrocoagulation followed by electrosorption process. Following are the general reactions takes place in electrocoagulation processes,

(i) When iron is used as the electrode, the reactions are as follows:

At the cathode: $2H_2O + 2e^- \longrightarrow H_2(g) + 2OH^-$ (1)

At the anode: $4Fe(s) \longrightarrow 4Fe^{2+}(aq) + 8e^-$ (2)

In the solution: $Fe(s) + 10H_2O(l) + O_2(g) \longrightarrow Fe(OH)_3(s) + 4H_2(g)$ (3)

(ii) When aluminum is used as electrode, the reactions are as follows:

At the cathode: $2H_2O + 2e^- \longrightarrow H_2(g) + 2OH^-$ (4)

At the anode: $Al \longrightarrow Al^{3+} + 3e^-$ (5)

In the solution: $Al^{3+}(aq) + 3H_2O \longrightarrow Al(OH)_3 \downarrow + 3H^+(aq)$ (6)

This method is characterized by reduced sludge production, a minimum requirement of chemicals, and ease of operation. There are several publications related to electrocoagulation for the removal of fluoride, iron, chromium and phosphate from drinking water [21, 24-26], but there are limited work on removal of boron by electrocoagulation and its adsorption and kinetics studies. This article presents the results of the laboratory scale as well as scale-up studies on the removal of boron using aluminium alloy and stainless steel as anode and cathode respectively. To optimize the maximum removal efficiency of boron, different parameters like effect of current density, initial boron concentration, temperature, pH and effect of co-existing ions like carbonate, phosphate, silicate, fluoride and arsenic were studied. The adsorption kinetics of electrocoagulants is analyzed by using first and second order kinetic models. The equilibrium adsorption models of Langmuir, Freundlich and Dubinin-Radushkevich were also studied and the activation energy is calculated to study the nature of the adsorption.

Materials and Methods

1. Cell Construction and Electrolysis

The electrolytic cell (Figure 1) consists of a 1.0 – L Plexiglas vessel that was fitted with a polycarbonate cell cover with slots to introduce the anode, cathode, pH sensor, a thermometer and electrolytes. Aluminum alloy (consisting of Zn (1-4%), In (0.006-0.025%), Fe (0.15%), Si (0-.15%) (CECRI, (CSIR), India), with surface area of 0.2 dm^2 acted as the anode and stainless steel (SS 304; SAIL, India; consisting of <0.08% C, 17.5-20% Cr, 8-11% Ni, <2% Mn, <1% Si, <0.045% P, <0.03% S) as cathode was placed with an inter electrode distance of 0.005m. The temperature of the electrolyte has been controlled to the desired value with a variation of ± 2 K by adjusting the rate of flow of thermostatically controlled water through an external glass-cooling spiral. A regulated direct current (DC) was supplied from a rectifier (10 A, 0-25 V; Aplab model).

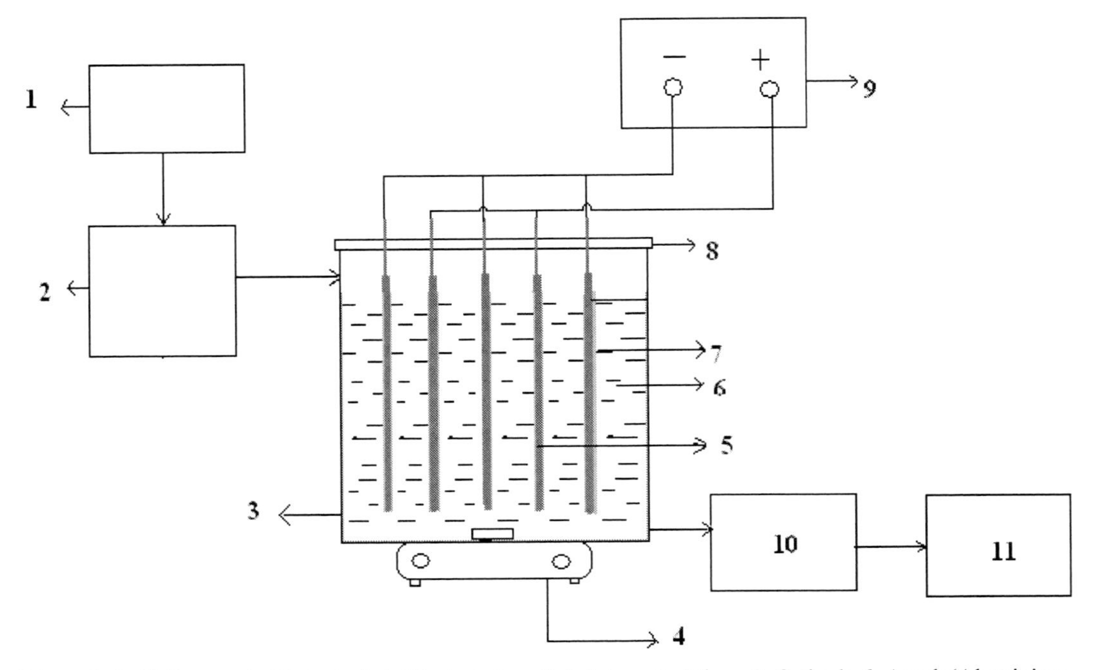

Figure 1. 1.pH Sensor, 2. Water tank 3. Electrolytic cell 4. Magnetic Stirrer 5. Cathode 6. Anode(Aluminium alloy) 7. Electrolyte 8. PVC cover 9. Power supply 10. Filter 11. Treated water tank.

The boron as boric acid (H_3BO_4) (Analar Reagent) was dissolved in deionised water for the required concentration (3-7 mg L^{-1} as boron). The solution of 0.90 L was used for each experiment as the electrolyte. The pH of the electrolyte was adjusted, if required, with 1 M HCl or 1 M NaOH solutions (AR Grade) before adsorption experiments. Temperature studies were carried at varying temperature (313-343 K) to determine the type of reaction. To study the effect of co-existing ions, in the removal of boron, sodium salts (Analar Grade) of phosphate (5-50 mg L^{-1}), silicate (5-15 mg L^{-1}), carbonate (5-250 mg L^{-1}), fluoride (0.2-5 mg L^{-1}) and arsenic (0-5 mg L^{-1}) was added to the electrolyte.

2. Analytical Method

The concentration of boron was analyzed by UV-Visible Spectrophotometer (MERCK, Pharo 300, Germany). The SEM of electrocoagulant (aluminum hydroxide) was analyzed with a Scanning Electron Microscope (SEM) made by Hitachi (model s-3000h). The Fourier transform infrared spectrum of aluminum hydroxide was obtained using Nexus 670 FTIR spectrometer made by Thermo Electron Corporation, USA. The concentration of carbonate, fluoride, silicate, arsenic and phosphate were determined using UV-Visible Spectrophotometer (MERCK, Pharo 300, Germany).

RESULT AND DISCUSSION

1. Effect of Current Density

Among the various operating variables, current density is an important factor which strongly influences the performance of electrocoagulation.The amount of boron removal depends upon the quantity of adsorbent (Aluminum hydroxide) generated, which is related to the time and current density. The amount of adsorbent $[Al(OH)_3]$ was determined from the Faraday law [29],

$$E_c = ItM/ZF \tag{7}$$

where I is current in A, t is the time (s), M is the molecular weight, Z is the electron involved, and F is the Faraday constant (96485.3 coulomb $mole^{-1}$). With the increase in current density the amount of aluminium hydroxide also increases. To investigate the effect of current density on the boron removal, a series of experiments were carried out by solutions containing a constant pollutants loading of 5mg L^{-1}, at a pH 7.0, with current density being varied from 0.1 to 0.5A dm^{-2}. Figure 2 shows the plot between removal efficiency of boron with respect to current density. From the figure it is found that, beyond 0.2A dm^{-2} the removal efficiency remains almost constant for higher current densities. So, further studies were carried out at 0.2 A dm^{-2}.

2. Effect of pH

The pH is one of the important factors affecting the performance of electrochemical process. To examine this effect, a series of experiments were carried out using 5 mg L^{-1} boron containing solutions, with an initial pH varying in the range 2 to 12 at 0.2 A.dm^{-2}.

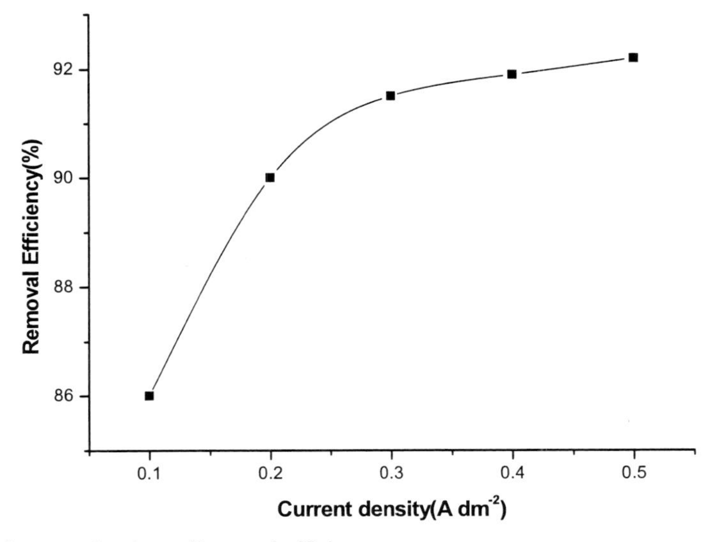

Figure 2. Effect of current density vs Removal efficiency.

From the results it is found that boron removal efficiency increased with the pH up to 7.0 and then decreased. The maximum removal efficiency for the removal of boron is 90% at pH 7 and the minimum efficiency is 72% at pH 12.

The decrease of removal efficiency at more acidic and alkaline pH was observed by many investigators and was attributed to an amphoteric behavior of $Al(OH)_3$ which leads to soluble Al^{3+} cations (at acidic pH) and to monomeric anions $Al(OH)^{4-}$ (at alkaline pH). It is well known that these soluble species are not useful for water treatment. When the initial pH was kept in neutral, all the aluminum produced at the anode formed polymeric species ($Al_{13}O_4(OH)_{24}{}^{7+}$) and precipitated $Al(OH)_3$ leading to more removal efficiency. Apart from the above, when solution pH was alkaline, borate ions in solution was dominantly $B(OH)_4^-$ form. When solution pH was acidic, borate ions in solution was dominantly $B(OH)_3$ form. The highest boron removal efficiency was obtained at pH 7.0 because boron was at $B(OH)_3$ form and the formation $Al(OH)_3$ was a quite high at this pH.

3. Effect of Inter Electrode Distance

To optimize the inter-electrode distance of electrodes, various experiments were conducted from 0.5 - 1.0 cm keeping all other conditions constant. Table 1 shows the cell voltage and removal efficiencies obtained for different inter-electrode distances. It was found that when the value of distance was increased from 0.5 to 1.0 cm, the boron removal efficiency decreased from 90 to 69 %. The cell voltage was increases with increase in distance and leading to higher energy consumption. Hence, the optimum distance for electrode was 0.5cm was fixed for further studies.

Table 1. Effect of Interelectrode distance at current density 0.2A dm^{-2}, concentration 5mg L^{-1} and at pH7

Interelectrode distance (cm)	Voltage (V)	Removal efficiency (%)	Energy consumption kWhr/kL
0.5	2.88	90	1.177
0.6	3.06	83	1.248
0.7	3.15	78	1.467
0.8	3.31	75	1.553
0.9	3.79	71.5	1.719
1.0	4.0	69	1.893

4. Effect of Initial Boron Concentration

In order to evaluate the effect of initial boron concentration, experiments were conducted at varying initial concentration from 3-7 mg L^{-1}. Figure 3 shows that the uptake of boron ($mg.g^{-1}$) increased with increase in boron concentration and remained nearly constant after equilibrium time. The equilibrium time was found to be 60 min for all concentration studied. The amount of boron adsorbed (q_e) increased from 1.8820 to 5.2144 mg g^{-1} as the concentration increased from 3-7 mg L^{-1}. The figure also shows that the adsorption is the rapid in the initial stages and remains almost constant with the progress of the adsorption. The plots are single, smooth and continuous curves leading to saturation, suggesting the possible monolayer coverage to boron on the surface of the adsorbent [30].

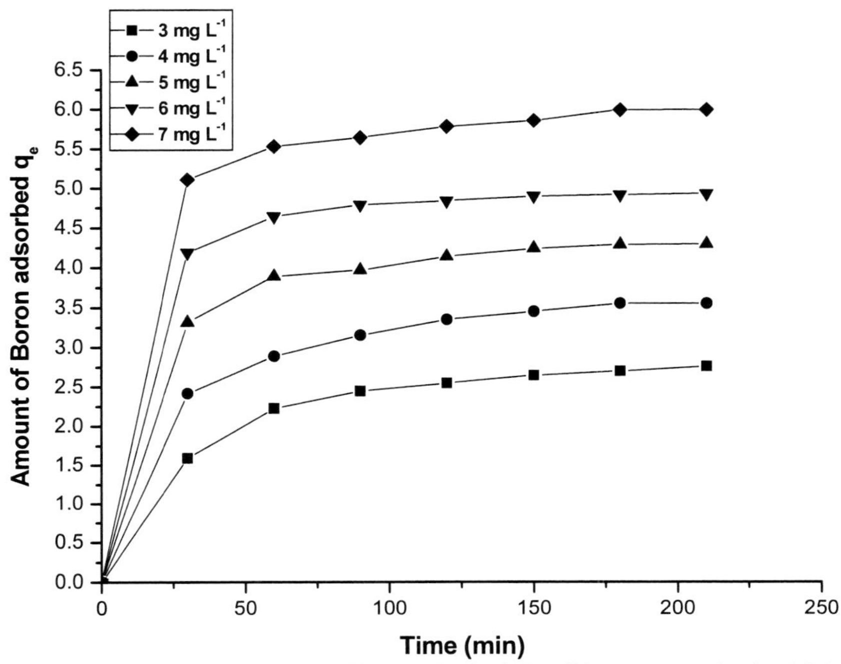

Figure 3. Effect of agitation time and amount of boron adsorbed. Conditions, current density 0.2 A dm^{-2}; pH of the electrolyte: 7.0; temperature:305K.

5. Effect of Coexisting Ions

5.1. Carbonate

Effect of carbonate on boron removal was evaluated by increasing the carbonate concentration from 2 to 250 mg L^{-1} in the electrolyte. The removal efficiencies are 90, 89.2, 49.2, 45.8, 22 and 12% for the carbonate ion concentration of 0, 2, 5, 65, 150 and 250 mg L^{-1} respectively. From the results it is found that the removal efficiency of the boron is not affected by the presence of carbonate below 2 mg L^{-1}. Significant reduction in removal efficiency was observed above 5 mg L^{-1} of carbonate concentration is due to the passivation of anode resulting, the hindering of the dissolution process of anode.

5.2. Phosphate

The concentration of phosphate ion was increased from 2 to 50 mg L^{-1}, the contaminant range of phosphate in the ground water. The removal efficiency for boron was 90, 89.5, 58, 47 and 32% for 0, 2, 5, 25 and 50 mg L^{-1} of phosphate ion respectively. There is no change in removal efficiency of boron below 2 mg L^{-1} of phosphate in the water. At higher concentrations (at and above 5mg L^{-1}) of phosphate, the removal efficiency decreases drastically. This is due to the preferential adsorption of phosphate over boron as the concentration of phosphate increase.

5.3. Arsenic

From the results it is found that the efficiency decreased from 90, 86.3, 77.1, 69 and 35% by increasing the concentration of arsenate from 0, 0.2, 0.5, 2 and 5 mg L^{-1}. Like phosphate ion, this is due to the preferential adsorption of arsenic over boron as the concentration of arsenate increases. So, when arsenic ions are present in the water to be treated arsenic ions compete greatly with boron ions for the binding sites.

5.4. Silicate

From the results it is found that no significant change in boron removal was observed, when the silicate concentration was increased from 0 to 2 mg L^{-1}. The respective efficiencies for 0, 2, 5, 10 and 15 mg L^{-1} of silicate are 90, 89, 70, 60 and 41%.

The removal of boron decreased with increasing silicate concentration from 2 to 15 mg L^{-1}. In addition to preferential adsorption, silicate can interact with aluminum hydroxide to form soluble and highly dispersed colloids that are not removed by normal filtration.

5.5. Fluoride

The efficiency was found to decrease with the increase in concentration of fluoride. The respective removal efficiencies were 90, 81.5, 79, 60.4, to 33% for 0, 0.2, 0.5, 2.0 and 5.0mg/L. Like other ion, the decrease in removal efficiency is due to the preferential adsorption of fluoride over boron as the concentration of fluoride increases. So, when fluoride ions are present in the water to be treated fluoride ions compete greatly with boron ions for the binding sites.

6. Adsorption Kinetics

In order to establish kinetic of boron adsorption, adsorption kinetics of aluminium alloy was investigated by using first order, second order kinetic models, Elovich and intraparticle diffusion.

6.1. First Order Lagergren Model

The first order Lagergren model is generally expressed as follows [31]

$$dq_t/dt = k_1 (q_e-q_t) \qquad (8)$$

where q_e and q_t are the adsorption capacities at equilibrium and at time t (min) respectively, and $k_1(min^{-1})$ is a rate constant. Eq. (8) can be linearized for use in the kinetic analysis of experimental analysis by applying boundary conditions t=0 to t=t and q_t = 0 to q_t = q_t as follows,

$$\log (q_e-q_t)=\log (q_e)-k_1t / 2.303 \qquad (9)$$

The values of log (q_e-q_t) were linearly correlated with t. The plot of log (q_e-q_t) vs t should give the linear relationship from which k_1 and q_e can be determined by the slope and intercept of the respectively. Table 2 shows the computed results of first order kinetics.

Table 2. Comparison between the experimental and calculated q_e values for different initial boron concentration in first order and second order adsorption isotherm at room temperature

Concentration ($mg\ L^{-1}$)	q_e (exp)	First order adsorption			Second order adsorption		
		q_e(Cal)	$K_1 \times 10^4$ ($min\ mg^{-1}$)	R^2	q_e(Cal)	$K_2 \times 10^4$ ($min\ mg^{-1}$)	R^2
3	1.8820	26.31	-0.0037	0.8234	2.3440	0.0444	0.9988
4	2.5961	28.42	-0.0060	0.8176	3.2214	0.0551	0.9947
5	3.6652	29.77	-0.0072	0.8341	4.1262	0.0623	0.9899
6	4.4426	30.43	-0.0077	0.9044	5.0011	0.0740	0.9941
7	5.2144	31.42	-0.0081	0.8046	6.1429	0.0799	0.9932

6.2. Second Order Lagergren Model

The lagregren second order kinetic model is expressed as [32]

$$dq_t/dt = k_2 (q_e-q_t)^2 \quad (10)$$

Where k_2 is the rate constant of second order adsorption. The integrated form of Eq. (10) with the boundary condition t=0 to >0 (q= 0 to >0) is

$$1/(q_e-q_t) = 1/q_e + k_2 t \quad (11)$$

Eq. (11) can be rearranged and linearized as,

$$t/q_t = 1/k_2 q_e^2 + t/q_e \quad (12)$$

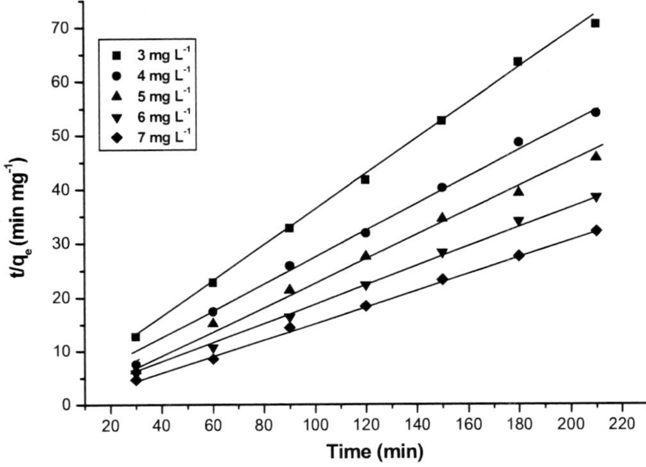

Figure 4. Second order kinetic model plot of different concentrations of boron. Conditions, current density: 0.2 A dm^{-2}; temperature: 305K; pH of the electrolyte:7.

The plot of t/q_t and time (t) (Figure 4) gave a linear relationship from which q_e and k_2 can be determined from the slope and intercept of the plot with high regression co-efficient. Table 2 shows the computed results of second order kinetics. The calculated q_e values well agree with experimental q_e values with high regression value.

6.3. Elovich Equation

The Elovich model equation is generally expressed as [33]

$$dq_t /dt = \alpha \exp(-\beta q_t) \tag{13}$$

the simplified form of Elovich Eq. (13) is

$$q_t = 1/\beta \log_e (\alpha \beta) + 1/ \beta \log_e(t) \tag{14}$$

where α is the initial adsorption rate (mg $g^{-1}.h^{-1}$) and, β is the desorption constant (g mg^{-1}). If boron adsorption fits the elovich model, a plot of q_t vs $\log_e(t)$ should yield a linear relationship with the slope of $(1/ \beta)$ and an intercept of $1/\beta \log_e (\alpha \beta)$. Table 3 depicts the results obtained from Elovich equation. Lower regression value shows the inapplicability of this model.

Table 3. Elovich model and Intra particle diffusion for different initial boron concentrations at temperature 305 K and pH 7

Elovich model			Intra particle diffusion		
A (mg/g.h)	β (g/mg)	R^2	k_{id} (l/h)	A (%/h)	R^2
13.36	54.38	0.9762	40.28	0.133	0.9914
4.69	39.26	0.9435	36.35	0.165	0.9770
1.26	20.22	0.9332	32.14	0.198	0.9681
1.03	19.34	0.9862	30.26	0.268	0.9763
0.82	12.36	0.9662	28.32	0.339	0.9798

6.4. Intra-Particle Diffusion

The intraparticle diffusion model is expressed [34, 35]

$$R = k_{id} (t)^{\alpha z} \tag{15}$$

A linearized form the Eq. (15) is followed by

$$\log R = \log k_{id} + a \log(t) \tag{16}$$

in which 'a' depicts the adsorption mechanism and k_{id} may be taken as the rate factor (percent of boron adsorbed per unit time) . Lower and higher value of k_{id} illustrates an enhancement in the rate of adsorption and better adsorption with improved bonding between pollutant and the adsorbent particles respectively. The results are presented in Table 3.

Table 2 and 3 depicts the computed results obtained from first order, second order, Elovich and intraparticle diffusion. From tables, it is found that the correlation coefficient decreases from second order, first order, intraparticle diffusion to Elovich model. This indicates that the adsorption follows the second order than the other models. Further, the calculated q_e values well agrees with the experimental q_e values for second order kinetics model concluding that the second order kinetics equation is the best fitting kinetic model.

7. Adsorption Isotherm

In order to investigate the adsorption isotherm Freundlich, Langmuir and Dubinin-Radushkevich (D-R) isotherm models were analyzed to quantify adsorption capacity of adsorbent. To determine the isotherms, the initial pH was kept at 7 and the concentration of boron used was in the range of 3-7 mg L^{-1}.

7.1. Freundlich Isotherm

According to Freundlich isotherm model, initially amount of adsorbed compounds increases rapidly, this increase slow down with the increasing surface coverage. The linearised in logarithmic form and the Freundlich constants can be expressed as [36, 37]

$$\log q_e = \log k_f + n \log C_e \quad (17)$$

where, k_f is the Freundlich constant related to adsorption capacity, n is the energy or intensity of adsorption, C_e is the equilibrium concentration of boron (mg L^{-1}). In testing the isotherm, the boron concentration used was 3-7 mg L^{-1} and at an initial pH 7, the adsorption data is plotted as log q_e versus log C_e and should result in a straight line with slope n and intercept k_f. The intercept and the slope are indicators of adsorption capacity and adsorption intensity, respectively. The value of n falling in the range of 1-10 indicates favorable sorption. The Freundlich constants k_f and n values are 0.3944 (mg g^{-1}) and 0.8126(L mg^{-1}) respectively. It has been reported that values of n lying between 0 and 10 indicate favorable adsorption. From the analysis of the results it is found that the Freundlich plots fit satisfactorily with the experimental data obtained in the present study. This is agreed with the results presented in the literature [38].

7.2. Langmuir Isotherm

An alternative equation was derived by Langmuir on the basis of a definite case of the nature of the process of adsorption from solution. The maximum adsorption occurs when molecules adsorbed on the surface of the adsorbent form a saturated layer. The linearized form of Langmuir adsorption isotherm model is, [39]

$$C_e/q_e = 1/q_m k_a + C_e/q_m \quad (18)$$

where C_e is the concentration of the boron solution (mg L^{-1}) at equilibrium, q_m is the adsorption capacity (Langmuir constant) and k_a is the energy of adsorption. Figure 5 shows the Langmuir plot with experimental data. Langmuir plot is a better fit with the experimental data compare to Freundlich plots. The value of the adsorption capacity q_m as found to be 73.24 mg g^{-1}, which is higher than that of other adsorbents studied.

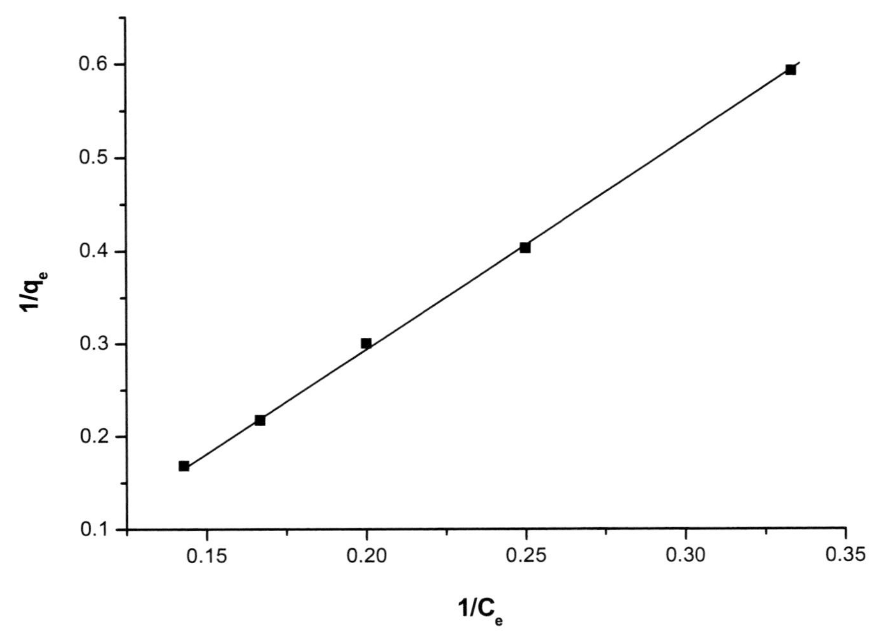

Figure 5. Langmuir plot ($1/C_e$ Vs $1/q_e$).

The essential characteristics of the Langmuir isotherm can be expressed as the dimensionless constant R_L. [40]

$$R_L=1/(1+bc_o) \quad (19)$$

where R_L is the equilibrium constant it indicates the type of adsorption, b, is the Langmuir constant. C_o is various concentration of boron solution. The R_L values between 0 and 1 indicate the favorable adsorption. The R_L values were found to be between 0 and 1 for all the concentration of boron studied.

7.3. Dubinin-Radushkevich Isotherm

Dubinin-Radushkevich isotherm assumes that characteristic sorption curve is related to the porous structure of the sorbent and apparent energy of adsorption. This model is given by

$$q_e = q_s \exp(-B\varepsilon^2) \quad (20)$$

where ε is polanyi potential, equal to RT $\ln(1+1/C_e)$, B is related to the free energy of sorption and q_s is the Dubinin-Radushkevich (D-R) isotherm constant [41]. The linearized form is,

$$\ln q_e = \ln q_s - 2B\ RT \ln[1+1/C_e] \quad (21)$$

The isotherm constants of q_s and B are obtained from the intercept and slope of the plot of ln q_e versus ε^2 [42] The constant B gives the mean free energy of adsorption per molecule of the adsorbate when it is transferred from the solid from infinity in the solution and the relation is given as,

$$E = [1/\sqrt{2B}] \quad (22)$$

The magnitude of E is useful for estimating the type of adsorption process. It was found to be 12.30 KJ mol^{-1}, which is too much smaller than the energy range of adsorption reaction, 8-16 KJ mol^{-1} [43]. So the type of adsorption of boron on Aluminum hydroxide was defined as chemical adsorption.

The correlation co-efficient values of different isotherm models are listed in Table 4. The Langmuir isotherm model has higher regression co-efficient ($R^2 = 0.999$) when compared to the other models. The value of R_L for the Langmuir isotherm was calculated between 0 and 1, indicating the favorable adsorption of boron.

Table 4. Constant parameters and correlation coefficient for different adsorption isotherm models for boron adsorption at 5 mg L^{-1}

Isotherm	Constants			
Langmuir	Q_o(mg g^{-1})	b(L mg $^{-1}$)	R_L	R^2
	73.24	0.0019	0.8942	0.9997
Freundlich	K_f(mg g^{-1})	n (L mg^{-1})		R^2
	0.3944	0.8126		0.9841
D-R	Q_s(x 10^3 mol g^{-1})	B(x 10^3 mol^2 KJ $^{-2}$)	E(KJ mol^{-1})	R^2
	0.422	0.561	12.30	0.8984

8. Thermodynamic Studies

To understand the effect of temperature on the adsorption process, thermodynamic parameters should be determined at various temperatures. The energy of activation for adsorption of boron can be determined by the second order rate constant is expressed in Arrhenius form.

$$\ln k_2 = \ln k_o - E/RT \quad (23)$$

where k_o is the constant of the equation (g mg^{-1}.min^{-1}), E is the energy of activation (J mol^{-1}), R is the gas constant (8.314 J mol^{-1} K^{-1}) and T is the temperature in K. Figure 6 shows that the rate constants vary with temperature according to equation (23). The activation energy (13.2 KJ mol^{-1}) is calculated from slope of the fitted equation. The free energy change is obtained using the following relationship

$$\Delta G = -RT\ln K_c \quad (24)$$

where ΔG is the free energy (KJ mol^{-1}), K_c is the equilibrium constant, R is the gas constant and T is the temperature in K. The K_c and ΔG values are presented in Table 5. From the table it is found that the negative value of ΔG indicates the spontaneous nature of adsorption. Other thermodynamic parameters such as entropy change (ΔS) and enthalpy change (ΔH) were determined using van't Hoff equation,

$$lnK_c = \frac{\Delta S}{R} - \frac{\Delta H}{RT} \quad (25)$$

The enthalpy change (ΔH = 2.9661 kJ mol^{-1}) and entropy change (ΔS=13.445J mol^{-1} K^{-1}) were obtained from the slope and intercept of the van't Hoff linear plots of lnk_c versus 1/T (Figure 7).

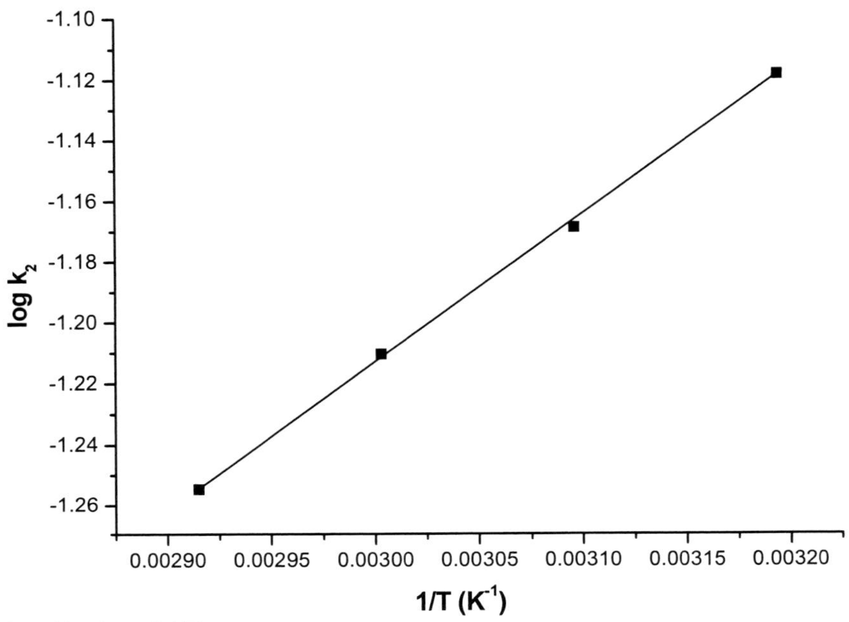

Figure 6. Plot of log k_2 and 1/T.

Table 5. Thermodynamics parameters for adsorption of boron

Temperature (K)	K_c	ΔG^o (J mol^{-1})	ΔH^o (KJ mol^{-1})	ΔS^o (J mol^{-1} K^{-1})
313	1.0063	-106.20		
323	1.1254	-210.75	2.9661	13.445
333	1.1499	-355.17		
343	1.1674	-457.18		

Positive value of enthalpy change (ΔH) indicates that the adsorption process is endothermic in nature, and the negative value of change in internal energy (ΔG) show the spontaneous adsorption of boron on the adsorbent. Positive values of entropy change show the increased randomness of the solution interface during the adsorption of boron on the adsorbent (Table 5). Enhancement of adsorption capacity of electrocoagulant (Aluminum hydroxide) at higher temperatures may be attributed to the enlargement of pore size and or activation of the adsorbent surface. Using Lagergren rate equation, first order rate constants and correlation co-efficient were calculated for different temperatures (305-343 K). The calculated 'q_e' values obtained from the second order kinetics agrees with the experimental 'q_e' values better than the first order kinetics model, indicating adsorption following second order kinetics.

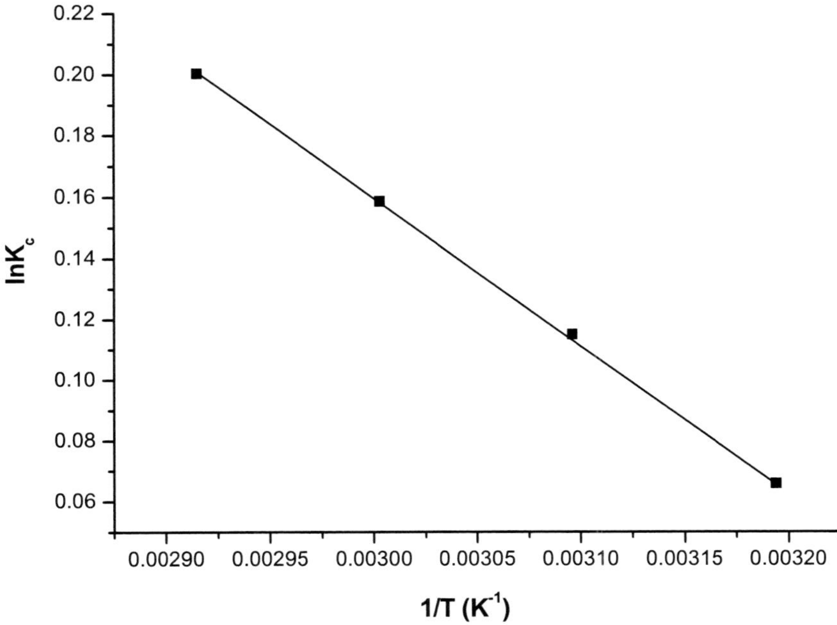

Figure 7. Plot of ln K_c and 1/T.

Table 6 depicts the computed results obtained from first and second order kinetic models. The diffusion co-efficient (D) for intraparticle transport of boron species into the adsorbent particles has been calculated at different temperature by,

$$t_{1/2} = 0.03 x r_o^2 / D \tag{26}$$

where $t_{1/2}$ is the time of half adsorption (s), r_o is the radius of the adsorbent particle (cm), D is the diffusion co-efficient in $cm^2\ s^{-1}$. For all chemisorption system the diffusivity co-efficient should be 10^{-5} to 10^{-13} $cm^2\ s^{-1}$ [44]. In the present work, D is found to be in the range of 10^{-10} $cm^2 s^{-1}$. The pore diffusion coefficient (D) values for various temperatures and different initial concentrations of boron are presented in Table 7 respectively.

Table 6. Comparison between the experimental and calculated q_e values for the boron concentration of 5 mg L^{-1} in first and second order adsorption kinetics

Concentration (mg L^{-1})	q_e (exp)	First order adsorption			Second order adsorption		
		q_e(Cal)	K_1 x 10^4 (min mg^{-1})	R^2	q_e(Cal)	K_2 x 10^4 (min mg^{-1})	R^2
313	3.2571	28.66	-0.0033	0.8001	3.4216	0.0856	0.9994
323	3.5211	30.42	-0.0042	0.7622	3.6441	0.0669	0.9996
333	3.7662	33.55	-0.0068	0.8846	3.9148	0.0601	0.9946
343	3.9033	35.01	-0.0089	0.6477	4.1265	0.0554	0.9849

Table 7. Pore diffusion coefficients for the adsorption of boron at temperature 305K

Concentration (mg L^{-1})	Pore diffusion Constant $D \times 10^{-9}$ ($cm^2\ s^{-1}$)
3	2.47
4	1.22
5	0.88
6	0.59
7	0.41

Temperature (K)	Pore diffusion Constant $D \times 10^{-9}$ ($cm^2\ s^{-1}$)
313	0.51
323	0.83
333	1.87
343	2.35

9. Monopolar and Bipolar Configuration

The influences of selected parameters on removal of boron by monopolar process obtained are compared with results obtained by bipolar electrocoagulation process. The removal efficiency of boron was almost same for both monopolar and bipolar experiments. So it can be concluded there is no significant difference in $Al(OH)_3$ mass ratio between monopolar and bipolar system. Figure 8 shows the significant removal efficiency of monopolar and bipolar plate on varying time.

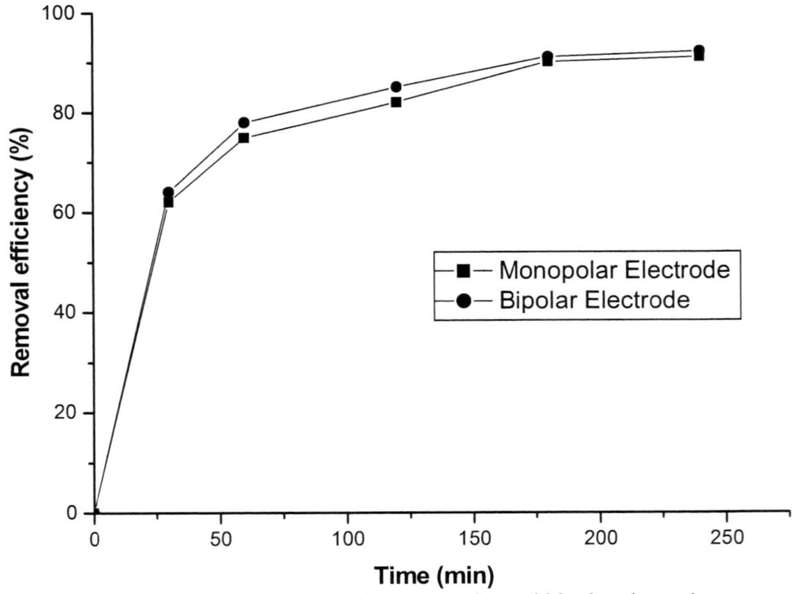

Figure 8. Plot between time vs removal efficiency for monopolar and bipolar electrodes at current density of 0.2 A dm^{-2}, concentration of 5 mg L^{-1} temperature of 303K, pH of 7.00.

10. Process Scale-Up

On the basis of results obtained on the laboratory scale, a large capacity cell was designed, fabricated and operated for the removal of boron from drinking water. The solution of 8.5 L was used for each experiment as the electrolyte. A cell [0.35(length) X0.25 (width) X 0.25 m (height)] was fitted with PVC cover having suitable holes to introduce anode, cathode, thermometer and the electrolyte acted as the cell. A aluminium anode [0.17 (width) X 0.18 m (height)] was used. Stainless steel plates of same dimension as that of anode were used as cathode. The cell was operated at $0.2 A.dm^{-2}$ and the electrolyte pH of 7.0. The results showed that the maximum removal efficiency of 89.5% was achieved at a current density of $0.2 A.dm^{-2}$ and a pH of 7 using aluminium as the anode and stainless steel as the cathode. The results were consistent with the results obtained from the laboratory scale, showing the robustness of the process.

11. Material Characterization

11.1. SEM Studies

SEM images of aluminium electrode, before and after, electrocoagulation of arsenate electrolyte was obtained to compare the surface texture. Figure 9 (a) shows the original aluminium plate surface prior to its use in electrocoagulation experiments.

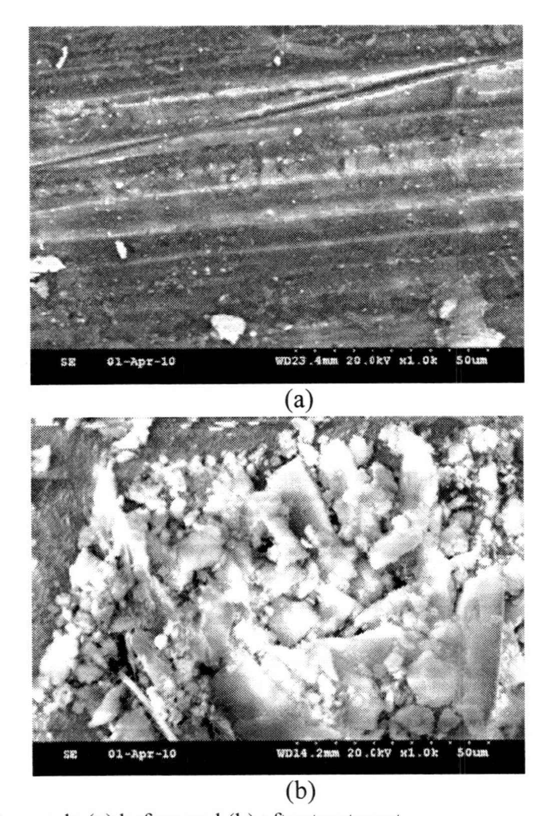

Figure 9. SEM image of the anode (a) before and (b) after treatment.

The surface of the electrode is uniform. Figure 9 (b) shows the SEM of the same electrode after several cycles of use in electrocoagulation experiments. The electrode surface is now found to be rough, with a number of dents.

These dents are formed around the nucleus of the active sites where the electrode dissolution results in the production of aluminium hydroxides.

The formation of a large number of dents may be attributed to the anode material consumption at active sites due to the generation of oxygen at its surface.

11.2. FTIR Studies

Figure 10 presents the FT-IR spectrum of boron - aluminum hydroxide. The sharp and strong peak at 3226.50 cm^{-1} is due to the O–H stretching vibration in the $Al(OH)_3$ structures.

The 1644.70 cm^{-1} peak indicates the bent vibration of H–O–H. The strong peak at 1077.48 cm^{-1} is assigned to the Al–O-H bending. B-O vibration at 708 cm^{-1} also observed [45,46].

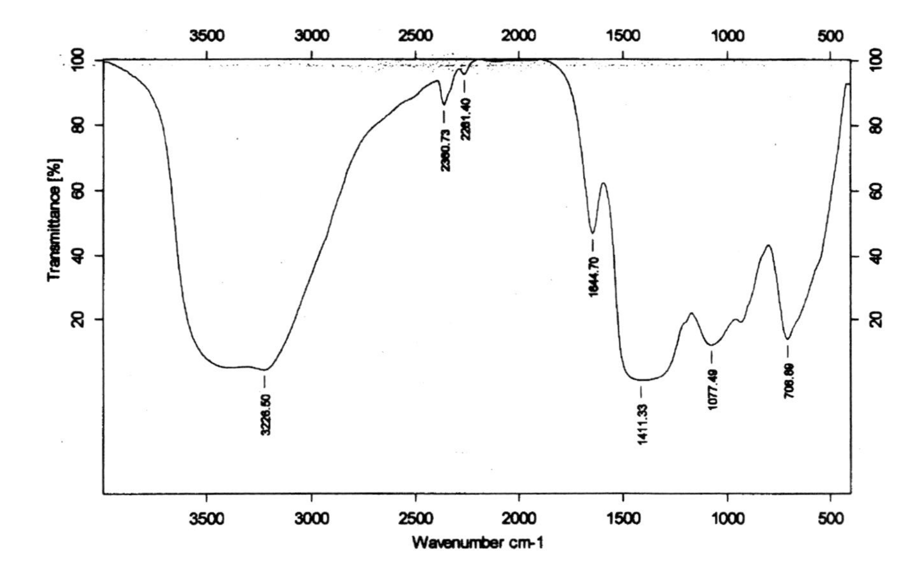

Figure 10. FTIR spectrum of boron-adsorbed aluminium hydroxide.

CONCLUSION

The results showed that the optimized removal efficiency of 90 % was achieved at a current density of 0.2A dm^{-2} and pH of 7.0 using aluminum alloy as anode and stainless steel as cathode with an interelectrode distance of 0.5 cm. The aluminum hydroxide generated in the cell remove the boron present in the water and made it for drinking. The results indicate that the process can be scaled up to higher capacity. The adsorption of boron preferably fitting the Langmuir adsorption isotherm suggests monolayer coverage of adsorbed molecules. The adsorption process follows second order kinetics. Temperature studies showed that adsorption was endothermic and spontaneous in nature.

Acknowledgments

The authors wish to express their gratitude to the Director, Central Electrochemical Research Institute, Karaikudi to publish this chapter.

References

[1] Peter A, Distribution of boron in the environment. *Bio. Trace Element Res.* 66:131- 143 (1998).

[2] Senkal BK and Bicak N, Polymer supported iminodipropylene glycol functions for removal of boron. *React Funct Polym* 55: 27–33(2003)

[3] Kistler RB and Helvacı C, "Boron and Borates". Industrial Minerals and Rocks (Donald D. Carr editor) 6 th Edition (Society of Mining, Metalurgy and Exploration, Inc.): 171–186 (1994).

[4] Bryjak M, Wolska J and Kabay N, Removal of boron from seawater by adsorption–membrane hybrid process: implementation and challenges. *Desalinaton* 223:57-62 (2008).

[5] *Yurdakoc M, Seki Y, Karahan S and Yurdakoc K,* Kinetic and thermodynamic studies of boron removal by Siral 5, Siral 40, and Siral 80. *J. Colloid Interface Sci. 286: 431-440(2005).*

[6] *Baek KW, Song SH, Kang SH, Rhee YW, Lee CS, Lee BJ, Hundson S and Hwang TS,* Adsorption Kinetics of Boron by Anion Exchange Resin in Packed Column Bed. *J. Ind. Eng. Chem. 13: 452-456 (2007).*

[7] Sartaj M and Fernandes L, Adsorption of boron from landfill leachate by peat and the effect of environmental factors. *J. Environ. Eng. Sci.* 4: 19–28 (2005).

[8] Bouguerra W, Mnif A, Hamroouni B and Dhahbi M, Boron removal by adsorption onto activated alumina and by reverse osmosis. *Desalination* 223:31-37(2008).

[9] Xu Y and Jiang JQ, Technologies for boron removal. *Ind. Eng. Chem. Res.* 47:16-24 (2008).

[10] Kabay N, Arar O,Acar F, Ghazal A,Yuksel U and Yuksel M, Removal of boron From water by electrodialysis: effect of feed characteristics and interfering ions. *Desalination* 223: 63-72 (2008).

[11] Guidelines for Drinking-Water Quality 2nd edn, WHO(1998).

[12] Jiang JQ, Xu Y, Quill K, Simon J and Shettle K, Laboratory Study of Boron Removal by Mg/Al Double-Layered Hydroxides. *Ind. Eng. Chem. Res.* 46: 4577 (2007).

[13] EnnilKose T and Ozturk N, Boron removal from aqueous solutions by ion-exchange resin: Column sorption–elution, *J. Hazard Mat.* 152: 744-749 (2008)

[14] Bick A and Ora G,Post-treatment design of seawater reverse osmosis plants: boron removal technology selection for potable water production and environmental *Desalination,* 178: 233-246 (2005).

[15] Srivastava VC, Mall ID and Mishra IM, Characterization of mesoporous rice husk ash (RHA) and adsorption kinetics of metal ions from aqueous solution onto RHA. *J. Hazard Mater* B134: 257- 267(2006).

[16] Inglezakis VJ, Loizidou MD and Grigoropoulou HP, on exchange of Pb2+, Cu2+, Fe3+, and Cr3+ on natural clinoptilolite: selectivity determination and influence of acidity on metal uptake. *J. Colloid Interf. Sci.* 261: 49-54(2003)
[17] Hazef AI, Manharawy MS and Khedr MA, RO membrane removal of unreacted chromium from spent tanning effluent. A pilot-scale study, Part 2. *Desalination* 144: 237-242(2002).
[18] Vik EA, Carlson DA, Eikum AS and Gjessing ET, Electrocoagulation of potable water. *Water Res.* 18:1355-1360(1984).
[19] Thella K, Verma B, Srivastava CV and Srivastava KK, Electrocoagulation study for the removal of arsenic and chromium from aqueous solution. *J. Environ. Sci. Health Part A* 43: 554-562(2008).
[20] Sahin S, A mathematical relationship for the explanation of ion exchange for boron adsorption. *Desalination* 143 :35–43(2003).
[21] Vasudevan S, Lakshmi J and Sozhan G, Studies on the Mg-Al-Zn – alloy as anode for the removal of fluoride from drinking water in electrolcoagulation process. *Clean* 37: 372 – 378 (2009).
[22] Chen X, Chen G and Yue PL, Modeling the electrolysis voltage of electrocoagulation process using aluminum electrodes. *Chem Eng. Sci.* 57: 2449- 2455 (2002).
[23] Chen G, Electrochemical technologies in wastewater treatment. *Sep. Purif. Technol.* 38: 11-41(2004).
[24] Vasudevan S, Lakshmi J and Vanathi R, Electrochemical coagulation for chromium removal: Process optimization, Kinetics, Isotherm and Sludge characterization. *Clean* 38: 9 – 16(2010).
[25] Vasudevan S, Lakshmi J and Sozhan G, Studies on the removal of iron from drinking water by electrocoagulation – A clean process, *Clean*, 37: 45-51(2009).
[26] Vasudevan S, Lakshmi J, Jayaraj J and Sozhan G, Remediation of phosphate-contaminated water by electrocoagualtion with aluminium, aluminum alloy and mild steel anodes. *J. Hazard. Mater.* 164: 1480-1486(2009).
[27] Yilmaz EA, Boncukcuoglu R, Muhtar Kocakerim M, Tolga Yilmaz M and Paluluoglu C, Boron removal from geothermal waters by electrocoagulation. *J. Hazard. Mater* 153: 146–151(2008).
[28] Yilmaz EA, Boncukcuoglu R and Muhtar Kocakerim M, An empirical model for parameters affecting energy consumption in boron removal from boron-containing wastewaters by electrocoagulation. *J. Hazard. Mater* 144: 101–107 (2007).
[29] Golder AK, Samantha AN and Ray, Removal of phosphate from aqueous solutions using calcined metal hydroxides sludge waste generated from electrocoagulation. *Sep. Purif. Technol.* 52: 102-109 (2006).
[30] Namasivayam C and Prathap K, Recycling Fe(III)/Cr(III) hydroxide, an industrial solid waste for the removal of phosphate from water. *J. Hazard. Mater* 123B: 127-134 (2005).
[31] Singh KK, Rastogi R and Hasan SH, Removal of cadmium from wastewater using agriculture waste rice polish. *J. Hazard. Mater* 121A: 51-58 (2005).
[32] Mckay G and Ho YS, The sorption of lead(II) ions on peat. *Water Res.* 33: 578-584 (1999).
[33] Oke IA, Olarinoye NO and Adewusi SRA, Adsorption kinetics for arsenic removal from aqueous solutions by untreated powdered eggshell. *Adsorption.,*14: 73-83 (2008).

[34] Weber Jr WJ and Morris JC, Kinetics of adsorption on carbon from solutions. *J. Sanit. Div. Am. Soc. Civ. Eng.* 89: 31-60 (1963).
[35] Allen SJ, Mckay G and Khader KHY, Intraparticle diffusion of basic dye during adsorption onto sphagnum peat. *Environ. Pollut.* 56: 39-50 (1989).
[36] Gasser MS, Morad GH and Aly HF, Batch kinetics and thermodynamics of chromium ions removal from waste solutions using synthetic adsorbents. *J. Hazard. Mater* 142: 118-129 (2007).
[37] Uber FH, Die adsorption in losungen. *Z Phy Chem.* 57: 387-470 (1906).
[38] Giles CH, MacEwan TH, Nakhwa SN and Smith D, Studies in adsorption. Part XI. A system of classification of solution adsorption isotherms, and its use in diagnosis of adsorption mechanisms and in measurement of specific surface areas of solids *J. Chem. Soc.* 12: 3973-3993 (1960).
[39] Langmuir I, The adsorption of gases on plane surfaces of glass, mica and platinum. *J. Am. Chem. Soc.* 40: 1361-1403 (1918).
[40] Michelson LD, Gideon PG, Pace EG and Kutal LH, US Department of Industry, *Office of Water Research and Technology Bulletin* Washington DC, 1975.
[41] Tan IAW, Hameed BH and Ahmed AL, Equilibrium and kinetic studies on basic dye adsorption by oil palm fibre activated carbon. *Chem. Eng. J.* 127: 111-119 (2007).
[42] Demiral H, Demiral I, Tumsek F and Karacbacakoglu B, Adsorption of chromium(VI) from aqueous solution by activated carbon derived from olive bagasse and applicability of different adsorption models. *Chem. Eng. J.* 144: 188-196(2008).
[43] Oguz E, Adsorption characteristics and kinetics of the Cr(VI) on the Thuja oriantalis. *Colloid Surf. A* 252: 121-128(2005).
[44] Yang XY and Al-Duri B, Application of branched pore diffusion model in the adsorption of reactive dyes on activated carbon. *Chem. Eng. J.* 83: 15-23(2001).
[45] Hernandez-Moreno MJ, Ulibarri MA, Renon JL and Serna CJ, IR characteristics of hydrotalcitelike compounds. *Phys. Chem. Miner* 12:34-38(1985).
[46] Weir CE and Lippincott ER, Infrared studies of aragonite. *J. Res. Natl. Bur. Standards*, 65A: 173-184 (1961).

In: Aluminum Alloys
Editor: Erik L. Persson

ISBN: 978-1-61122-311-8

Chapter 5

SIGNIFICANCE OF ALUMINIUM ALLOYS FOR CATHODIC PROTECTION TECHNOLOGY

*I. Gurrappa and I. V. S. Yashwanth**
Defence Metallurgical Research Laboratory
Kanchanbagh PO, Hyderabad-500 058, India

ABSTRACT

The present chapter provides a review on a variety of aluminium alloy sacrificial anodes used for protection against corrosion of structures submerged in a marine environment. The superiority of aluminium alloy anodes over other sacrificial anodes and their characterisation are explained in detail. The performance of anodes depends not only on the alloy composition but also on the treatment / homogenization technique that is to be used for their preparation. This aspect is reviewed critically as both these factors play an important role in enhancing the life of anodes significantly. The activation mechanism of aluminium alloy anodes based on the surface free energy concept is an advanced technique. This concept is explained after critically reviewing it and pointing out the disadvantages of mechanisms proposed by earlier investigators. It is also mentioned that the surface free energy phenomenon is a novel, simple and non-electrochemical tool for understanding their working mechanism, as well as for the development of efficient sacrificial anodes. Finally, the necessity of innovation of a smart anode to protect the structures against corrosion both in marine and fresh / river water environments, the critical steps involved in their development is emphasized.

1. INTRODUCTION

Development of novel materials that are light weight, smart and possess physical and engineering properties superior to the existing materials are constantly necessitated for continued technical advances in a variety of fields. In recent times, the design, development, processing and characterization of newer materials have been greatly aided by novel

* Email: igp1@rediffmail.com

approaches to materials design and synthesis that are based on intelligent and unified understanding of the processing - structure properties - performance relationships for a wide range of materials. Excellent corrosion resistance is of paramount importance for most of the applications. However, for cathodic protection of structures using galvanic system / sacrificial anodes, the characteristics are different. Here the anodes have to undergo uniform dissolution to provide sufficient current to maintain the required protection potential for a designed period. Aluminium alloy anodes are the most reliable and proved high performance systems for protection of marine structures, pipelines, offshore structures, submarines etc. The availability of aluminium alloys with requisite combination of alloying elements largely dictates the life of structures. Several attempts have been made by the researchers in the field to develop suitable high performance aluminium alloy anodes with necessary surface modifications, suitable treatments and alloying elements.

Cathodic protection is more reliable, effective and economic method for protection of a variety of pipelines, tanks, marine structures including ships hulls and submarines against corrosion. Cathodic protection works primarily by depressing the natural corrosion potential of the structure to be protected to a value where it does not corrode. It can be obtained either by using sacrificial anodes or by using impressed current cathodic protection (ICCP) system (Figure 1). The primary advantage of using sacrificial anodes is that the investment is low and they can be used in preference to an ICCP system when the current requirement is low. In addition, sacrificial anode protection is the most economical method for short term protection. On the basis of experience, it has been established that the corrosion of structural steel stops when it's potential is -800 mV against silver/silver chloride or +250 mV against zinc or -850mV against copper/copper sulphate reference electrode.

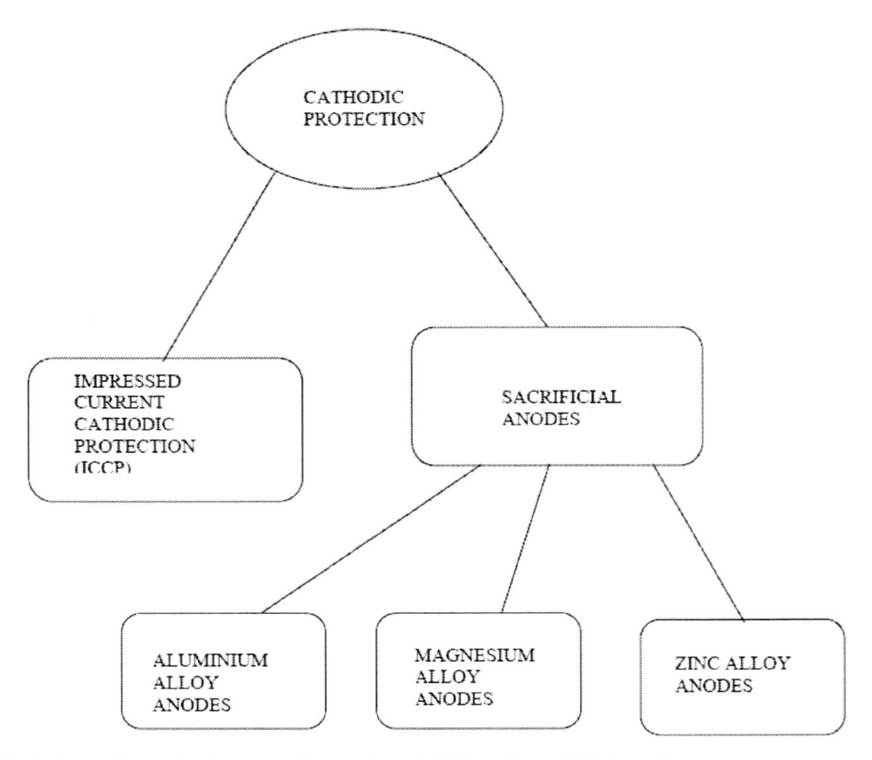

Figure 1. Techniques for cathodic protection and availability of sacrificial anodes.

This potential is the most significant measurement with respect to corrosion control. Table I provides the potential of structural steel under different conditions (over protection potential, optimum potential and corroding potential) with respect to three different reference electrodes which are normally in use for monitoring the effectiveness of cathodic protection or to determine the status of the structural steel in a given environment. For measurement of soil to pipe potentials, $Cu/CuSO_4$ electrode is the most suitable while high purity zinc and Ag/AgCl reference electrodes are ideal in marine environment. It is necessary to maintain the potential of the structures (pipelines / tanks/ ship hulls etc.) in a protective range because under protection or over protection are not useful. Under protection can not protect the structure completely from corrosion while over protection results in generating more hydrogen at the surface of structure (cathode) leading to either blistering or disbonding of organic coatings applied on the steel surface or hydrogen embrittlement of the steel (loss of ductility through absorption of hydrogen) or both. Apart from this, over protection causes wastage of electric power and increased consumption of anodes.

Table I. Corroding, optimum and overprotection w.r.t. $Cu/CuSO_4$, Ag/AgCl and Zn reference electrodes

Reference Electrode	Corroding potential	Optimum potential	Overprotection potential
Copper/Copper sulphate	-0.65 or more	-0.85 to –1.0 V	-1.05 or higher
Silver/Silver Chloride	-0.60 V or more	-0.80 to -0.95 V	-1.00 V or higher
Zinc	+0.45 V or higher	+0.150 to 0.250 V	+0.10 or less

Therefore, it is normal practice to design the cathodic protection system for a given structure to maintain the protective potentials during the designing period. The next section describes the basic principle of protecting the structures by using sacrificial anodes, different types of sacrificial anodes, merits and demerits of each type.

2. Sacrificial Anodes

The principle involved in the development of sacrificial anodes for cathodic protection purpose is that galvanic current flows when two dissimilar metals are electrically shorted in the conducting environment. The noble metal gets protected with consuming the less noble metal. Aluminium, Zinc and Magnesium alloys are more base materials than mild steel / carbon steel, which are the structural materials for the pipelines, offshore structures, ships, submarines etc and therefore produce galvanic current when coupled with the steel in the seawater resulting in their sacrificial dissolution for protecting the steel. As mentioned above, sacrificial anodes are often used in preference to ICCP system when the current requirements are low and in relatively low resistivity environments. Capital investment will generally be lower and it is often the most economical method for short time protection.

The salient features of Magnesium, Zinc and Aluminium alloy anodes are discussed below:

2.1. Magnesium Alloy Anodes

They have a highly negative free corrosion potential and thereby dissolve too vigorously in seawater. Therefore, magnesium alloy anodes are restricted to the protection of pipelines in the soil or the structures in estuarine waters where the resistivity is high enough to limit the effectiveness of zinc or aluminium alloy anodes. They have also used for protection of ships when they enter into high resistivity waters such as river water. The protection of storage tanks containing fresh or brackish water is another suitable application. These anodes are helpful in protecting condenser boxes in cooling water systems. The current efficiency, potential and life of magnesium alloy anodes are as follows:

i. The current efficiency is about 50% only
ii. Anodes have high driving voltage
iii. Areas in the vicinity of anodes get overprotected
iv. The life range is from 6 months to 20 months depending on the system design

2.2. Zinc Alloy Anodes

An important property of zinc alloy anodes is that the capacity of the anode is virtually unaffected by the operating current density. Therefore, zinc alloy anodes are generally used for protection of pipelines, where the effects of burial together with a high-duty coating tend to give low anode current densities. Zinc anodes can be successfully used for protection of structures in water or mud resistivity upto 1,000 ohms-cm. The driving voltage of zinc alloy anodes diminishes with increasing temperature such that they are virtually ineffective at temperatures of above 60^0 C. Therefore, zinc alloy anodes have limited application and can not be used for protection of structures where the temperature is high. These anodes are suitable for protecting cooling water systems, gas pipelines, storage tanks etc where the soil resistivity is upto 1000 ohms-cm. The following are the properties of zinc alloy anodes:

i. The efficiency is about 95%
ii. Anodes are too heavy and the structure gets appreciably loaded with dead weight. However, there may not be a problem with the pipelines buried in the soil.
iii. High purity zinc (99.99%) is essential for reliable performance

2.3. Aluminium Alloy Anodes

These anodes became popular because of their superiority over magnesium and zinc alloy anodes in respect of

a) low cost
b) long life
c) high energy capability and
d) light weight

These anodes are most ideal not only for the protection of weight limited structures such as submarines and weapon packages but also for all types of structures including offshore platforms and ships. In fact, the importance of aluminium as a sacrificial anode material was recognised in the early 1960s. However, unalloyed aluminium is not useful as sacrificial anode due to the formation of an impervious Al_2O_3 layer. Developmental work was therefore initiated during the same period to produce an aluminium alloy anode with suitable alloying elements such as zinc, mercury, indium, tin, magnesium, bismuth, cadmium etc to improve the cathodic protection properties of unalloyed aluminium. Schrieber and Reding [1] developed an alloy (Al-Zn-Hg) for the first time with mercury as an activator. This anode had shown a current efficiency in seawater exceeding 90% and current capacity of 2750 A-hr/kg.

This alloy became popular soon after this investigation and marketed under the trade name "Galvalum". However, mercury is an environmentally controversial element and some aluminium anode users prefer not to use anodes containing mercury. This aspect made Cathodic Protection Engineers find an alternative to mercury and successfully developed anodes with indium, bismuth, tin etc. as activators. Later work was concentrated on the electrochemical characteristics of various combinations of alloying elements and their optimum concentrations required for specific applications. The compositions used in commercial scale now-a-days generally contain 2 to 5% zinc and one of the elements such as indium, tin and bismuth as an activator. Based on the activating elements, the aluminium alloy anodes can be subdivided into three types:

i. Tin activated aluminium alloy anodes
ii. Bismuth activated aluminium alloy anodes and
iii. Indium activated aluminium alloy anodes

2.3.1. Tin Activated Aluminium Alloy Anodes

These anodes were developed by Keir et al [2] and have been in use for protection of marine structures. Its composition is (Al + 5%Zn + 0.12%Sn). The reported anode efficiency is 70%. The effectiveness of tin is attributed to its ability to enter the surface oxide film as Sn^{4+} ions and thereby creating additional cation vacancies. However, these anodes require controlled heat treatment for their performance and better efficiency and as such are not cost effective. Gurrappa et al [3] successfully avoided the costly heat treatment process by the controlled addition of small amount of bismuth to tin activated aluminium alloy anodes. The effect of bismuth in improving the polarisation behaviour of basic composition (Al + 5% Zn + 0.25% Sn) was studied in detail (Figure 2) [3]. It was reported that bismuth helps in expanding the aluminium matrix and thereby increases the tin solubility in aluminium, which is the reason for enhanced anode capacity comparable to that of heat-treated anodes. The results also show that anode capacities depend on various compositions of tin activated aluminium alloy anodes. Table II provides closed circuit potential of various compositions of tin activated aluminium alloy anodes including the heat-treated alloy. The anode capacity / efficiency is closely related to its closed circuit potential i.e. the lower the closed circuit potential, the higher the anode capacity of an alloy anode. The anode capacities of different tin activated aluminium alloy anodes are shown in Figure 3.

Table II. Characteristics of tin activated aluminium alloy anodes

S. No.	Anode Composition	Closed Circuit Potential (V)(SCE)
1.	Al + 5%Zn + 0.25%Sn	-0.973
2.	Al + 5%Zn + 0.25%Sn + 0.1%Bi	-0.977
3.	Al + 5%Zn + 0.25%Sn + 0.2%Bi	-0.982
4.	Al + 5%Zn + 0.25%Sn + 0.3%Bi	-0.996
5.	Al + 5%Zn + 0.25%Sn + 0.4%Bi	-0.978
6.	Al + 5%Zn + 0.25%Sn + 0.5%Bi	-0.974
7.	Al + 5%Zn + 0.25%Sn*	-0.993

* Heat-treated at 620^0 C for 24 hours and then water quenched.

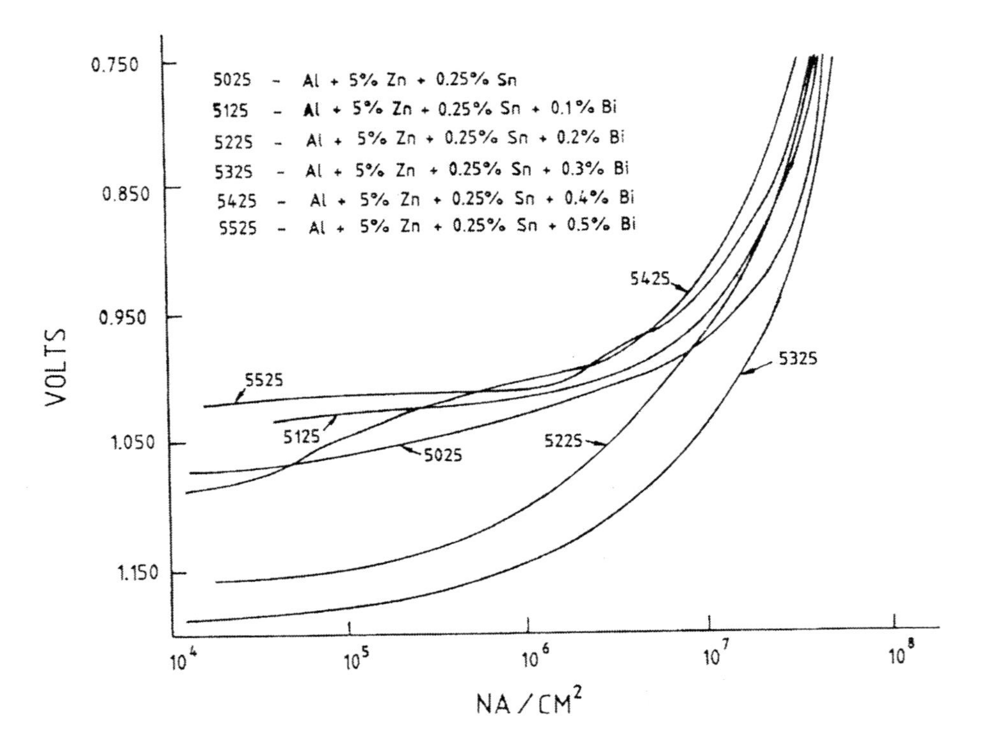

Figure 2. Anodic polarization characteristics of Al + 5% Zn + 0.25% Sn + varying percentages of bismuth.

There are different views on the role of tin and assumed that creation of additional cation vacancies by entering tin as quadravalent Sn^{4+} is responsible for improving the cathodic protection properties. However, no experimental evidence is available on the subject. Recently, the mechanism involved in increasing the anode efficiency before and after heat treatment of tin activated aluminium alloy anodes was explained on the basis of surface free energy concept which was evaluated by determining the Young's Modulus and atomic spacing of the anode materials experimentally [4]. It was clearly shown that decrease in surface free energy with the addition of tin to (Al + 5% Zn) and after the heat treatment is the principal reason for increasing the cathodic protection properties of these alloys.

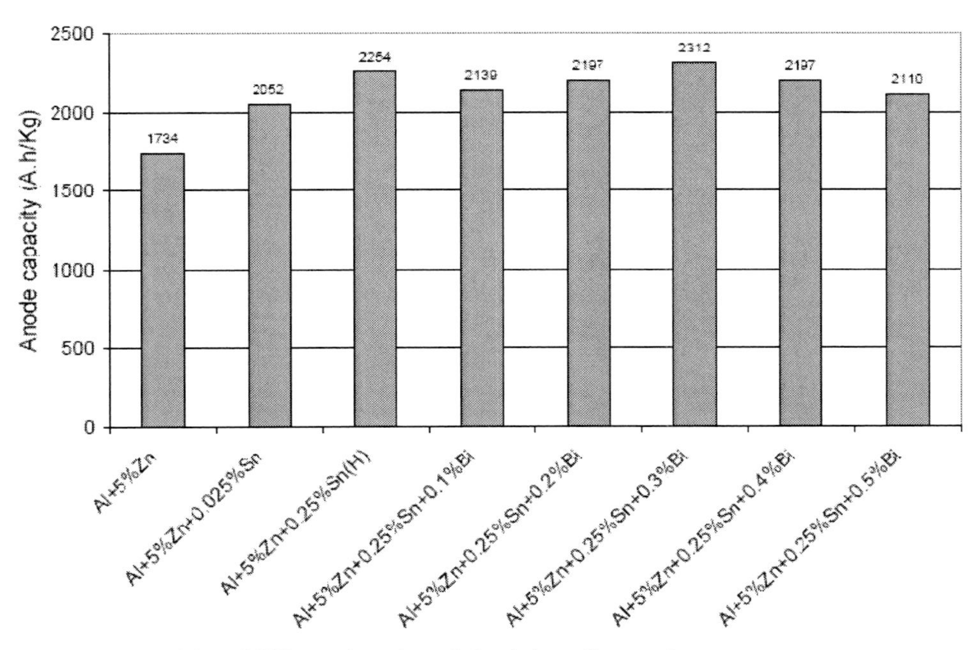

Figure 3. Anode capacities of different tin activated aluminium alloy anodes.

2.3.2. Bismuth Activated Aluminium Alloy Anodes

Murai et al [5] were reported that 0.005 to 1.0% bismuth addition is helpful in enhancing the anode efficiency of (Al + 5%Zn) from 65 to 71%. Bismuth addition to tin activated aluminium alloy anodes is advantageous in hot water boiler applications [6]. But the details are not available since this was patented. For seawater applications, suitable addition of bismuth was identified and the anode efficiency of (Al+ 5%Zn + 0.2%Bi) was improved to 83% by the addition of 2% magnesium [7]. It was reported that expansion of aluminium lattice and grain refinement after the addition of 2% magnesium are the responsible factors in increasing the efficiency. It was also shown that (Al+ 5%Zn+0.2%Bi+ 2%Mg) is superior to other ternary aluminium alloys containing different percentages of bismuth.

Table III. Characteristics of Bismuth activated aluminium alloy anodes

S.No	Anode Composition	Closed Circuit Potential (V) (SCE)
1.	Al + 5% Zn + 0.05% Bi	-0.899
2.	Al + 5% Zn + 0.1% Bi	-0.901
3.	Al + 5% Zn + 0.15% Bi	-0.903
4.	Al + 5% Zn + 0.2% Bi	-0.910
5.	Al + 5% Zn + 0.25% Bi	-0.905
6.	Al + 5% Zn + 0.2% Bi + 1% Mg	-0.923
7.	Al + 5% Zn + 0.2% Bi + 2% Mg	-0.937
8.	Al + 5% Zn + 0.2% Bi + 3% Mg	-0.920
9.	Al + 5% Zn + 0.2% Bi + 4% Mg	-0.918

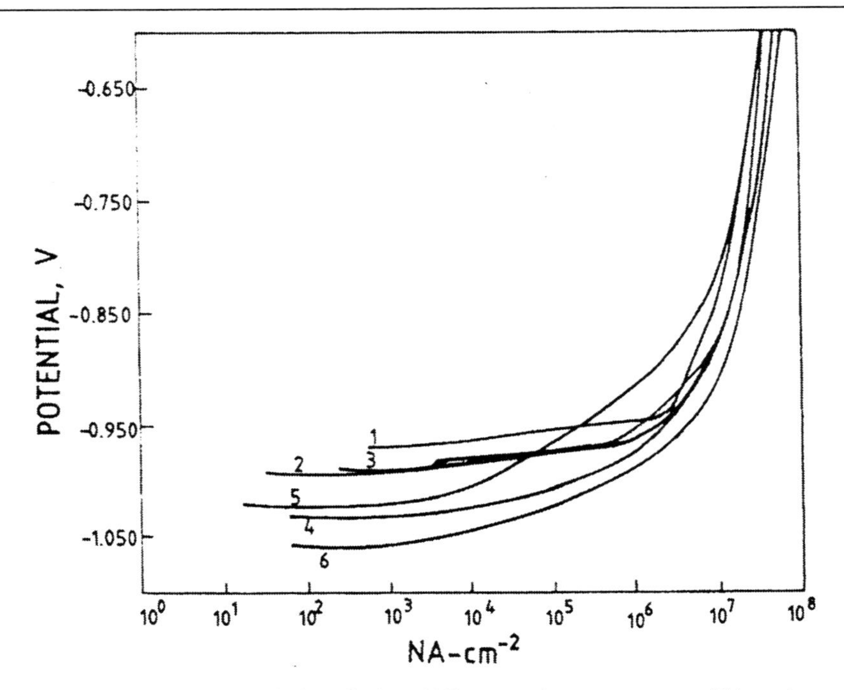

Figure 4. Anodic polarization characteristics of Al + 5% Zn + varying percentages of bismuth.

Figure 4 illustrates the superior performance of (Al + 5%Zn + 0.2%Bi + 2%Mg) alloy to other ternary aluminium alloys containing different percentages of bismuth. The properties of different bismuth activated aluminium alloy anodes are presented in Table III and their anode capacities are demonstrated in Figure 5.

2.3.3. Indium Activated Aluminium Alloy Anodes

Sakano et al [8] developed indium activated aluminium alloy anodes and their composition is (Al + 5%Zn + 0.03%In). They exhibit anode efficiency of 80%. These anodes do not require

a) any heat treatment
b) no reported pollution problem
c) exhibits maximum efficiency and
d) works satisfactorily in seawater mud in addition to excellent performance in seawater.

The reported underlying mechanism is that once indium is dissolved, it is immediately reduced at localised sites and thus promotes activation. It is important to note that presence of chloride ions in solution is essential for indium to activate aluminium [9-12]. The reduced indium ions produce a highly polarizing condition, which promote chloride migration and adsorption, thereby enhancing aluminium dissolution by forming complex chlorides. Reboul et al proposed a three-step mechanism for the activation of aluminium by indium [10] as well as mercury [13]. However, experimentally established mechanisms have not been reported in the literature to the best of authors' knowledge.

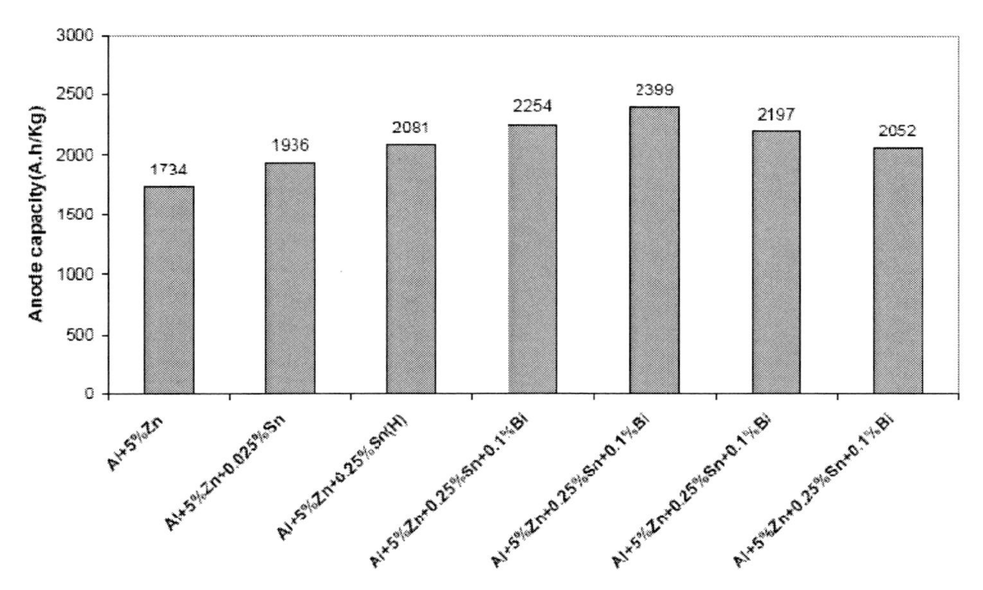

Figure 5. Anode capacities of various bismuth activated aluminium alloy anodes.

The anode efficiency was improved to 90% by the addition of 2% magnesium to (Al +5%Zn +0.03%In (Figure 6) [14]. Tamada and Tamura [15] were confirmed that magnesium addition to (Al + 5%Zn + 0.03% In) is helpful in improving the cathodic protection properties. Homogenisation of (Al+5%Zn+0.03%In) [16] and (Al+5%Zn+0.03%In+2%Mg) [17] also exhibits improved anode performance. Homogenisation of (Al+5%Zn+0.03% In+2%Mg) followed by water quenching resulted improved Ecorr and corrosion current compared to as cast and furnace cooled anodes [18]. Further, homogenisation of (Al + 5%Zn + 0.03%In + 2%Mg) alloy anode followed by water quenching helps to improve Ecorr and corrosion current compared to as cast and furnace cooled anodes (Figure 7).

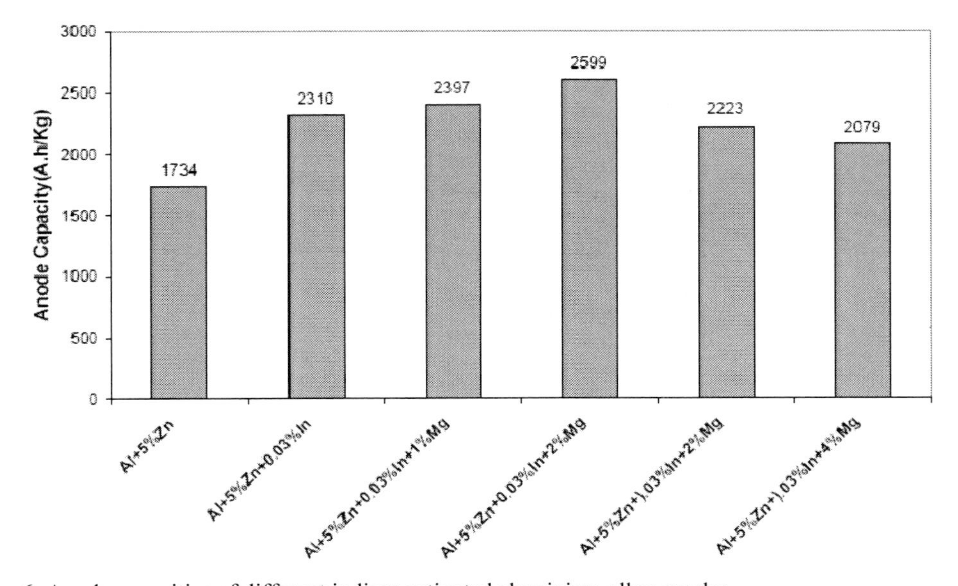

Figure 6. Anode capacities of different indium activated aluminium alloy anodes.

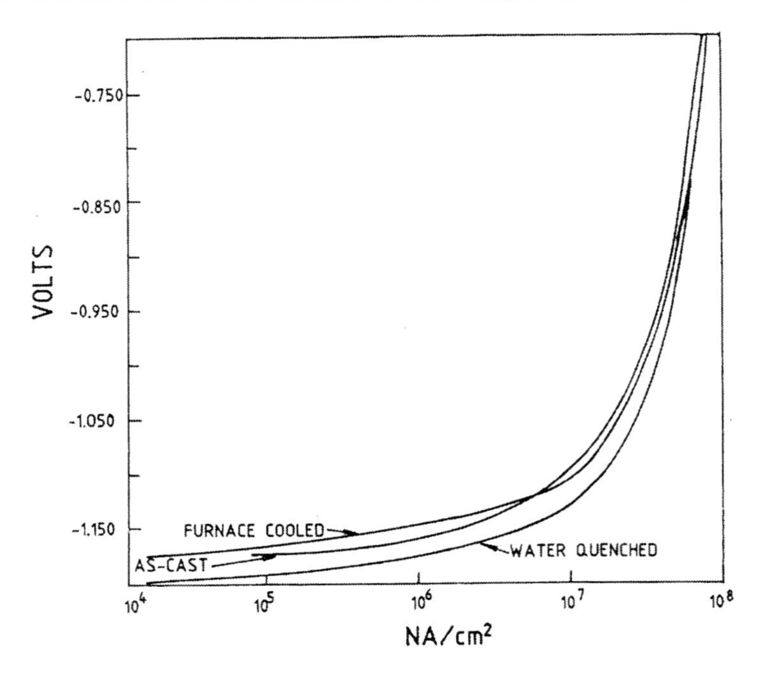

Figure 7. Anodic polarization characteristics of Al + 5% Zn + 0.03% In + 2% Mg with different treatments.

Table IV. Atomic spacing, Young's Modulus, Surface free energy and anode capacity of indium activated aluminium alloy anodes

S. No.	Anode Composition	Atomic spacing "nm"	Young's Modulus GNm^{-2}	Surface Free Energy $10^{-2}Jm^{-2}$	Anode capacity A.h / Kg
1.	Pure Aluminium	0.4048 ± 0.0066	62.455 ± 1.271	126.40	-
2.	Al+5%Zn	0.4023 ± 0.0020	60.689 ± 2.074	122.00	-
3.	Al+5%Zn+0.03%In	0.4038 ± 0.0013	56.265 ± 1.084	113.60	2110
4.	Al+5%Zn+0.03%In+1 %Mg	0.4036 ± 0.0015	53.748 ± 0.431	108.40	2129
5.	Al+5%Zn+0.03%In+2 %Mg	0.4032 ± 0.0016	52.371 ± 0.378	105.50	2348
6.	Al+5%Zn+0.03%In+3 %Mg	0.4047 ± 0.0014	59.262 ± 0.147	119.90	2066
7.	Al+5%Zn+0.03%In+4 %Mg	0.4053 ± 0.0013	64.678 ± 1.178	131.00	1811

With a view to understand the role of different alloying elements and various homogenisation treatments in exhibiting different cathodic protection characteristics, the metallurgical aspects of these anodes were studied in detail and it was experimentally [14] shown that decrease in surface free energy after the addition of zinc, indium and 2% magnesium, is the principal reason to improve the cathodic protection properties of pure aluminium, (Al + 5%Zn) and (Al + 5%Zn + 0.03% In) alloys respectively. This process appears to be a novel technique because this technique was not reported earlier and at the same time, it is very simple and no electrochemical reactions are involved. Therefore, it seems to be highly useful not only in evaluating the existing anode materials but also for the innovation of newer anode materials for cathodic protection of marine structures. The experimentally measured Young's Modulus, atomic spacing and evaluated surface free energies of different indium activated aluminium alloy anodes are tabulated in Table IV.

Scanning electron microphotographs of corroded (Al + 5%Zn + 0.03% In) and (Al + 5%Zn + 0.03%In + 2%Mg) alloy anodes are shown in Figure 8. It clearly explains that the composition containing 2% magnesium undergoes uniform and smooth dissolution and thus exhibits excellent cathodic protection properties over the alloy containing no magnesium.

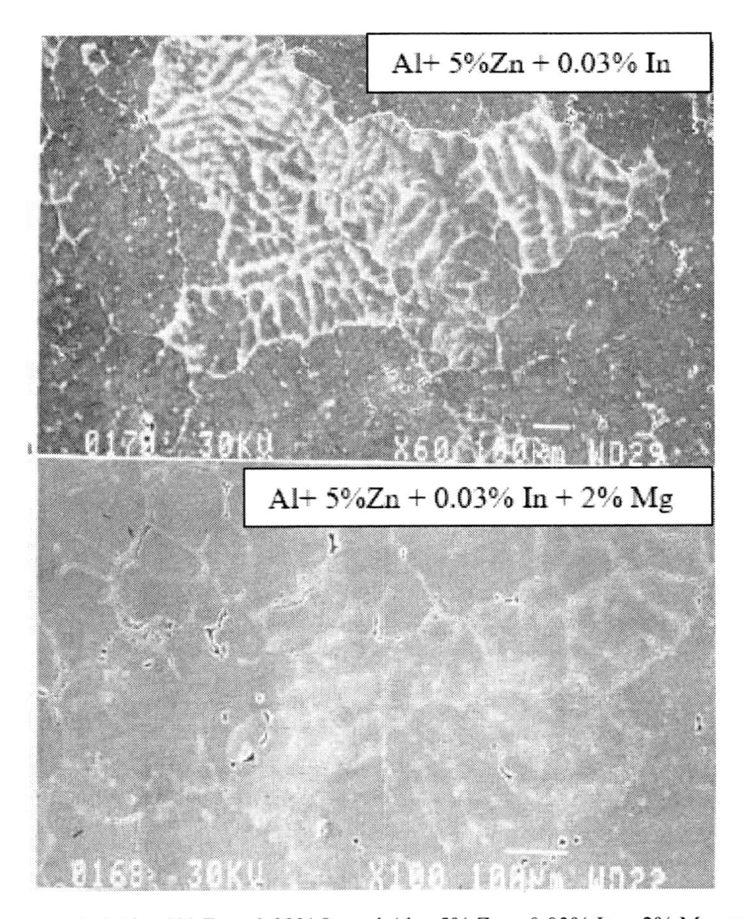

Figure 8. SEM of corroded Al + 5% Zn + 0.03% In and Al + 5% Zn + 0.03% In + 2% Mg anodes in seawater.

The developed anode is found to work satisfactorily in highly conducting media such as seawater and therefore, can be used for protection of marine structures in seawater environment. However, it can not provide protection for the structures that immersed in low conductivity waters. This aspect necessitated the development of aluminium alloy anodes that can protect the structures in low conductivity waters. Gurrappa et al [18] have developed an hanging anode with the addition of small amount of bismuth to (Al +5%Zn + 0.03%In) and showed that it can be an alternative to magnesium alloy hanging anode, having moderate driving voltage and thus avoiding the over protection problem which is being faced with magnesium alloy hanging anodes.

3. ACTIVATION MECHANISMS

As mentioned earlier, the cathodic protection properties of an alloy anode depend both on the alloying elements as well as on their concentration in the solid solution of aluminium. The elements which are present in solid solution are only effective in activating aluminium and thereby enhancing the cathodic protection properties. The presences of activating elements outside the matrix of aluminium are not useful for activating the aluminium and are protected by the aluminium since they are cathodic to it. Therefore, identification of a suitable element and its optimum composition are the real challenges to the cathodic protection Engineer. It is also important to understand the activation mechanism in order to develop highly efficient sacrificial anodes which can protect the structure for a long time with minimum energy.

Different mechanisms were proposed by earlier investigators based on some assumptions; these are described and commented upon below.

3.1. Keir's Mechanism

After observing the dramatic effect of tin on aluminium (among the IV group elements) in shifting the potential of aluminium to a more negative value, Keir et al {12] proposed a mechanism as follows:

> "Tin present in aluminium alloy is able to enter the surface oxide film as Sn^{4+} ions and create additional cation vacancies. As a result ionic resistance of the film reduces, thereby improving the cathodic protection properties of the aluminium alloy"

Comments

This mechanism cannot be applied to mercury activated or indium activated aluminium alloys because In^{3+} and Hg^{2+} cannot reduce the ionic resistance of oxide film by creating additional cation vacancies.

3.2. Reboul's Mechanism

Reboul et al [21] were proposed a three step mechanism for activation of aluminium alloy anodes as follows:

Step 1: Aluminium and the alloying elements present in the solid solution of aluminium (Zn,Hg, In, Sn) oxidize upon galvanic coupling and produce cations in the electrolyte

$$Al\,(M) \quad = \quad Al^{3+} \quad + \quad M^{n+}$$

Step 2: The cations that are produced during step 1 plate back onto the aluminium surface, since the alloying elements are cathodic to aluminium according to the electrochemical exchange reaction

$$Al + M^{n+} = Al^{3+} + M$$

Step 3: In this step, which occurs simultaneously with Step 2, the aluminium oxide film is separated locally.

Comments

This mechanism can generally explain the activation of aluminium by certain alloying elements. However, it cannot explain the reason or underlying mechanism why does each ternary alloying element (for example Hg, In, Sn) exhibit a different anode efficiency or current capacity when the zinc concentration is almost the same in all the alloys and what way the anode efficiency be correlated? It is also a fact that solution treatment results in improving the cathodic protection characteristics of certain alloys. This aspect also cannot be explained based on this mechanism.

3.3. Gurrappa's Mechanism

It is therefore essential to develop a mechanism which can be used to explain both the influence of different alloying elements and solution treatment on the cathodic protection properties of aluminium alloy anodes. This will not only help in understanding the working mechanism of an anode, but also in the development of efficient sacrificial anodes.

Gurrappa strongly feel that both alloying elements and solution treatment change the metallurgy of aluminium. Therefore, it is appropriate to study the metallurgical aspects of aluminium and evaluate a parameter based on which the cathodic protection characteristics can be correlated. Accordingly, experiments were conducted for indium activated aluminium alloy anodes, with the addition of various percentages of magnesium as well as for tin activated aluminium alloy anodes with and without solution treatment.

Figure 9 provides the results of different compositions of indium activated aluminium alloy anodes and tin activated aluminium alloy anodes. The surface free energy was evaluated by determining Young's modulus and atomic spacing of the anode materials experimentally and by using the following formula:

$$F = 4.99478 \times 10^{-11} \times E \times a$$

where F = surface free energy

E = Young's modulus and

A = atomic spacing

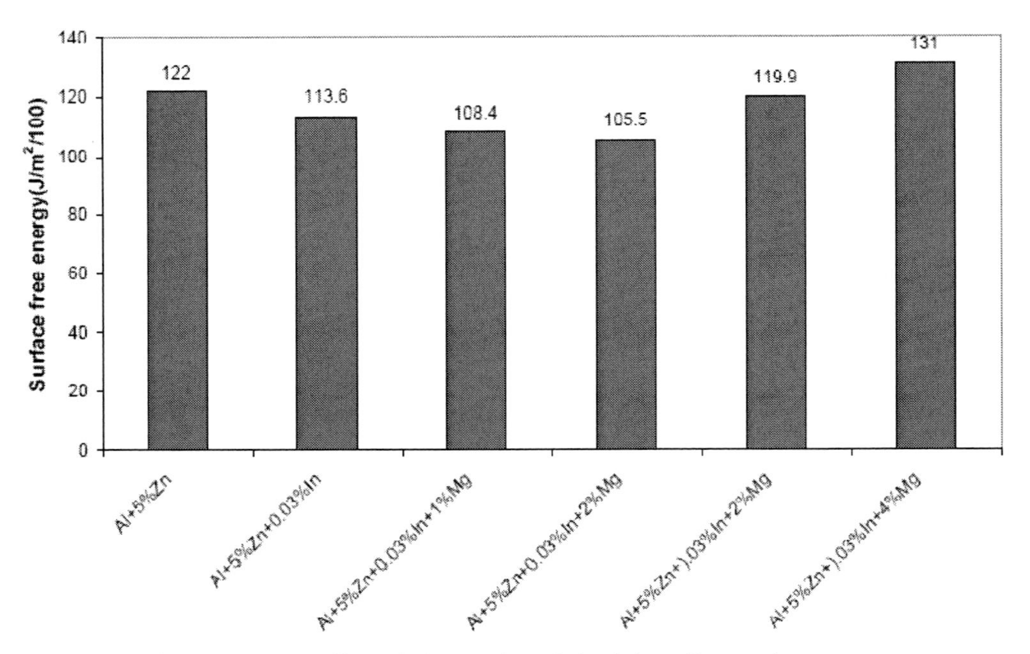

Figure 9. Surface free energies of different indium activated aluminium alloy anodes.

The lower surface free energy of an alloy, the lower is the film thickness of Al2O3 and weaker is the bonding between the metal and the surface oxide film. Therefore, the efficiency of an alloy improves as the oxide film can easily be ruptured by chloride ions present in the electrolyte allowing the alloy to be available for uniform dissolution. This criterion was applied to tin activated, mercury activated and indium activated aluminium alloy anodes and correlated most satisfactorily [4, 14, 26] with the results. It also explains the effect of different alloying elements and different efficiencies / capacities and the effect of solution treatment.

In fact, in the authors' opinion, this concept is invaluable in explaining the working mechanism of different aluminium alloy anodes and also in developing sacrificial anodes for the purpose of cathodic protection. It is also useful in understanding the metallurgical changes taking place in the activation mechanism. It is a novel, non-electrochemical and simple tool for a metallurgist to adopt and to use in the development of an efficient sacrificial anode.

4. Advantages of Sacrificial Anodes

1. They operate independent of electric supply
2. They are relatively simple to install. Additional anodes can always be fitted if adequate protection is not achieved
3. There is no electrical hazard to the divers
4. No control to be exercised
5. Incorrect fitting problem can be eliminated
6. Minimum maintenance costs and
7. Installation costs are low

5. Design and Development of Smart Anodes

Though the aluminium alloy anodes work satisfactorily in marine environment, their utility in fresh / river water environments is restricted. Therefore, there is a need to innovated smart sacrificial anodes. The designed and developed smart anodes should be optimised compositionally with appropriate nanostructured materials or completely with nanostructured materials that promote appropriate surface for dissolution depending on the surrounding environmental conditions that helps to enhance the efficiency of anodes, which in turn possible to achieve more than designed life of structures. It is imperative to mention that advanced marine systems / structures essential necessitate such anodes for enhanced efficiency.

6. Economics of Cathodic Protection

When more than thousands of pipelines and million tons of structures including cooling water systems is in service in the world today, the means of application of anti-corrosive devices on buried pipelines and structures and subsequent saving of an enormous amount of money cannot be over emphasized. The estimated annual loss in the year 2000 due to corrosion is about Rs. 30,000 crores to India while for United States of America is 275 billion US$. For example, in a particular cooling water system or a gas pipeline, a single leak from a pipe can cause numerous losses and may include

i. Efficiency loss
ii. Expensive repairs
iii. Service interruption
iv. Contamination of water / gas and thereby changing the chemistry of water / gas etc

Besides this, once a section of pipeline system starts developing leaks, experiences have shown that further leaks will develop at a continuously increasing rate.

For a giant petrochemical complex of worth Rs.5000 crores located at a very corrosive belt of eastern India, CP system of worth Rs.5.35 crores was installed sometime back, which was about 0.1% of the total cost. The system was installed in the installation stage itself. It was anticipated that CP system in conjunction with protective coatings will prevent occurrence of corrosion of the plant cooling, fire water lines and the bottom plates of A/G hydrocarbon storage tanks. As a result, hindrance to plant operation and safety due to leak formation at these structures are eliminated completely even though the soil is highly corrosive. In essence, the total cost of cathodic protection system is about 0.1% of total cost of the plant, which is more economical in installing the system during commissioning of the equipment. Therefore, it is recommended to install the CP system during the installation stage of the plants which helps in enhancing the productivity by elimination of downtimes and thereby increasing the profitability.

SUMMARY

i. Aluminium alloy anodes are superior to zinc and magnesium alloy anodes
ii. The performance of aluminium alloy anodes depends on the alloying elements and their concentration in the aluminium solid solution
iii. Among the aluminium alloy anodes available, indium activated aluminium alloy anodes work satisfactorily in marine and marine muds.
iv. The surface free energy evaluation phenomenon is a novel tool for developing efficient sacrificial anodes as well as for understanding the metallurgy involved after the addition of alloying elements and after solution treatment. Therefore, the surface free energy concept can be successfully adopted in explaining the activation mechanism of aluminium by the alloying elements as well as solution treatments.
v. Magnesium addition to (Al + 5% Zn + 0.03% In) and (Al + 5% Zn + 0.25% Bi) helps in improving anode capacity / efficiency.
vi. Successful avoidance of costly heat treatment is possible by the controlled addition of bismuth to tin activated aluminium alloy anodes.
vii. An anode's efficiency / capacity is closely related to its closed circuit potential i.e. the lower the closed circuit potential, the higher the anode capacity of an alloy anode.
viii. For obtaining maximum efficiency when a sacrificial anode is coupled to a steel structure, it is important that the anode has to undergo uniform dissolution / corrosion.
ix. Design and development of smart anodes which work intelligently depending on the surrounding environmental conditions is highly essential and future research should be focused on this subject.

ACKNOWLEDGMENTS

Defence research and Development Organisation is gratefully acknowledged for financial assistance.

REFERENCES

[1] C.F. Schrieber and J.T. Reding, *Mater. Protec.*, 6 (1967) 33.
[2] 2. D. S. Keir, M. J. Pryor and P. R. Sperry, *J. Electrochem. Soc.*, 114 (1977) 777.
[3] I. Gurrappa and J. A. Karnik, *Corr. Prev. and Control*, 41 (1994) 117.
[4] I. Gurrappa, *Corr. Prev. and Control*, 40 (1993) 111.
[5] T. Murai, C. Miura and Y. Tamura, *Bull. Bismuth Institute*, 1977, Third Quarter, 17.
[6] U. S. Patent, 1968, No.3.368958.
[7] A. G. Kulkarni, I. Gurrappa and J. A. Karnik, *Bull. Electrochem.*, 7 (1991) 549.
[8] T. Sakano, K. Toda and M. Hanada, *Mater. Protec.*, 5 (1966) 45.
[9] J. U. Chavarin, *Corrosion*, 47 (1991) 472.
[10] M. C. Reboul, P.H. Gimenez and J. J. Rameau, Corrosion, 40 (1984) 366-371.
[11] C. B. Breslin and W. M. Carroll, *Corr. Sci.*, 34 (1993) 327.

[12] W. M. Carroll and C. B. Breslin, *Corr. Sci.,* 33 (1992) 1161.
[13] M. C. Reboul and M.C. Delatte, *Mater. Perform.*, 19 (1980) 35.
[14] A. G. Kulkarni and I. Gurrappa, *Brit. Corr. J.*, 28 (1993) 67.
[15] A. Tamada and Y. Tamura, *Corr. Sci.,* 34 (1993) 261.
[16] J. C. Lin and H. C. Shih, *J. Electrochem. Soc.*, 134 (1987) 817.
[17] I. Gurrappa and J. A. Karnik, *SAEST,* 28 (1993) 118.
[18] I. Gurrappa and J.A.Karnik, *Corr. Prev. and Control*, 43 (1996) 77.
[19] J.A.Jakobs, *Materials Perform.*, 20 (1981) 17.
[20] E.W.Dreyman, *Materials Protection and Perform.*, 11 (1972) 17.
[21] R. Babosian, *Materials Perform.*, 16 (1977) 20.
[22] A.Kumar and M.D. Armstrong, *Materials Perform.*, 27 (1988) 19.
[23] R.J. Kessler and R.G. Powers, *Materials Perform.*, 28 (1989) 24.
[24] A.K.Sinha, A.K.Mitra, U.C.Bhakta and S.K.Sanyal, CORRCORN 1999, Vol I, p 837.
[25] V. Ashworth and C.J.L. Booker (ed.), Cathodic Protection, Ellis Horwood, Chichester, UK, 1986.
[26] I.Gurrappa, Proceedings of All India Convention on Cooling Water Treatment, December 2001, Hyderabad, p B1-B16.
[27] J.N. Wanklyn and N.J.M. Wilkins, *Br. Corros. J.*, 20 (1985) 161.

In: Aluminum Alloys
Editor: Erik L. Persson

ISBN: 978-1-61122-311-8

Chapter 6

PERSPECTIVES OF ALUMINUM AS AN ENERGY CARRIER

H. Z. Wang and D. Y. C. Leung*

Department of Mechanical Engineering,
The University of Hong Kong, Pokfulam Road,
Hong Kong

ABSTRACT

In this paper, a brief overview and perspectives of the development of aluminum-derived energy are given. Aluminum has received a lot of attention particularly in recent years for energy applications because of its outstanding properties including high electrochemical equivalent (2980Ah kg^{-1}), highly negative voltage (-1.66V versus SHE in neutral electrolyte; -2.33V versus SHE in alkaline electrolyte) as well as high energy density (29.5 MJ kg^{-1}). Other attractive features of aluminum are its abundance and recyclability.

All these features make aluminum and its alloys promising candidates to meet the urgent demand of advanced energy carrier arising from energy and environmental crises. Through proper activation, aluminum and its alloys are capable of reacting with water to generate hydrogen or being directly used as a "fuel" to produce electricity through aluminum-air semi-fuel cells.

The hydrogen produced from aluminum-water reaction can be delivered to fuel cells or internal combustion engine for further energy conversion.

The technical characteristics and development challenges of the above-mentioned two aluminum energy conversion routes, i.e. aluminum-assisted hydrogen production and aluminum-air semi-fuel cells, are briefly reviewed.

In spite of the existence of challenges, this study indicates that aluminum and its alloys may serve as an excellent energy carrier over the near to long-term.

---

* Corresponding author. Tel.: +852 2859 7911; Fax: +852 2858 5415; *E-mail:* ycleung@hku.hk

1. INTRODUCTION

In response to increasing concerns about energy crisis and climate change, considerable efforts have been devoted to developing new generation of energy carriers to replace the present fossil fuels or fossil-derived energy carriers.

Aluminum, which possesses a number of outstanding properties, has received ever increasing interest, especially in recent years. However, most attention has been focused on technical aspects of aluminum energy conversion. There is a lack of thorough thinking on the following issues:

i. Why to choose aluminum? Is it a near-term or long-term solution?
ii. How to make maximum use of the aluminum energy?
iii. What will be the challenges supposing a long-term reliance on aluminum energy?

The present paper aims to find answers to these questions. In the following sections, we start from the motivation to use aluminum energy by introducing the present energy situation and aluminum features. Two typical aluminum energy conversion routes, i.e. aluminum-assisted hydrogen production and aluminum-air semi-fuel cells, are then briefly reviewed with an emphasis on their characteristics and development potentials. Concepts of combined production of multiple energies from aluminum are also discussed. Finally, major challenges faced by future development of the aluminum energy are suggested.

2. ENERGY CARRIER BACKGROUND

2.1. Advanced Energy Carrier: Transition to Future Renewable Energy

The present energy economy that heavily relies on fossil fuels (i.e. natural gas, petroleum and coal) confronts us with series of global issues such as energy security risk, resource depletion, environmental pollution and climate change. Particularly in recent years, dramatically growing concerns about fossil fuel-induced carbon dioxide (CO_2) emissions and consequential global warming spurred the establishments of renewable energy policy targets in many countries to promote the transition to renewable energy [1-4]. However, due to high economic and technological barriers, the global energy demand for quite a long period is most likely to be met by mix of primary energy resources of fossil fuels and renewable ones. Even at the early stage of the transition, fossil fuels will keep to dominate the energy supplies [5]. The fact of continuous fossil fuel use together with the urgency of climate stabilization not only necessitates the deployment of CO_2 abatement technologies such as efficient fossil-fuel combustion and CO_2 capture and sequestration (CCS), but also makes the development of new energy carriers an immediate priority for the gradual switchover to a renewable energy future. Energy carrier, an intermediate medium for storing and delivering energy, is an indispensable component in the future energy-supply chain between intermittent and site-specific renewable energy sources and end-use applications. However, its importance in helping mitigate the present fossil fuel issues also should never be overlooked. The use of carbon-free energy carrier to replace the present gasoline or other fossil-derived carriers for end-uses, particularly transportation, can provide an immediate local emission reduction. From a long-term point of view, this move may facilitate a more economical centralized implementation of CO_2 abatement technologies such as CCS at central locations (e.g. fossil-fired power plants) instead of largely dispersed end-use locations. In addition, introducing alternative energy carriers produced from various energy sources into the current energy chain encourages the diversity of world's energy mix and thus enhances the global energy security. Despite the long lead-time for the transition to a renewable energy future, it is undoubted that an eventual CO_2 emission reduction should only be achieved with the joint actions of renewable energy sources and advanced energy carriers.

2.2. Options of Alternative Energy Carriers

To date, lots of alternative energy carriers have been proposed towards a renewable energy future such as hydrogen, methanol, ethanol, biodiesel, lithium and even silicon [6-11]. Among them, hydrogen and lithium represent the most promising choices. A concept of hydrogen economy has been raised due to a number of attractive properties of hydrogen. And fuel cell technology is thought to be the most efficient way to convert the hydrogen energy into usable electricity [12]. Nevertheless, not to mention the technical and cost issues related to fuel cells, hydrogen itself faces great challenges impeding its widespread application at the present stage, of which the storage problem stands out from all. At ambient conditions, hydrogen is a gas with extremely low volumetric energy density. Conventional physical storage methods including compression and liquefaction are not very sensible options particularly for mobile applications, since 100MPa or higher pressure is needed to compress the hydrogen for a car traveling of 500 kilometers into a reasonable volume while the liquefaction requires a temperature below 23K [13]. Among all existing storage techniques, chemisorptions, in which hydrogen reacts with adsorption materials to form hydrides and consequently reduces the volume, provide an impressive long-term solution to the on-board hydrogen storage. In spite of an improved volumetric hydrogen capacity, these techniques still suffer unresolved problems of high material costs, low gravimetric capacity, high operating temperatures and slow kinetics [13-14]. Alternatively, on-site hydrogen generation could be employed instead of storage. Reforming hydrocarbons is the most mature and widely-used hydrogen production method. It is able to offer kinetics favorable for on-board automotive applications [9]. Unfortunately, the reforming process involves the use of fossil fuels and produces CO_2, and thus departs from the global low-carbon target. In brief, a better solution to either on-board storage or on-site generation is needed before hydrogen can fulfill our expectations.

As for lithium, it has received considerable concerns owing to its high electropositivity (-3.04V versus a standard hydrogen electrode (SHE)) and attractive electrochemical equivalent (3860 $Ah(kg\text{-}Li)^{-1}$) [15]. Like hydrogen fuel cells, lithium batteries are also hot choices of power sources for future electric vehicles and energy plants. However, again, apart from the difficulties in scaling up the lithium battery technology [16], the sustainability of lithium remains a debate. It is hard to tell whether the world's limited and unevenly distributed lithium reserves can be solely relied on to meet the yearly increasing demand from electric vehicles in the future. While 230 billion tons of lithium in seawater is potentially a large resource, its low concentration in the seawater (0.1-0.2ppm) makes the extraction of lithium unrealistically expensive at the present stage [15,17].

From the above, it is seen that lots of challenging issues should be addressed before hydrogen or lithium become truly competitive choices. Thus, there is also a desire to seek a new energy carrier for our use in short, medium and long term. In this paper, we discuss aluminum as a potential energy carrier.

3. Properties of Aluminum as an Attractive Energy Carrier

For all non-fossil derived energy carrier candidates, aluminum possesses a gravimetric energy density of 29.5MJ kg^{-1} that is preceded only by hydrogen and lithium as compared in Figure 1 [18]. From a volume standpoint, the energy density of aluminum ranks no.1. The highest volumetric energy density makes aluminum particularly favorable for vehicles and small-scale portable devices in which space is limited.

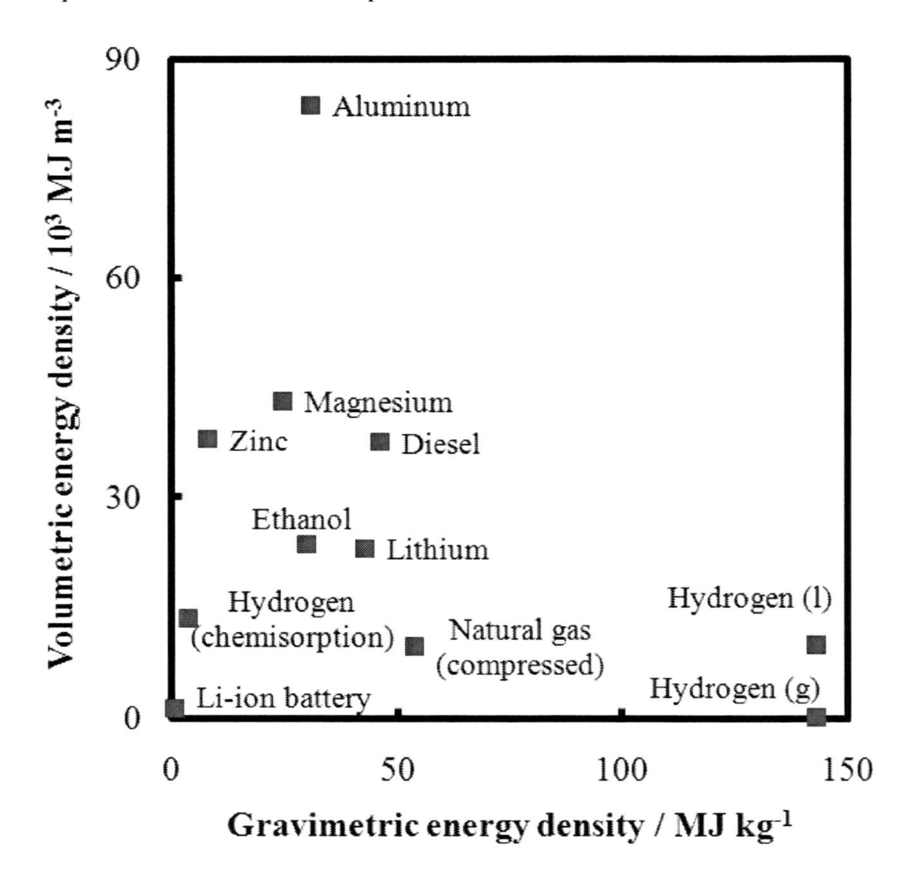

Figure 1. Energy density of common energy carriers.

Sustainability could be another advantage of aluminum as a consideration of an energy carrier. Aluminum is the most abundant metallic element (8.3% by weight) and the third most abundant element (after oxygen and silicon) on the earth. The identified reserves of bauxite

($AlO_x(OH)_{3-2x}$), a main ore of aluminum, reach 55 billion tons [19]. Taking into account other potentially exploitable aluminum resource forms [19], aluminum should be virtually inexhaustible like hydrogen as compared in Figure 2. Even if the immediate availability is considered, the world's aluminum reserves are 4 times higher than Mg, 25 times higher than Zn, and 2000 times higher than Li as shown in Figure 2. Based on a conservative estimation of 30kg of aluminum required to power a 40kWh full electric car (considering the product recovery) and assumption of 30% share of aluminum-based energy in future car energy supplies, a yearly increase of 0.45 million tons in aluminum production from ore will arise from the transformation of 50 million vehicles produced every year into full electric ones. Such an increase is tolerable compared with current production capacity of 50 million tons per year. In contrast, a similar estimation process yields a use-up of lithium reserves in two decades. Aluminum in theory can be 100% regenerated without any loss of qualities from $Al(OH)_3$ (or Al_2O_3), the only by-product of aluminum-energy conversion processes. Figure 3 depicts a conceptual aluminum energy cycle, in which aluminum is produced with energy from renewable sources or carbon-sequestered fossil fuel plants, and then delivered to high-density end-users in the urban area with by-products collected back to a recycling center for further recovery of fresh aluminum. This cycle is conceptually green and self-sustained without pollutant emissions and additional aluminum ore consumptions.

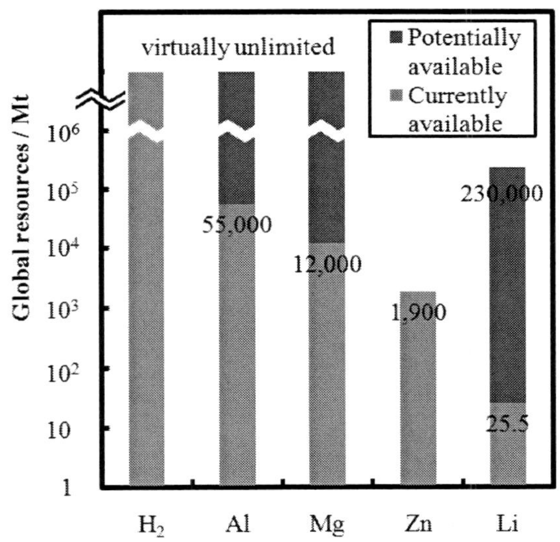

Figure 2. Global resources of selected energy carriers.

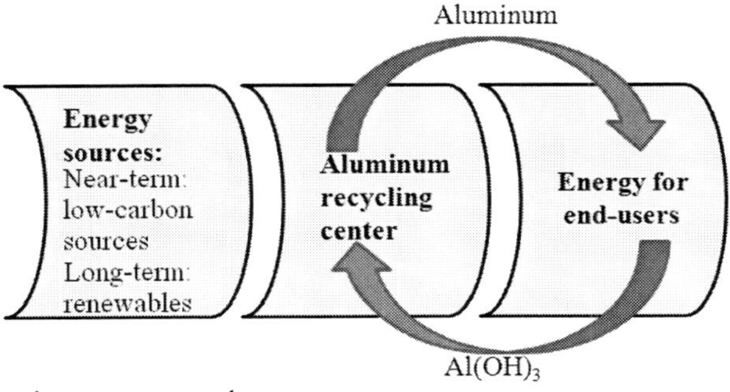

Figure 3. Conceptual aluminum energy cycle.

4. Applications of Aluminum as an Energy Carrier

4.1. Aluminum as an Intermediate Energy Carrier for Hydrogen Production

Due to its reactivity, aluminum is able to extract hydrogen from water. Typical aluminum-water reactions are as follows:

$$2Al + 6H_2O \rightarrow 2Al(OH)_3 + 3H_2 \tag{1}$$

$$2Al + 3H_2O \rightarrow Al_2O_3 + 3H_2 \tag{2}$$

The first reaction takes place at room temperatures, while the second one is initiated at temperatures above 753K [20]. These reactions are believed to provide possible routes for emission-free on-site hydrogen production, which may act as an interim solution to the on-board storage of hydrogen and therefore serve as a bridge to hydrogen economy. Therefore, in what follows, we discuss aluminum-water reactions in terms of their hydrogen capacity, kinetics (i.e. hydrogen generation rate), energy efficiency and cost, which are important indicators of the performance of on-site hydrogen systems.

4.1.1. Hydrogen Capacity

Assuming a consumption of pure aluminum with 100% hydrogen yield, the theoretical gravimetric hydrogen capacity is calculated to be 3.7wt.% for reaction (1) and 5.6wt.% for reaction (2). If a full recovery of water produced from the driven fuel cell is considered, the theoretical gravimetric hydrogen capacity will increase up to 5.6wt.% for reaction (1) and 11.1wt.% for reaction (2). The above indicates that the aluminum-water systems have potential to approach the DOE revised target of 5.5wt.% on system-level capacities for hydrogen storage in 2015 [21]. However, it was reported that most existing activation schemes required to promote the reaction lead to a serious reduction of the material hydrogen capacity [20]. Effects of several activation schemes on material hydrogen capacity from reaction (1) are listed in Table 1. Different activation schemes will be reviewed along with the aluminum-water reaction kinetics in the below sub-section.

Table 1. Hydrogen capacities for different activation methods [20]

Promoter schemes	Gravimetric capacity (wt.%)	Volumetric capacity (H_2 L^{-1})
Hydroxide promoter	2.5	36
Oxide promoter	2.5	40.6
NaCl salt promoter	2.8	39
Ga20/Al80 alloy	3.0	37

4.1.2. Kinetics

In spite of its high electropositivity, aluminum is normally well protected by an oxide layer formed on their surface due to their strong affinity for oxygen, which shifts its corrosion potential by nearly 1V in the positive direction. This blocks the aluminum-water reaction under usual conditions. Existing approaches to improve aluminum activity in water can be divided into three categories: mechanical activation, chemical activation and electrochemical methods [22]. Freshly exposed metal surface is known to have a relatively higher chemical activity. Uehara et al. [23] observed bubble release when cutting, drilling or grinding of aluminum and its alloys in water, by which the fresh metal surface was kept exposed in water. Although a highest volume of hydrogen generated per unit volume of metal removal was found in the case of grinding, the reaction stopped immediately after the machining stopped. High-energy ball milling, which produces fine aluminum particles with defects and exposed fresh surfaces, is an alternative method to facilitate continuous generation of hydrogen. However, properly-controlled milling time is required since prolonged milling would induce a reduction in powder surface area and the oxidation of powders, both of which increase the corrosion resistance of metals and therefore inhibit the hydrogen generating reaction [22, 24]. As a supplement to the mechanical activation, chemical activation aims at improving the aluminum-water reaction kinetics through either modifying the composition of aluminum alloys, or introducing additives into the aqueous solution. Enhanced activity of aluminum alloyed with Zn, Ca, Ga, Bi, In or Sn has been found [25-27], but these particular alloys have obvious drawbacks that they are not readily available and unstable. It was reported that their chemical activity can only be well retained upon storage at liquid nitrogen temperature [25]. Alternatively, effects of solution additives include potassium hydroxide, sodium hydroxide, sodium chloride, sodium aluminate, sodium stannate as well as aluminum hydroxide on hydrogen generation kinetics have been investigated [28-31]. From published data to date, sodium stannate demonstrated the best activation performance under the same experimental condition [30]. As for electrochemical method, it involves the formation of corrosion cells by coupling aluminum with more cathodic metals. In the presence of electrolytes, aluminum dissolution is accentuated via the electrochemical process while the hydrogen evolution mainly occurs at the cathodic side. However, for corrosion cells, the stored chemical energy cannot be utilized but dissipate in the form of heat. Current trend in further improving aluminum-water reaction kinetics is to combine two or three of the above activation strategies together [32-33]. By using ball-milled aluminum containing Ga-In gallams, a hydrogen generation rate of ~3750ml min^{-1} $(g\text{-}Al)^{-1}$ was achieved [33], which is among the fastest kinetics in the literatures to date. While recent molecular dynamics simulations [34] have suggested that the use of nanotechnology could help achieve an enormous enhancement of aluminum-water reaction rates compared with existing activation strategies, no laboratory experimentation has been implemented to realize the theory yet.

4.1.3. Efficiency and Costs

Excluding the energy consumption and cost arising from various activation schemes, we roughly estimate the energy efficiency and hydrogen costs involved in aluminum-water reaction based on different aluminum production routes. The present aluminum is classified as primary aluminum if it is produced from ore and as secondary aluminum if it is produced from recycled scrap material [35]. Considering today's state-of-the-art technologies, producing 1kg of primary aluminum requires energy of 76.5MJ and producing 1kg of secondary aluminum requires energy of 1.4MJ [35]. If a hydrogen yield of 100% is reached, total 9kg of aluminum will be consumed to generate 1kg of hydrogen from aluminum-water reactions. Calculated from the lower heating value of hydrogen (119.93MJ kg^{-1}) together with the U.S. electricity price in 2007 (US$0.0572 kWh^{-1}) [36-37], the hydrogen production using primary aluminum yields an efficiency of 17.4% and cost of US$10.9 $(kg\text{-}H_2)^{-1}$ that is obviously not competitive to other commonly-used hydrogen production methods such as hydrocarbon reforming and water electrolysis. However, if secondary aluminum is used for hydrogen production, much higher output-input energy ratio and a low cost of US$0.2 $(kg\text{-}H_2)^{-1}$ can be achieved due to the low energy inputs for aluminum scrap recycling. From the foregoing estimations, aluminum used for hydrogen production is encouraged to shift to secondary aluminum or aluminum scrap if possible.

4.2. Aluminum-Air Semi Fuel Cells

In addition to a water splitting agent for hydrogen production, aluminum has been an attractive anode material for a long time because of its outstanding electrochemical properties including highly negative voltage (-1.66V versus SHE in neutral electrolyte; -2.33V versus SHE in alkaline electrolyte) and high electrochemical equivalent (2980 $Ah(kg\text{-}Al)^{-1}$). Although aluminum is inferior to lithium in these properties, its abundance, safety and readiness for fabrication make it a potential alternative to lithium anode for future batteries [38]. An overview of aluminum batteries has been completed by Li and Bjerrum [39]. Among these batteries, aluminum-air batteries exhibit the highest theoretical energy density and have received most attention. An aluminum-air single cell, composed of a consumable aluminum anode and oxygen diffusion cathode with oxygen fed from atmosphere (Figure 4), is normally classified into a category of semi-fuel cell. Its discharging reaction is expressed as

$$4Al + 6H_2O + 3O_2 \rightarrow 4Al(OH)_3 \qquad E^0 = 2.73V \quad (3)$$

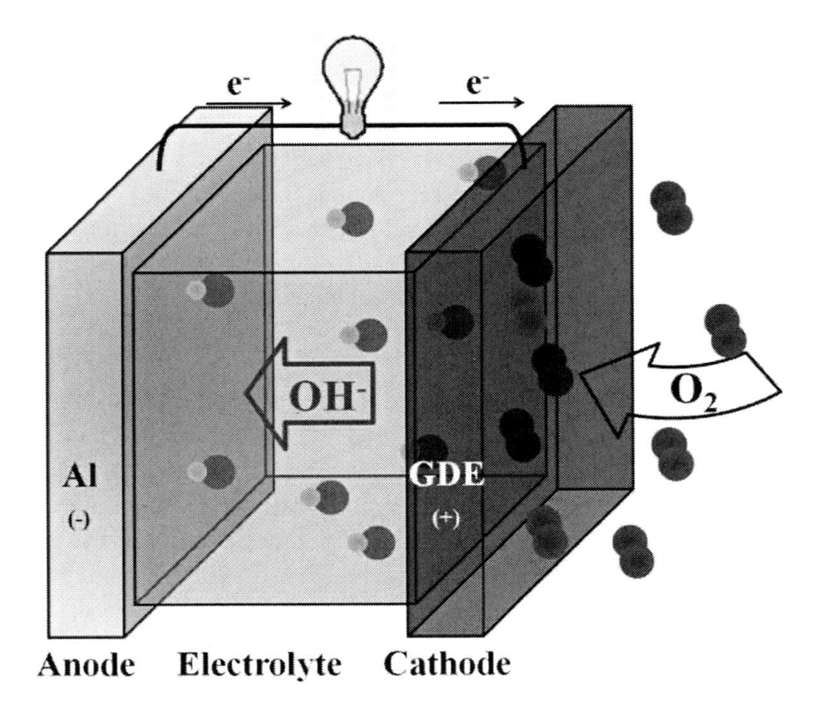

Figure 4. Illustration of an aluminum-air single cell.

An aluminum-air cell can work with either neutral or alkaline aqueous electrolytes. Electrochemical behaviors of gel electrolyte-based aluminum-air cells also have been reported [40]. In contrast to the foregoing aluminum-assisted hydrogen production, the aluminum-air cell directly converts aluminum-stored energy into usable electricity, which must be more efficient than using the converted hydrogen to produce energy.

However, the aluminum anode suffers from the following problems: i) the passivation of aluminum anode surface by oxide layer that decreases the reversible anode potential and thus decreases the cell voltage, and ii) the parasitic aluminum corrosion that results in less than 100% utilization of anode metal as well as hydrogen evolution.

The improvement of anode behaviors has been attempted either by doping high purity grade aluminum with trace elements [41-44] or by introducing corrosion inhibitors into electrolyte [45-48]. Considering that the above strategies may increase the battery production cost, Zhuk et al. [49] studied the feasibility of the alternative utilization of the parasitically produced hydrogen and proposed a concept of electricity-hydrogen co-generation system, which will be discussed in the following sub-sections.

4.3. Concept of Combined Production

To tap the full potential of aluminum energy, several concepts of multiple energy generation from aluminum were raised. Systems for both i) combined production of hydrogen and electricity and ii) combined production of hydrogen and heat have been developed.

4.3.1. Combined Production of Hydrogen and Electrical Energy

Aiming at an economical approach to increase the energy efficiency of aluminum-air cells, Zhuk et al. [49] proposed the use of aluminum anodes in alkaline aluminum-air cells for the hydrogen-electricity co-generation. Such a combined production was realized by collecting and utilizing the hydrogen formed from the self-corrosion of aluminum anode during discharge. The feasibility of this concept has been further confirmed through their experiments, which shows that the energy efficiency with more corrodible commercial-grade aluminum alloys is comparable to that with the special anode alloys if the energy stored in the parasitically released hydrogen is also taken into account.

Another type of co-generation set-up is so-called aluminum-water hydrogen storage cell, invented and patented by Li and Wang [50]. This generator, different from the foregoing design, is composed of aluminum anodes and hydrogen evolution cathodes such as Ni, and therefore hydrogen is mainly produced from the water reduction on the cathode, which serves as a current-generating reaction in the mean time. While both types of co-generators hold certain promises in exploring aluminum energy, the existing studies on this topic are far from sufficient. For a better performance of an aluminum-based hydrogen-electricity co-generator, a thorough understanding of the interaction between the two production processes is necessary.

4.3.2. Combined Production of Hydrogen and Heat

Inspired by the largely negative enthalpy change arising from aluminum-water reaction (2), Franzoni et al. [51] proposed a system for combined production of hydrogen and heat based on reaction (2). In such a system, pressurized hydrogen was produced to power a fuel cell, while the reaction heat was used to vaporize the water into high pressure steam to be fed in a turbine. A grinding tool was used to pulverize the solid aluminum rod and to remove the aluminum oxide layer that inhibits the aluminum oxidation with water. It was evaluated that the energy conversion efficiency of this combined production system ranges between 62% and 85%, which are comparable with the state-of-the-art superheated steam cycle as well as heat and power plants that use standard fuels [51].

5. Future Concerns of Aluminum Energy Utilization

Besides the technical concerns regarding different aluminum energy conversion routes, the challenges faced by a long-term reliance on aluminum energy are discussed below.

5.1. Aluminum Recovery

Aluminum recovery from by-products is a very important step in the aluminum energy cycle as indicated in Figure 3. It is this process that converts the primary energy into the secondary energy stored in aluminum for subsequent use. Here are two major concerns about this process, energy efficiency and environmental impact. Almost all existing commercial aluminum reduction facilities are based on Hall-Héroult carbon anode reduction process as depicted in Figure 5. If considering the state-of-the-art Hall-Héroult reduction cells, the total energy required for recovering 1kg of aluminum from $Al(OH)_3$ exceeds 61.3MJ with the inclusion of energy consumptions in carbon anode production and calcination that transforms $Al(OH)_3$ to Al_2O_3 [35]. In view of aluminum exergy of 29.5MJ kg^{-1}, the energy conversion efficiency is merely 48%. In addition, pollutant emission is another problem associated with the existing Hall-Héroult process. CO_2 and CO are directly involved as reaction products from a Hall-Héroult cell. It is estimated that a release of 1.52kg of carbon dioxide equivalents could be induced by the recovery of 1kg aluminum using the existing process [35]. Therefore, targeting a sustainable future, alternative recovery technologies, which are efficient, clean, and economical, are necessary for a long-term development of the aluminum energy.

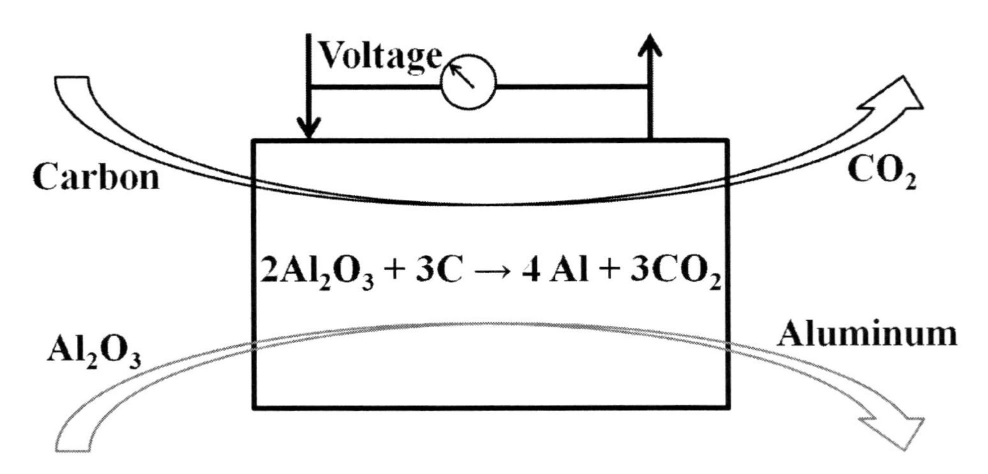

Figure 5. Hall-Héroult carbon anode reduction process for aluminum recovery.

5.2. Recharge Concerns

The rechargeability of the aluminum-air battery would be another crucial issue related to the future reliance on aluminum energy. Because of its high charging potential, the aluminum-air battery cannot be electrically recharged in an aqueous system, where water is preferentially electrolyzed [52]. Currently, aluminum-air batteries rely on mechanical recharging by replacing the discharged aluminum anode with a new one. From a near-term viewpoint, it is an advantage rather than a drawback as mechanical recharging can be more time-efficient than conventional electrical recharging method. Take a conventional battery electric vehicle for instance, a complete recharging takes from 30 minutes up to many hours or even a whole day depending on the infrastructure and battery size [53]. In comparison, electric vehicles powered by aluminum-air batteries with ~10-minute refueling time are more acceptable by consumers [54]. From a long-term viewpoint, however, electrical recharging should be the trend. The main reason is that it not only eliminates the heavy transport of anode metal and by-products between the recycling center and urban areas but also meets the demand of future energy systems relying on multiple energy carriers-electric vehicles powered by different batteries can share the same recharging infrastructure. Nevertheless, there has been no report on the electrically rechargeable (secondary) aluminum-air batteries. To realize the long-term survival of aluminum-air batteries for the electric vehicle application, efforts on developing secondary aluminum-air batteries are inevitable.

CONCLUSION

This paper considers aluminum and its alloys as an energy carrier. In addition to its high energy density, aluminum is abundant, safe, easily storable, which makes it more readily usable compared to other energy carriers of choice such as hydrogen and lithium. Extensively-studied routes for aluminum energy conversion include reacting aluminum with water to produce hydrogen and using aluminum as an anode for battery discharge. For both routes, activation is the key technology as aluminum is easily passivated by the oxide layer. Added to the above two routes, concept of combined generation multiple kinds of energies from aluminum has also been raised for a full exploitation of aluminum energy. While aluminum along with these energy conversion techniques holds promise for mitigating current fossil fuel issues, a long-term dependence on aluminum energy is still facing great challenges. With further efforts on solving these challenges, aluminum is expected to occupy a place in the renewable energy future.

ACKNOWLEDGMENT

The authors would like to acknowledge the support of the CRCG and ICEE of the University of Hong Kong in providing funding for this study.

REFERENCES

[1] Office of the Renewable Energy Regulator of Australian government (2009). MRET - the Basics. http://www.orer.gov.au/publications/ pubs/mret-thebasics-0709.pdf.

[2] Commission of the European Communities (2007). Renewable Energy Road Map. http://ec.europa.eu/energy/energy_policy/ doc/03_renewable_energy_roadmap_en.pdf.

[3] US Department of Energy. http://www1.eere.energy.gov/femp/ regulations/ requirements_by_subject.html.

[4] Information Office of the State Council of the People's Republic of China (2007). China's Energy Conditions and Policies. http://en.ndrc.gov.cn/policyrelease/ P020071227502260511798.pdf.

[5] Kramer, G.J.; Haigh, M. Nature. 2009, 462, 568-569.

[6] Ni, M.; Leung, M.K.H.; Sumathy, K.; Leung, D.Y.C. *Int. J. Hydrogen Energy*. 2006, 31, 1401 – 1412.

[7] Ni, M.; Leung, M.K.H.; Leung, D.Y.C.; Sumathy, K. Renew. *Sust. Energ. Rev.*. 2007, 11, 401–425.

[8] Olah, G.A. *Catal. Lett.*. 2004, 93, 1-2.

[9] Xuan, J.; Leung, M.K.H.; Leung, D.Y.C.; Ni, M. Renew. *Sust. Energ. Rev.*. 2009, 13, 1301–1313.

[10] Bruce, P.G. *Solid State Ionics*. 2008, 179, 752-760.

[11] Auner, N.; Holl, S. *Energy*. 2006, 31, 1395-1402.

[12] Marbán, G.; Valdés-Solís, T. *Int. J. Hydrogen Energy*. 2007, 32, 1625 – 1637.

[13] Schlapbach, L. *Nature*. 2009, 460, 809-811.

[14] Zhou, L. Renew. *Sust. Energ. Rev.*. 2005, 9, 395-408.

[15] Tarascon, J.M. *Nat.Chem.*. 2010, 2, 510.

[16] Scrosati, B.; Garche, J. *J. Power Sources*. 2010, 195, 2419-2430.

[17] Onishi, K.; Nakamura, T.; Nishihama, S.; Yoshizuka, K. Ind. Eng. Chem. Res.. 2010, 49, 6554-6558.

[18] Wikipedia. Energy Density. http://en.wikipedia.org/wiki/ Energy_density.

[19] US Geological Survey. Mineral Commodity Summaries 2010: U.S. Geological Survey; 2010.

[20] Petrovic, J.; Thomas, G. Reaction of Aluminum with Water to Produce Hydrogen; white paper for US Department of Energy, 2008.

[21] US Department of Energy. Targets for Onboard Hydrogen Storage Systems for Light-Duty Vehicles; 2009.

[22] Wang, H.Z.; Leung, D.Y.C.; Leung, M.K.H.; Ni, M. *Sust. Energ. Rev.*. 2009, 13, 845-853.

[23] Uehara, K.; Takeshita, H.; Kotaka, H. *J. Mater Process Technol.*. 2002, 127, 174–177.

[24] Czech, E.; Troczynski, T. *Int. J. Hydrogen Energy*. 2010, 35, 1029-1037.

[25] Kravchenko, O.V.; Semenenko, K.N.; Bulychev, B.M.; Kalmykov, K.B. *J. Alloys Compd.*. 2005, 397, 58–62.

[26] Fan, M.Q.; Xu, F.; Sun, L.X. *Int. J. Hydrogen Energy*. 2007, 32, 2809–2815.

[27] Parmuzina, A.V; Kravchenko O.V. *Int. J. Hydrogen Energy*. 2008, 33, 3073 – 3076.

[28] Soler, L.; Macanás, J.; Muñoz, M.; Casado, J. *J. Power Sources*. 2007,169(1),144–149.

[29] Soler, L.; Candela, A.M.; Macanás, J.; Muñoz, M.; Casado, J. *J. Power Sources*. 2009, 192, 21–26.
[30] Soler, L.; Candela, A.M.; Macanás, J.; Muñoz, M.; Casado, J. *Int. J. Hydrogen Energy*. 2010, 35, 1038-1048.
[31] Soler, L.; Candela, A.M.; Macanás, J.; Muñoz, M.; Casado, J. *Int. J. Hydrogen Energy*. 2009, 34, 8511-8518.
[32] Alinejad, B.; Mahmoodi, K. *Int. J. Hydrogen energy*. 2009, 34, 7934-7938.
[33] Ilyukhina, A.V.; Kravchenko O.V.; Bulychev, B.M.; Shkolnikov, E.I. *Int. J. Hydrogen Energy*. 2010, 35, 1905-1910.
[34] Shimojo, F.; Ohmura, S.; Kalia, R.K.; Nakano, A.; Vashishta, P. *Phys. Rev. Lett.*. 2010, 104, 126102-1.
[35] US Department of Energy. US Energy Requirements for Aluminum Production: Historical Perspective, Theoretical Limits and Current Practices; 2007.
[36] US Department of Energy. Hydrogen Properties. http://www1.eere.energy.gov/hydrogenandfuelcells/tech_validation/pdfs/fcm01r0.pdf, 2001.
[37] US Department of Energy. Energy Information Administration, Form EIA-861, Annual Electric Power Industry Report. http://www.eia.doe.gov/cneaf/electricity/wholesale/wholesalet2.xls.
[38] Armand, M.; Tarascon, J.M. *Nature*. 2008, 451, 652-657.
[39] Li, Q.; Bjerrum, N.J. *J. Power Sources*. 2002, 110, 1-10.
[40] Mohamad, A.A. *Corros. Sci.*. 2008, 50, 3475–3479.
[41] Tang, Y.; Lu, L.; Roesky, H.W.; Wang, L.; Huang, B. *J. Power Sources*. 2004, 138, 313–318.
[42] Ferrando, W.A. *J. Power Sources*. 2004, 130, 309–314.
[43] Nestoridi, M.; Pletcher, D.; Wood, R.J.K.; Wang, S.; Jones, R.L.; Stokes, K.R.; Wilcock, I. *J. Power Sources*. 2008, 178, 445–455.
[44] Cao, D.; Wu, L.; Sun, Y.; Wang, G.; Lv, Y. *J. Power Sources*. 2008, 177, 624–630.
[45] Doche, M.L.; Novel-Cattin, F.; Durand, R.; Rameau, J.J. *J. Power Sources*. 1997, 65, 197-205.
[46] Abdel-Gaber, A.M.; Khamis, E.; Abo-ElDahab, H.; Adeel, Sh. *Mater. Chem. Phys.*. 2008, 109, 297–305.
[47] Wang, J.M.; Wang, J.B.; Shao, H.B.; Zeng, X.X.; Zhang, J.Q.; Cao, C.N. *Mater. Corros.*. 2009, 60, 977-981.
[48] Ma, Z.; Zuo, L.; Pang, X.; Zeng, S. *Trans. Nonferrous Met. Soc. China.* 2009, 19, 160-165.
[49] Zhuk, A.Z.; Sheindlin, A.E.; Kleymenov, B.V. *J. Power Sources*. 2006, 157, 921-926.
[50] Li, Z.; Wang, W. Aluminum-water hydrogen storage cell. Chinese Patent 02148850.9, 2003 (in Chinese).
[51] Franzoni, F.; Milani M.; Montorsi, L., Golovitchev, V. *Int. J. Hydrogen Energy*. 2010, 35, 1548–1559.
[52] Hamlen, R.P.; Atwater, T.B. In Handbook of Batteries; Linden, D.; Reddy, T.B.; Ed.; McGraw-Hill: NY, 2001; pp 38.2.
[53] Eberle, U.; Helmolt, R. *Energy Environ. Sci.*. 2010, 3, 689-699.
[54] Littauer, E.L.; Cooper, J.F. In Handbook of Batteries and Fuel Cells; Linden, D.; Ed.; McGraw-Hill: NY, 1984; pp 30-19.

In: Aluminum Alloys
Editor: Erik L. Persson

ISBN: 978-1-61122-311-8

Chapter 7

LASER WELDING OF ALUMINIUM ALLOYS

***R. M. Miranda*[*,1] *and L. Quintino*[2]**

[1]UNIDEMI, Departamento de Engenharia Mecânica e Industrial, Faculdade de Ciências e Tecnologia, FCT,
Universidade Nova de Lisboa, 2829-516 Caparica, Portugal
[2]IST-UTL Instituto Superior Técnico, Av. Rovisco Pais,
Lisboa, Portugal

ABSTRACT

Aluminium alloys are largely used in transportation industry, as automotive and aeronautics where the high strength-to-weight ratio is valuable, but also in corrosive environments due to the excellent corrosion resistance they exhibit, as in tanks, pressure vessels and packaging. Despite their wide acceptance in industry, they have limited weldability. Arc welding processes as tungsten inert gas (TIG) and metal inert gas (MIG) have been used as well as electron beam and laser welding. For high productivity and demanding manufacturing, laser welding is an alternative process to consider.

Laser welding is a precise high energy density process that reduces the fusion and heat affected zones, minimising some of the major weldability problems. However, the high reflectivity of aluminium to carbon dioxide laser has restricted its use. Infrared solid state lasers, as Nd:YAG, both disc and rod, and the recent high power fiber lasers showed to be more adequate for laser welding of Al alloys.

This chapter is structured as follows: Laser welding process is discussed focusing on the effect of operating parameters on the weld quality, welding modes (keyhole and conduction) and process control. An overview of the weldability problems of major groups of industrial aluminium alloys is given. Recent and emerging techniques are presented and discussed.

[*] Email: rmiranda@fct.unl.pt,

1. Introduction

A simple search in scientific databases revealed the existence of about 8,000 references on welding of aluminium alloys and about 2,000 on laser welding. This clearly shows the intense activity the scientific community devotes to this material due to its highly interest for industry. Till the end of the 90s, carbon dioxide lasers were investigated followed by solid state lasers as neodymium doped with yttrium aluminium garnet (Nd:YAG) in both continuous wave and pulsed wave modes and more recently fiber lasers. Hybrid lasers emerged in 2006 involving laser plus an arc welding process, MIG or TIG.

Research works are mostly classified in materials science and engineering. The former group of references focus on the weldability problems of weldable alloys series as the AA5XXX and AA6XXX in both similar and dissimilar welds, amongst these and to steels. Major issues targeted by researchers include thermo-mechanical modelling and simulation, residual stresses and distortions, fatigue resistance and crack growth mechanisms, and weld corrosion.

High power lasers have been investigated for welding applications where sheet metal and thin sheet is involved. Laser welding has a few distinctive characteristics which represent advantages when welding aluminium as: low heat input which minimise distortions and property degradation due to fusion welding, high welding speeds improving productivity and reducing processing times and costs.

Additionally, the flexibility of laser systems has to be considered since it allows to feed several workstations, be remotely controlled, weld with filler material to correct the chemical composition of the weld metal, prevent welding problems and improve welded joint strength.

2. Laser Welding

The basic principle of laser welding is that a laser radiation beam produced in a resonator cavity (the laser) is focused on the material surface through an optical system. The effectiveness of the beam coupling depends on the wavelength of the beam radiation and the optical properties of the material including the surface condition.

The absorption of the laser beam increases if the energy is transferred to the material so that the temperature increases according to classical thermal conduction theory. If the energy absorbed is below a threshold value, the material conducts the heat into adjacent regions leading to a wide and low penetrant molten pool. This is a conduction laser welding mode observed in highly reflective materials as Al or low power lasers. The resulting weld has a shape very similar to arc welding as shown in Figure 1a.

If the energy absorbed by the material is above the threshold value that is roughly the energy required to melt a unit volume of material, it evaporates forming a highly absorbent plasma which transmits the heat into depth. This column of plasma is sustained by surrounding molten metal. The resultant weld has a small aspect ratio, that is, the weld width to the weld penetration ratio is less than one. Figure 1b shows a schetch of the keyhole laser welding mode.

The molten metal motion is ruled by the forces exerted on the pool, namely the Marangoni effect as depicted in Figure 1a.

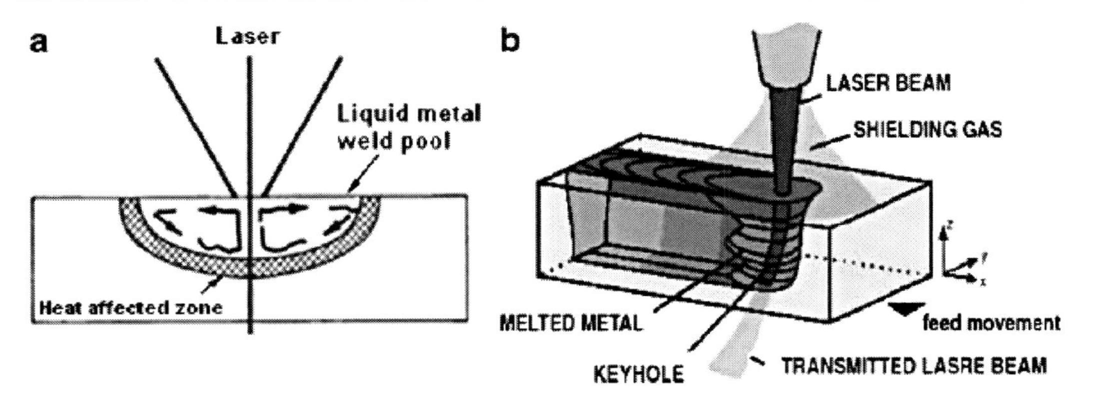

Figure 1. Schema of conduction (a) and keyhole (b) laser welding modes.

Among the operating factors influencing the shape and properties of the molten pool and, consequently the properties of the weld, absolute and relative factors can be identified. The absolute factors are related to:

- laser beam: wavelength, power, power distribution, operating mode, spot size
- materials: chemical composition, thermo-physical properties, microstructure, geometry and dimensions
- shielding gas: composition, flow rate, flow configuration
- filler material: chemical composition, thermo-physical properties, type and dimensions.

The relative factors are:

- relative position of laser beam, welding materials, shielding gas and filler material
- welding speed, filler material and feeding rate.

Table 1 shows the laser wavelength and beam quality of several commercialised laser systems.

Table 1. Beam quality for several laser systems

Laser Type	Typical Beam quality mm.mrad	Wavelength Nm
CO_2	3.7	10.600
Lamp pumped Nd:YAG	25	1060
Diode pumped Nd:YAG	12	1060
Yb Fibre laser (IPG YLR7000)	18.5	1070
Yb Fibre laser (IPG YLR 10000)	11.6	1070
Yb Fibre laser (IPG YLR 17000)	11.7	1070
Thin disc Yb YAG	7	1030

The high reflectivity of Al limits the amount of energy transferred or absorbed by the material and thus the welding mode preferentially observed. Moreover, during welding there is a danger of reflection of the beam from the workpiece surface back into the optical system, damaging the welding head and, eventually, the laser light cable. In extreme cases, the reflected radiation can also enter the resonator damaging it due to excessive thermal load, which is due to excessive power density inside the laser active element.

In order to protect the laser resonator and the welding head, it is advised to incline the head about 5 to 15° relatively to the beam axis, so that the reflected beam does not enter the optical system.

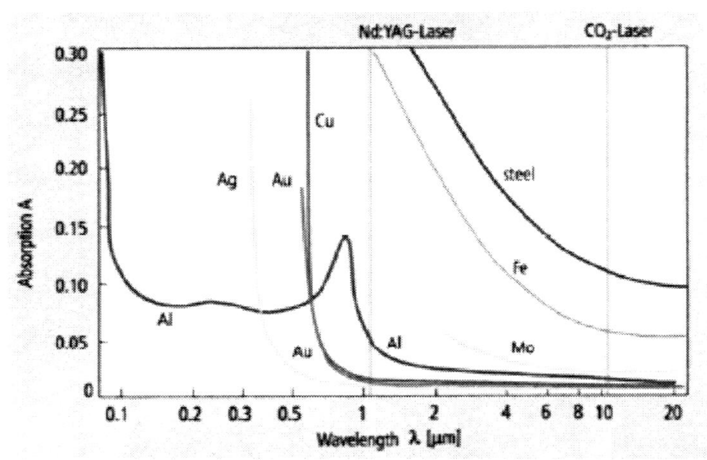

Figure 2. Absorption of a number of metals as a function of laser radiation wavelength.

Figure 2 shows the absorption of several metals as a function of the laser beam wavelength.

2.1. CO_2 and Nd:YAG Laser

The two most industrially used lasers are carbon dioxide and Nd/YAG lasers. The former is a gas laser emitting at a wavelength of 10.6 micron with higher output powers, efficiencies around 20% (higher than Nd:YAG lasers), with a good beam quality, easy to focus but requiring manipulation systems complex and robust since the light cannot be transmitted via optical fiber. The fact these lasers have high output powers (up to 50 kW) allows for keyhole welding modes in thick Al plate, despite the poor absorption to its wavelength.

Nd:YAG lasers are solid state lasers emitting at 1.06 micron in a range of wavelengths where the metallic materials are most absorbent, but have poor energy conversion efficiency (around 5%), limited output powers and high running costs specially due to maintenance. The possibility to be transported via optical fiber improves the system flexibility, but the most relevant advantage is the high absorption of Al alloys to its laser wavelength. That is, for the same weld penetration the required power is lower for Nd:YAG laser than for CO_2 laser.

Since these can work in pulsed wave mode they can be very precise, adequate for small components with limited distortions.

Diode lasers and diode pumped lasers have been the subject of innovative developments in the last decades. They use the same principles as a diode, so they are classified as solid state lasers and in recent years several developments have been performed by manufacturers.

The ressonator cavity is made of two coated bars to have the desired optical properties. They have good energy efficiencies in the range of 30 a 50% and can be transported via an optical fiber. Together with the low implementation area they are easily integrated in manufacturing lines. Diode lasers are particularly adequate for laser welding of Al since they emit in the range of 800 e 900 nm (less than Nd/YAG lasers).

2.2. Fiber Laser

Most recently high power diode pumped fiber lasers were developed with attractive characteristics for materials processing applications.

Multi-kilowatt fibre lasers have now been introduced for materials processing. These new lasers have multiple advantages, including: high efficiency, compared to lamp or diode pumped rod lasers; compact design, which simplifies installation; good beam quality, due to the use of small diameter fibers, and thus small beam focus diameter; and a robust setup for mobile applications.

Multi-kilowatt power levels are achieved by combining the output from several single mode fibre lasers. The lifetime of the pumping diodes is expected to exceed the lifetime of other diode pumped lasers [2], which leads to low costs of ownership.

High power fiber lasers can be used for deep penetration welding in a diversity of materials since the low wavelength that characterizes these lasers allows its absorption by almost all metals and alloys, and the fiber delivery system provides the necessary flexibility on the positioning of the beam.

The active medium of a fibre is a core of the fiber doped with a rare earth. Most commonly, this is a single-mode fiber laser made of silica. The pump beam is launched longitudinally along the fiber length and it may be guided by either the core itself, as occurs for the single-mode lasers or by an inner cladding around this core (double-clad fiber laser). Figure 3 shows a schema of these lasers [4].

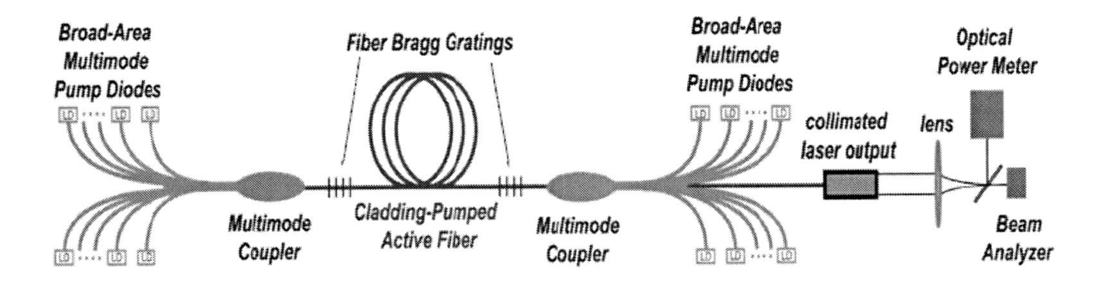

Figure 3. Fiber laser pumping schematic.

Fiber laser technology offers several benefits to the industrial user. The "footprint" area and maintenance needs are reduced in comparison with other laser because there is no need to replace flash lamps or diodes. The high electrical efficiency greatly reduces operating costs. Better beam quality, with very low divergence, allows the user to produce spot diameters substantially smaller than conventional lasers producing longer working distances. When all factors are accounted for, namely, floor space, chillers and maintenance, they are more cost effective than equivalent power rod-type Nd:YAG lasers. Figure 4 [8] shows calculated operating costs over an 8 year period, indicating that for intensive use, fibre lasers are projected to have a lower operating cost than either CO_2 or Nd:YAG lasers.

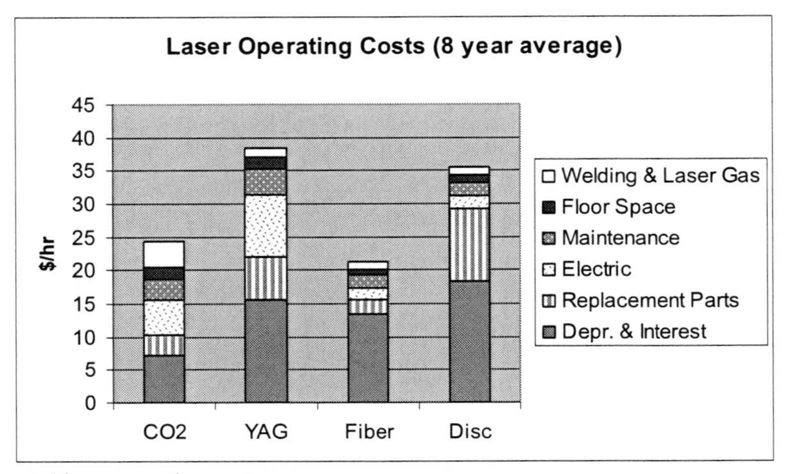

Figure 4. Estimated laser operating costs.

2.3. Hybrid Welding

In hybrid welding, a laser beam and an arc welding process interact in a common area where the produced molten pool is due to the synergic effect of both the plasma and the electric arc as shown in Figure 5. The use of hybrid processes has several advantages:

- It allows wider gaps between plates requiring less precise positioning systems and rougher surface preparation
- Higher welding speeds. The laser and the arc interact increasing the efficiency of both individual processes
- Small heat affected zones
- Higher productivity
- Allows the addition of filler material improving the mechanical and chemical properties of the weld metal

The major disadvantages are the high investment cost. In fact, hybrid processes minimise the drawbacks of laser welding like: joint preparation and fit-up and costs and simultaneously, they reduce the disadvantages of arc welding processes as: high heat inputs, low welding speeds and environmental and health and safety problems to welders.

In aluminium alloys, hybrid processes have an additional advantage. If the torch is in front of the laser beam, it heats the material increasing its absortivity, thus less power is required to give a certain penetration. In dissimilar welding the arc welding process provides the filler material so the control of the overall process is simplified specially if a standard welding head for hybrid processes is used.

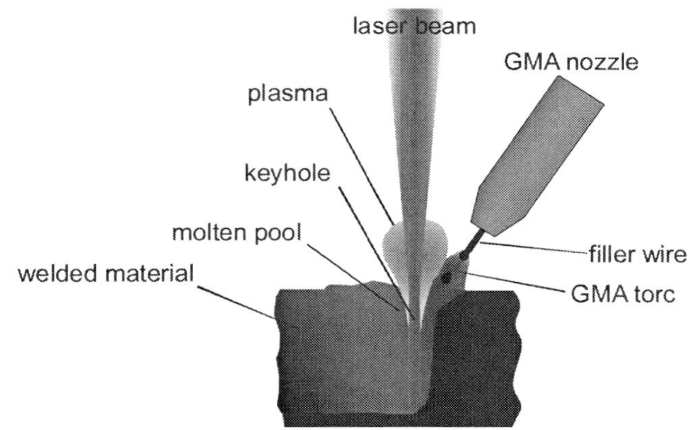

Figure 5. Schematic representation of a laser+MIG hybrid welding process.

Fillers have several functions: to bridge wide or irregular gaps in the joint, to reduce the amount of work required to prepare the seam, to add elements to the molten metal in order to improve weldability and modify the properties of the fusion zone including mechanical strength and corrosion resistance [3].

3. Weldability of Al Alloys

Aluminium alloys is a general designation of a wide range of aluminium based materials with different alloying elements introduced aiming at producing alloys with dedicated physical, chemical and mechanical properties. Usually they are classified according to the major alloying elements present, in series from which the 2xxx, 5xxx and 6xxx have major relevance as far as welding is concerned. The 2xxx series has Cu as the major alloying element, with minor additions of Mg and Li. The 5xxx series present Mg and the 6xxx have Si and Mg. The high strength 7xxx series, widely used in aeronautic industry, has Zn as major alloying element with additions of Cu, Mg, Cr and Zr and has limited weldability. Thermal and/or mechanical treatments shape the mechanical properties of the alloys.

From a weldability point of view, aluminium alloys have a set of characteristics that reduces its weldability and these are:

- Low melting point
- high affinity to oxygen, forming alumina, a refractory material with a melting point high above the melting point of pure Al
- high thermal conductivity (about three times that of carbon steel) that requires more heat input to melt a unit volume of material
- high thermal expansion coefficient (twice the one of carbon steel) that originates distortions and cracking
- the presence of volatile elements as magnesium or zinc in 7xxx series, which originates porosities in the weld metal
- when laser welding aluminium alloys, weldability problem occur due to mechanical and thermal treatments, as well as, to the physical properties, as these are:

Welding Problem	Alloy
Porosities	In almost all the alloys
Excessive material loss	In all alloys due to low melting point
Solidification cracking	High resistance alloys as: AA7025, AA2014, AA6061, AA5052
Softening in the heat affected zone	Cold worked alloys and heat treated alloys

The major cause of porosity formation is the hydrogen incorporation in the liquid. The solubility of hydrogen in liquid is higher than in the solid, thus porosities are formed as a result of the rejection of hydrogen from the solid. Any sources of hydrogen have to be minimised to prevent porosities, thus a good surface cleaning and a correct selection of shielding gas (type and mass flow) are important. Surface cleaning is preferably done by a chemical agent followed by mechanical brushing which increases the surface roughness and decreases the reflectivity to laser light. This is particularly valid when welding with CO_2 laser.

Inert gases of high purity are used for welding aluminium alloys in mass flow rates of about 20-25 l/min to prevent contact of the molten region with any contaminants including atmosphere.

Figure 6 shows an example of Nd:YAG laser welding of AA6061 alloy with porosity and micro cracking in the interface of the fusion zone of base material. Micro porosity was caused by hydrogen that was rejected from the liquid metal on solidifying. Potential hydrogen sources were moisture contaminants and hydrated oxides on the surface of both base materials and filler wires. Micro cracks appear in the AA6061 alloy interface due to the formation of low solidification temperature liquid induced by the microsegregation of magnesium and silicon of this alloy.

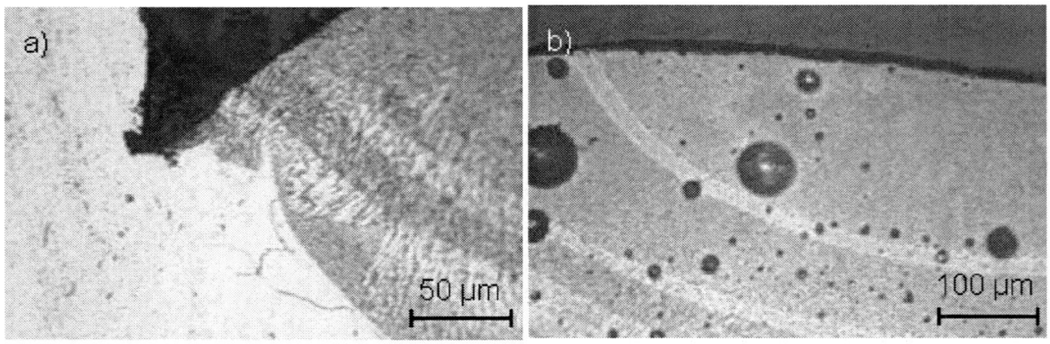

Figure 6. Microstructure of AA 6061. a) Interface with fusion zone in AA6061 alloy; b) Porosity in the fusion zone.

Solidification cracking is observed in the weld metal centreline as a result of aluminium high thermal expansion coefficient, high volume of molten metal and a large temperature range of the alloy.

In laser welding it is observed in the 5xxx and 6xxx series, in high strength alloys where the poor ductility of the fusion zone increases the stress concentration leading to cracking.

The weld depth to width ratio controls the volume of the weld material and thus the susceptibility to solidification cracking. Narrower welds are preferable.

The addition of filler materials allows correcting the chemical composition of the weld metal reducing both porosity and/or hot cracking. Controlling the weld shape and profile, namely reducing its convexity decreases the stress concentration and improves the high temperature behaviour of the weld metal. High welding speeds decrease the depth to width ratio thus preventing solidification cracking.

Softening in the heat affected zone reduces the mechanical strength of the joint, namely the tensile and yield strengths and elongation, due to dissolution of strengthening precipitates as Mg_2Si in 5xxx series alloys. A common procedure includes post weld heat treatment for reprecipitation of these particles.

Laser welding has an advantage compared to arc welding since the heat input is very low the extent of the heat affected zone is small and the time for structural modifications as precipitate dissolution is reduced. The rapid thermal cycle prevents these phenomena. Nevertheless, cold worked alloys as the 2xxx and 6xxx series are prone to softening in the heat affected zone.

Figure 7 shows a hardness profile of a dissimilar laser weld with Nd:YAG laser of two distinct 6xxx series alloys with several filler materials. It is evident the drop in hardness in the heat affected zone of both alloys more evident in the AA6082 alloy. In this case the loss of precipitates and the over ageing effect was identified as the main reason for hardness reduction.

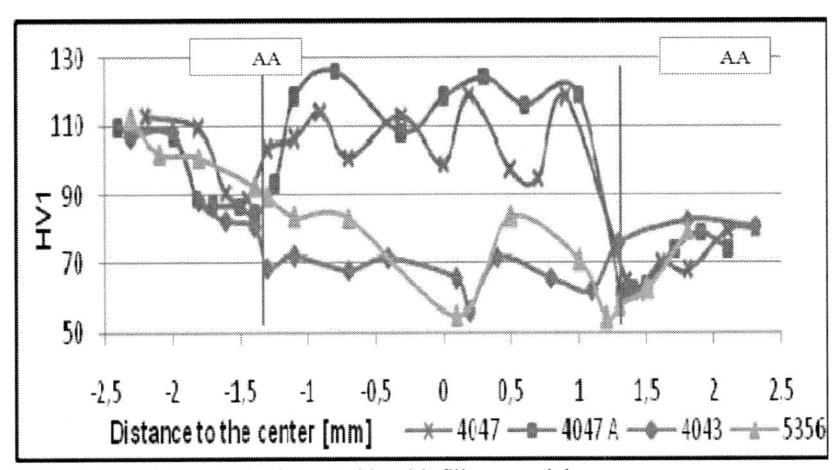

Figure 7. Hardness profile in dissimilar laser welds with filler material.

In the case of heat-treatable alloys, the loss of precipitates in the welds and over ageing in the HAZ has been identified as the main cause for hardness reduction. This degradation is caused by microstructural modifications associated with elevated temperatures experienced in this zone. For heat-treatable aluminium alloys, the HAZ is distinguished by dissolution or growth of precipitates. Postweld heat treatment can also be used to improve the strength of the HAZ in heat-treatable alloys.

This may involve complete postweld solution heat treating followed by ageing or postweld ageing only. Although the recovery of strength in the HAZ after postweld is lower than postweld solution heat treating and ageing, there are advantages to postweld ageing alone [3]. Aluminium-silicon alloys have good resistance to cracking, partially due to their abundance of liquid eutectic available for back-filling. However, their use should be avoided when welding high-magnesium alloys (>3 wt%) because of embrittlement from excessive Mg_2Si precipitation.

4. Applications

Important volume of industrial application of laser welding of aluminium can be found in shipbuilding and aero-space industry. Although the first aircraft made by Boeing was welded, mechanical fastening methods such as riveting have dominated because of the difficulties associated with fusion welding of the common aluminium aerospace alloys 2024 and 7075. Laser welding can reduce joining costs relative to adhesive bonding and riveting by about 20%.

EADS Airbus has invested heavily in laser welding as a replacement for riveting in non-critical applications. One application involves joining stiffening stringers to the skin of the fuselage. The damage-tolerant alloy 6013 is the base material and 4047 the filler material. The stringers are welded from two sides at 10 m/min, using two 2.5 kW CO_2 laser beams. The joint is designed such that the HAZ is contained in the stringer, and does not impinge on the skin.

The process was first used in series production of the Airbus 318, and after implemented in other aircraft types.

Laser welding of aluminium has important applications in automotive industry since aluminium started to be used not only for gear box, but also for structural parts.

The new concept of aluminium integrated car body involves the use of extruded high strength elements jointed in nodal points. Both laser welding and laser-GMA hybrid welding are used, besides classical arc welding and brazing processes.

Laser-GMA hybrid welding has much more volume of applications in industry when compared with laser welding. E.g., from about 500 km of weld in case of a cruiser, about 250 km are made by laser-GMA welding, but in this case the major part of the structure is made of naval steel.

Summary

In summary it can be concluded that:

Aluminium alloys are well disseminated in industry due to the particular characteristics they exhibit.

Weldability depends on the type of alloy, that is, on its chemical composition and previous heat and mechanical treatments. Several weldability problems have been identified and these occur specially in arc welding processes due to the high heat input of these processes.

Laser welding has significant advantages over arc welding though the high reflectivity of Al makes dioxide carbon lasers difficult to use. Solid state lasers and, in particular, fiber lasers have significant advantages. The recent hybrid processes are to be considered specially since the reflectivity to laser light is minimised and tolerance to fit up is very good.

REFERENCES

[1] K. R. Brown, M. S. Venie, and R. A. Woods, "The increasing use of aluminum in automotive applications," *Journal of Metals*, vol. 47, no. 7, pp. 20–23, 1995.

[2] W. van Haver, X. Stassart, J. Verwimp, B. De Meester, and A. Dhooge, "Friction stir welding and hybrid laser welding applied to 6056 alloy," *Welding in the World*, vol. 50, no. 11- 12, pp. 65–77, 2006.

[3] John F. Ready, "LIA Handbook of Laser Material Processing", Magnolia Publishing, Inc, 2001.

[4] W.M. Steen, "Laser Material processing", Ed. Springer Verlag, 3rd edition, 2003.

[5] Y. Arata, N. Abe, "Fundamental Phenomena CO_2 Laser Welding", Transactions of JWRI, Vol. 1 n° 1, 1985.

[6] M. Kutsuna, J. Suzuki, S. Kimura, S. Sugiyama, M. Yuhki, and H. Yamaoka, "CO2 laser welding of A2219, A5083 and A6063 aluminium alloys," *Welding in the World*, vol. 31, no. 2, pp. 126–135, 1993.

[7] J. Zhang, D. C. Weckman, and Y. Zhou, "Effects of temporal pulse shaping on cracking susceptibility of 6061-T6 aluminum Nd:YAG laser welds," *Welding Journal*, vol. 87, no. 1, pp. 18s– 30s, 2008.

[8] C. Thomy, T. Seefeld, F. Vollersten, "Application of high power fibre lasers in laser and MIG welding of steel and aluminium", pp 88-98 in "Proceedings of the IIW conference on benefits of new methods and tends in welding to economy, productivity and quality", 10-15 July, Prague, Czech Republic. Ed. L Junek.

[9] B. Shiner, "High power fiber lasers impact material processing", Industrial Laser Solutions, February 2003 , pp 9-11 (in *http://ils.pennnet.com*).

[10] S. Ream, "Laser welding efficiency and cost – CO_2, YAG, Fiber and disc lasers", ICALEO 2004.

[11] Abe, N., Hayashi, M.; Trends in laser arc combination welding methods, Welding Internacional 16, H.2, A.94-98, 2002.

[12] Kutsuna, M., Lin Z., Developments and applications of laser-arc hybrid welding, Doc. IIW-IV-939-07.

[13] Dilthey U., Olschok S., "Robotic fiber Laser GMA hybrid welding in shipbuilding", Doc IIW-IV-950-07.

[14] KatayamaS., Kawahito Y., Mizu Tani M., Understanding of laser and hybrid welding phenomena, Doc. IIW-IV-943-07.

[15] J. Polmear, "Metallurgy of the Light Metals", Edward Arnold Ltd, London 1981.

[16] U. Dilthey, A. Brandenburg, and F. Reich, "Investigation of the strength and quality of aluminium laser-mig-hybrid welded joints," *Welding in the World*, vol. 50, no. 7-8, pp. 7–10, 2006.

[17] Haboudoua, A., Peyrea, P., Vannes, A.B., Peix, G.; *Reduction of porosity content generated during Nd:YAG laser welding of A356 and AA5083 aluminium alloys*; Materials Science and Engineering A (2003) 363, 1, 40–52.

[18] H. Zhao and T. DebRoy, "Macroporosity free aluminum alloy weldments through numerical simulation of keyhole mode laser welding," *Journal of Applied Physics*, vol. 93, no. 12, pp. 10089–10096, 2003.

[19] Cicala, E., Duffet, G., Andrzejewski, H., Grevey, D., Ignat, S.; *Hot cracking in Al–Mg–Si alloy laser welding – operating parameters and their effects*; Materials Science and Engineering A (2005), 395, 1, 1–9.

[20] Paleocrassas, A. G. Feasibility Investigation of Laser Welding Aluminum Alloy 7075-T6 through the use of a 300 W, Single-Mode, Ytterbium Fiber Optic Laser, Degree of Master of Science, Graduate Faculty of North Carolina State University, 2005.

[21] Rao, Prasad, K., Ramanaiah, N., Viswanathan, N.; *Partially melted zone cracking in AA6061 welds*; Materials and Design (2008) 29, 1, 179–186.

[22] ASM Handbook; Properties and Selection: Nonferrous Alloys and Special-Purpose Materials; American Society for Metals (ASM); Vol. 2, 2004.

[23] E. Assunção, L. Quintino, R. M. Miranda, "Comparative Study of Laser Welding of Tailor Blanks for the Automotive Industry International Journal of Advanced Manufacturing Technology, (2010), 49,123–131.

[24] L. Quintino, A. Costa, R. Miranda, D. Yapp, V. Kumar, C.J. Kong, "Welding With high Power Fiber Lasers – A Preliminary Study", Materials and Design, 28, (4), 2007, 1231-1237.

[25] L. A. Pinto, L.Quintino, R. M. Miranda, P. Carr, "Laser welding of dissimilar aluminium alloys with filler materials, Doc. IIW – IV-978-09, Com IV – Power Beams, IIW Annual Meeting, 2009.

In: Aluminum Alloys
Editor: Erik L. Persson

ISBN: 978-1-61122-311-8

Chapter 8

MONITORING OF THE ALUMINUM ALLOY B95 (7075) ALARM STATES ARE CAUSED BY FATIGUE LOADS

S. Kh. Shamirzaev, J. K. Ziyovaddinov and Sh. B. Karimov
Physical Technical Institute of Uzbek Academy of Sciences,
Bodomzor Yuli 2-b, Tashkent, 700084,
Republic of Uzbekistan

ABSTRACT

The objects of our paper are aluminum alloy samples (AASs) contained the different amount of Cu, Mn, Mg, Si and Li. We are modeling the features of microstructure of potential relief of an Aluminum Alloy Sample (AAS) and studying its transformation under both imposed fatigue deformation and wetted by liquid metals (Ga; or Hg; Li ;In). Although fatigue in an AAS is characterized by permanent changes of its mechanical impedance, we illustrate the main ideas by using only the «time series» allied with effective internal friction Q^{-1}_{eff} of an AAS.

There is not a common law permits one in advance to determine both what kind of micro - structures and how many of micro - structures will be simultaneously belong to aluminum alloy (AA) after the N- number of cycles are loaded.. AAs like B-95 or 7075 are heterogeneous materials for which the more energy can be absorbed by selected micro-regions of a tested sample. So micro-crack in the space of AAS and alarm state of AAS arises. Each micro-regions will to contribute (the $Q^{-1}{}_{k}$ belong to k-th micro - region) to the effective internal friction – $Q^{-1}{}_{eff}$ accordance with fit statistic g_{k}. We find a number of micro-regions - L and series g_{k} and Q_{k} from the experimental data like as the internal friction $(Q^{-1})_{eff}$ versus both, the number of cycles - N and the deformation - ε. Series g_{k} and Q_{k} (k = 1,2,3,…,L) present the microstructures of AASs. So the monitoring of AASs alarm states was made.

In this paper also is presented the original technology to forecast fatigue damage of an aluminum alloy sample. Here was used the fatigue sensitive element (FSE). The various impurities in Aluminum implemented the various resistances to fatigue of an Aluminum Alloy Sample. In the other hand, we used multiphase heterogeneous mixtures (MHMs) which contents a variable volume of initial components. It is selected MHMs are using for produce FSEs. The correlation between output parameters of the MHMs and volume contents of initial components and form of their grains are given in our previous papers, The present paper is aimed to establish the correlation of the FSEs microstructures changes and corresponding changes of the aluminum alloy microstructures at imposing the same spectra deformation on both of them. A change of FSEs microstructure investigated by using their effective electrical resistance $R_{eff=}$ data.

Keywords: amplitude-dependent internal friction, fatigue, cracking process, alarm state, fatigue sensitive element.

INTRODUCTION

The microstructure of an AAS is reflected on the microstructure of its potential relief that is originated by both a residual deformation and an imposed fatigue deformation. In local regions of polycrystalline alloy, that undergoes fatigue loads, a wide spectrum of strongly exited states arises [2,3], These states can not be described by traditional methods of a perturbation theory, With the help of mechanics methods of damage one can study only the growth of macroscopic cracks using empirical - formula dependencies describing the growth of a crack.

The various impurities in Aluminum implemented the various resistance to fatigue of an Aluminum Alloy Samples. There is not a common low permits one in advance to determine both what kind of microstructures and haw many of microstructures will be simultaneously belong to AAS after a number cycles of loads.

The main objective of this paper is to present the experimental method for estimating the effective output parameters of AASs and their change under both the imposed fatigue deformation and the wetted by liquid metals. The output data of AASs depend on the microstructures they are possessed of. For example, effective internal friction – $Q^{-1}{}_{eff}$ of the AAS is formed as a sum of microstructures' internal friction – $Q^{-1}{}_{k}$ with fit statistics - g_{k}:

$$Q_{eff}^{-1} = \sum_{k=1}^{L} g_k * Q_k^{-1} \; ; \; \sum_{k=1}^{L} g_k = 1 \; ; \quad (k = 1,2,3, \ldots, L)$$

Damages accumulated at the initial stage of fatigue process change local parameters of the AASs [3,5,6]. The micro – structures are irreversibly changed if sufficiently high deformation is imposed on AASs. These processes become more intensive and more intricate under wetted by lick of liquid metals such as Gallium – Ga or Mercury – Hg, In this case the AAS become very brittle and reduce their lifetime to zero [4]. As the number of cycle's increases AAS's microstructures eventually begin to make change. As a result the alarm state of an AAS arises. Monitoring of the AASs' alarm states has to been developed. It is one of the main object of this paper.

New experimental techniques of g_{k} and $Q^{-1}{}_{k}t$ were developed. For simplicity sake assume that series g_{k} and $Q^{-1}{}_{k}$ present the microstructures of polycrystalline Aluminum Alloys [6]. The analysis of both new materials' micro - structures as well as existing materials' micro – structures by novel methods was made.

The ways of realization. After imposing a limited number of cycles of fatigue deformation that change the AAS's microstructures, the test cycle of deformation with selected value of amplitude are imposed on the AAS's.

The $Q^{-1}{}_{eff}$ - ε (ε - value of imposed deformation) curve, that was automatically recorded within test cycles of deformation, permits one to bring out the forthcoming structures of AAS and its features, that will take place at the near future cycles of imposed deformation. Reiteration of these processes also for $\mathbf{Q^{-1}{}_{eff}}$ - ε ($\mathbf{Q^{-1}{}_{eff}}$ – effective internal friction of AAS wetted by liquid metals – Hg, Ga, In or Li) curves give one the chance to do the monitoring of AASs' microstructures change,

Internal friction .Generally on a fatigue **SN** curve the numbers of cycles **N** are within 10^7 to 10^9. If an Aluminum alloy sample is wetted by liquid gallium – Ga, **N** tends to sharp decrease with the advance of cracking process. It is known [4] that a quantity of Ga leads to embrittlement of Aluminum alloy and decrease its durability under Strain because of cracking processes. The study of micro cracks, related with selected points on surface of a test sample, is essential circumstance for understanding the nature of fatigue. Another essential circumstance is the study of sample's internal friction (Q^{-1} (ε)). The internal friction relates to a number of structure-dependent phenomena which permit to judge about fine details of the real structure of a condensed hard matter. For some samples Q^{-1} (ε) curves possess a hysteretic dependence , which after Granato and Lucke [7] can be related with moving and damping of hesitating dislocation segments in the selected local regions of an AAS.

But there are ample samples of construction alloys' Q^{-1} (ε) curves, that meet difficulties under the quantitative comparison with the theory of dislocation damping. This theory was developed for the homogeneous media. The Aluminum Alloys are structurally heterogeneous. So there are overstrained micro regions with high speed evolution of damage processes. Those micro regions are transformed under the load and verged towards to unstable state. They store mechanical energy due to both the phonon mechanism [9] and a structural change which one can notice before micro-cracking processes.

Experimental Installation. The two-component oscillator (sample and piezoceramics), having an Eigen-frequency of about 18 kHz, was used by Mason [10] for study the amplitude dependent decrement on the vibration **S**train amplitude. High level of computer engineering was used by Mills D. and Bratina W. J., [8], Kardashov B.K. et al. [11], Schenck H. et al. [12] for studying the effect of the oscillation amplitude and temperature on the amplitude-dependent internal friction Q^{-1}(ε).We also are using the two-component oscillator (ZrTiPb-19 piezoceramics and the wetted by liquid Ga on the Aluminum Alloy Sample) having an Eigen-frequency of about 20 kHz, for studying the effect of the oscillation amplitude and cracking processes on the amplitude-dependent internal friction Q^{-1} (ε).

Figure 1 shows ZrTiPb-19 piezoceramics and the wetted by liquid Ga on the Aluminum Alloy Sample with a rigidly installed fatigue element (FE).

The FE is small weight (no more than 5g) and has film geometry. They are produced by thermal vacuum evaporation of the charge onto the polyamide support. The charge consists of a finely dispersed mixture of various initial components, e.g., a mixture of granular $Bi_{2-x}Te_{3+x}$ and $Sb_{2-y}Te_{3+y}$ with carbonyl iron [1].

An ample number of grains do not take part in fatigue process of both an AAS and a FE. In parallel with growth of the number of strong deformed grains in AAS as well as in FE, there is also the ample number of grains possessing their original state. The simple model of selected FE describes these situations are given in [2,3].

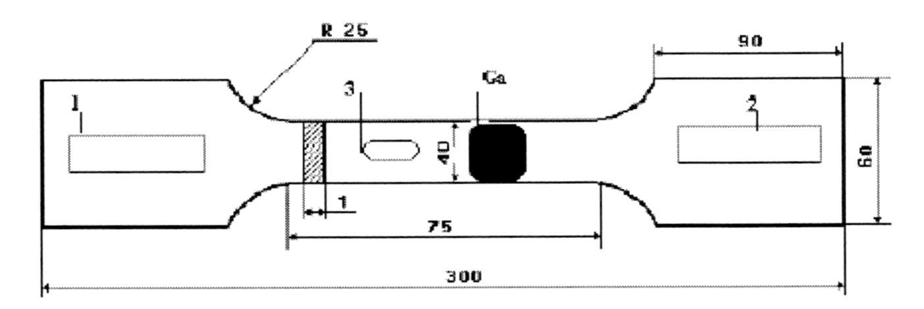

Figure 1. Shape and geometrical dimensions of the Aluminum Alloy Sample.. A drop of liquid Ga was imposed on this pattern. After 12 hours this drop became a slur, as pictured at the figure 1. 1 - ZrTiPb-19 – oscillator ; 2- ZrTiPb -19 – receiver; 3 - FE -fatigue element of the fatigue sensor.

Here is also next sufficient conclusion: For all kinds of imposed deformation (extension-compression; bending variations with different coefficients of asymmetry; various spectra of imposed deformations), at which the output parameters (electric resistance R_n for FE and internal friction Q_n^{-1} for AAS) of materials are recorded in equal time (τ) intervals, the next recurrent relations take place:

$$R_{n+1}[\sigma] = \mathbf{O}_n * R_n[\sigma] = B_n * R_n + (1-B_n) * M_n \; ; \tag{1}$$

$$Q^{-1}_{n+1}[\sigma] = \mathbf{O}_n * Q^{-1}_n[\sigma] = b_n * Q^{-1}_n + (1-b_n) * m_n \tag{1a}$$

Here: R_n is the electric resistance of FE measured under the same state of the environment; n is the ordinal number of measurement (n = 1,2,3,..., g); g is the maximum number of the measurement performed; $\mathbf{O}_n$ is the loading operator transferring the FE resistance from R_n-th state to R_{n+1} -th state; σ is a complicated parameter to characterize the type of imposed deformation ; the «time series» is composed of experimental data of the types:

$$R_1 , R_2 , \ldots , R_{n-1} , R_n , R_{n+1} , \ldots , R_q \; ; \tag{2}$$

$$(Q^{-1}_1)_{eff} , (Q^{-1}_2)_{eff}, \ldots , (Q^{-1}_{n-1})_{eff}, (Q^{-1}_n)_{eff}, (Q^{-1}_{n+1})_{eff}, .(Q^{-1}_q)_{eff}. \tag{2a}$$

After imposing a limit number of cycles of fatigue deformation, that changes both the AA's structure and the FE's structure, the test cycles of deformation with selected value of amplitudes are imposed on the sample. The Q^{-1} - ε (Q^{-1} – internal friction of AAS and ε - value of imposed deformation) curve, that was automatically recorded within test cycles of deformation, permits one to bring out the forthcoming structures of AAS and its features, that will take place at the near future cycles of imposed deformation. Reiteration of these processes gives one the chance to do the monitoring of fatigue features of different Aluminum Alloys Samples as well as of FEs (See [6]).

The general model. To gain a greater insight into why the Q^{-1} (ε) dependencies have the form as in Figure 2 (or Figure 3, before B point), assume that for every overstrained micro-regions the next formulae

$$\Delta Q_k(\varepsilon) = \begin{cases} G(m)^* \varepsilon \text{ if } \varepsilon \le \varepsilon_k \\ G(m)^* \varepsilon_k \text{ if } \varepsilon \ge \varepsilon_k \end{cases} \quad (3)$$

takes place; ε_k varies in value. A scale factor - G(m) is the same for each micro-region of the selected AAS and depends on the degree of asymmetry - m of the imposed cyclic deformation.

Each micro-region will be able to contribute to the $\Delta Q^{-1}{}_{eff}$ accordance with statistic g_k [3]:

$$\Delta Q^{-1}(\varepsilon) = \sum_{k=1}^{L} g_k * \Delta Q_k^{-1}(\varepsilon) \ ; \sum_{k=1}^{L} g_k = 1. \quad (4)$$

$\Delta Q^{-1}{}_k(\varepsilon;m, N)$ - variation of internal friction of a k-th micro-region depends on the amplitude - ε and degree of asymmetry - m of the imposed cyclic deformation. N is a number of cycles; L – number of micro-regions.

For the sake of convenience, assume that $\Delta Q_{max} = M_g(m)$ and introduce the next value Z_k:

$$Z_k = G(m)^* \varepsilon_k / M_g(m) \ ; \ (k = 1,2,3,..., L). \quad (5)$$

Z_k and g_k can be found from experimental data, Figure 2 (or Figure 3, before the B point), and formulae (4). Proceed as follows:

$$\Delta Q^{-1}(\varepsilon) = G^*\varepsilon^* \sum_{k=1}^{L} g_k = G^*\varepsilon; \qquad 0 \le \varepsilon \le \varepsilon_1$$

$$\Delta Q^{-1}(\varepsilon) = G^*\varepsilon_1{}^*g_1 + G^*\varepsilon^*\{1\text{-}g_1\}; \qquad \varepsilon_1 \le \varepsilon \le \varepsilon_2$$

$$\Delta Q^{-1}(\varepsilon) = G^*\varepsilon_1{}^*g_1 + G^*\varepsilon_2{}^*g_2 + G^*\varepsilon^*\{1\text{-}g_1 - g_2\}; \ \varepsilon_2 \le \varepsilon \le \varepsilon_3$$

(6)

$$\Delta Q^{-1}(\varepsilon) = G^* \sum_{k=1}^{L} \varepsilon_k * g_k \qquad \varepsilon_L \le \varepsilon$$

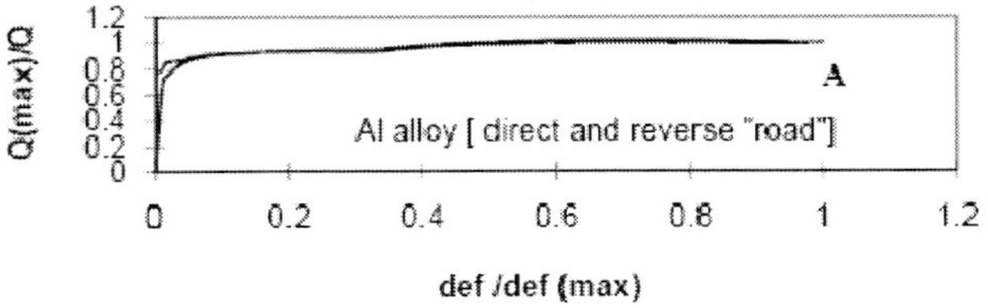

Q(max)/Q

1.2 1 0.8 0.6 0.4 0.2 0

A

Al alloy [direct and reverse "road"]

0 0.2 0.4 0.6 0.8 1 1.2

def /def (max)

Figure 2. Relative internal friction of Aluminum alloy Sample (B-95) versus the relative imposed deformation.

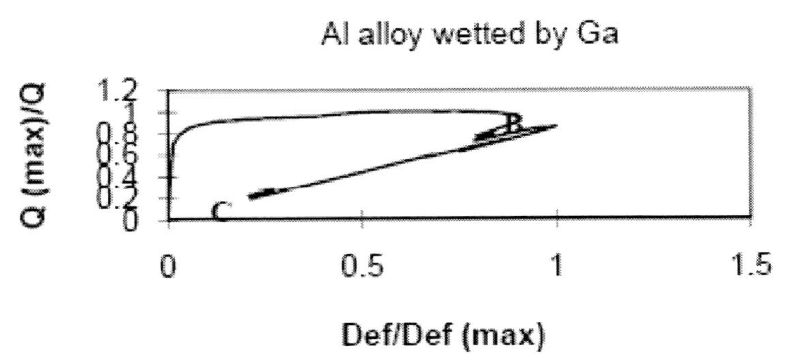

Figure 3. Relative internal friction of Aluminum Alloy's Sample(B-95) wetted by liquid gallium-Ga, versus the relative imposed deformation. This curve reflected the cracking process, beginning from B point. At the C point sample fractured.

Notice that $\varepsilon_1 < \varepsilon_2 <.....< \varepsilon_L$ and ε vary in each of interval of $[\varepsilon_k , \varepsilon_{k+1}]$. Let us take a derivative of ΔQ^{-1} (ε) with respect to ε in each of (6) equalities. Then each subsequent equality can be subtracted from the preceding one. As a result one finds all quantities of g_k, ε_k, as well as L; G(m) and M_g(m). Figure 4 presents the states of AAS after various numbers of cycles and value of the imposed deformation. Information entropy S:

$$\overrightarrow{u_2} \tag{7}$$

is a measure of the amount of disorder (different states) in AAS states. This model also admits control of the structural mutations in AAS and FE after N cycles of an imposed regular deformation [6].

The same procedure was used on to the $\mathbf{Q}^{-1}$ - ε (or $\mathbf{Q}^{-1}$ – N , $\mathbf{Q}^{-1}$ - wetted by liquid gallium - Ga internal friction of AASs) dependencies. Information entropy for this situation S = 1,374. As a result, one can in advance to bring out the forthcoming structures of FE and AAS and its features that will take place at the near future cycles of imposed deformation.

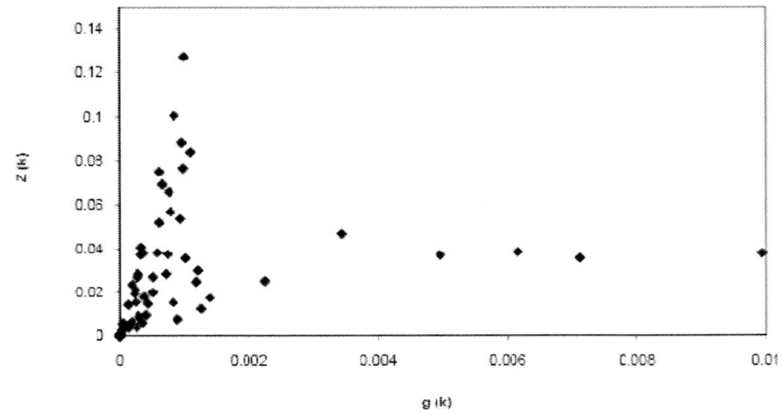

Figure 4. The state (gk&Zk) of Aluminum Alloy in the end (see point A on Figure 2) of the direct «road». Information entropy for this state S=0,950. The full diagram (gk&Zk) is inserted The greasy zero points of which are presented on the main figure.

High level of computer engineering was used by us to study the effect of the cracking processes, caused by liquid metals, on the amplitude-dependent internal friction $\bar{u}\ \vec{_3}$, The rate of sample mechanical irregularity was found to be different for the different time after wetted by liquid Ga.

In Figure 5 and 6 are presented the internal friction Q-1 of Aluminum Alloy Samples versus of both the N - cycles and the ε - deformations imposed on AASs (D16 – T)

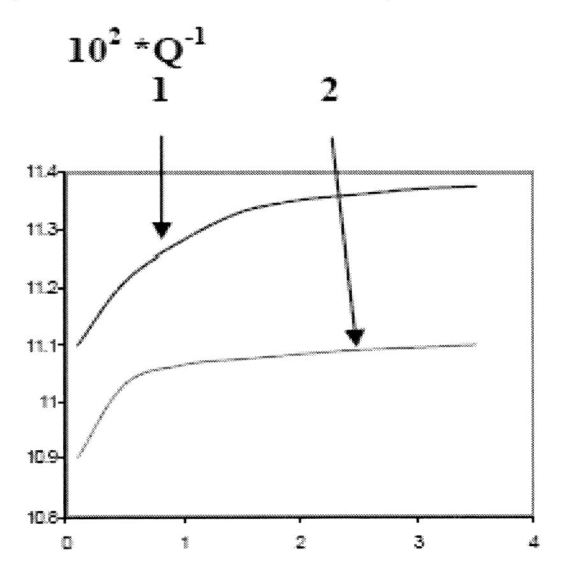

Figure 5. Internal friction Q-1 of AAS (D16 – T) versus N cycles of the imposed deformation coresponding both 32 kg/mm2 (1) and 26 kg / mm2 (2) loads. The asymmetry coefficient is unity (m =1).

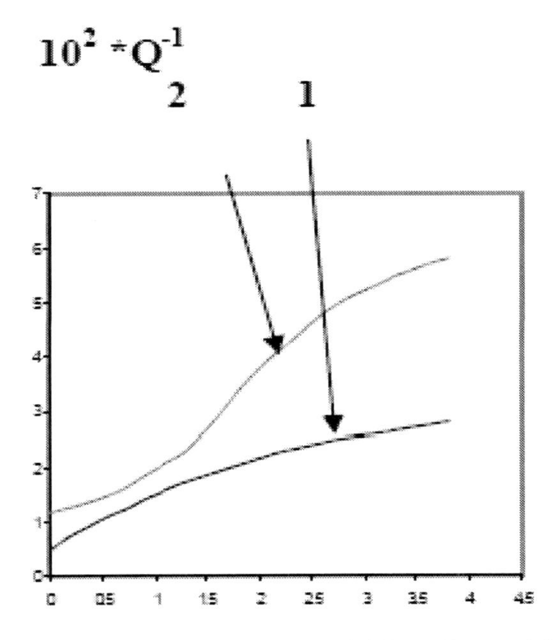

Figure 6. Internal friction **Q-1** of AAS (D16-T) versus imposed deformation. **1** - After imposed 5*104 cycles of 25 kg/mm 2 load. **2** - After imposed 5*106 cycles of 25 kg/mm2 load.

At the Figure 7 and Figure 8 are presented the R eff changes under simple and complicated mode of loads.

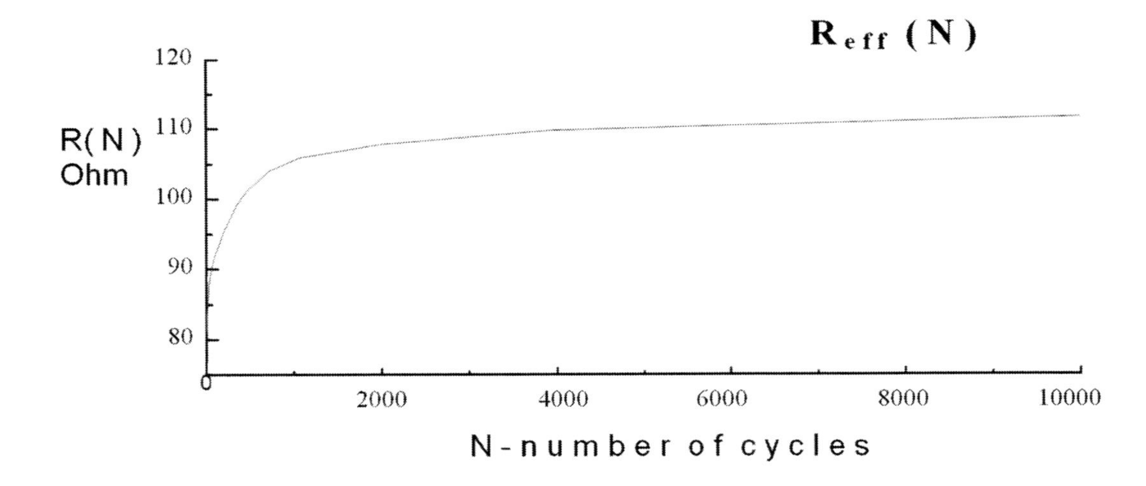

Figure 7. R_{ff} of SE versus N (simple mode of loads).

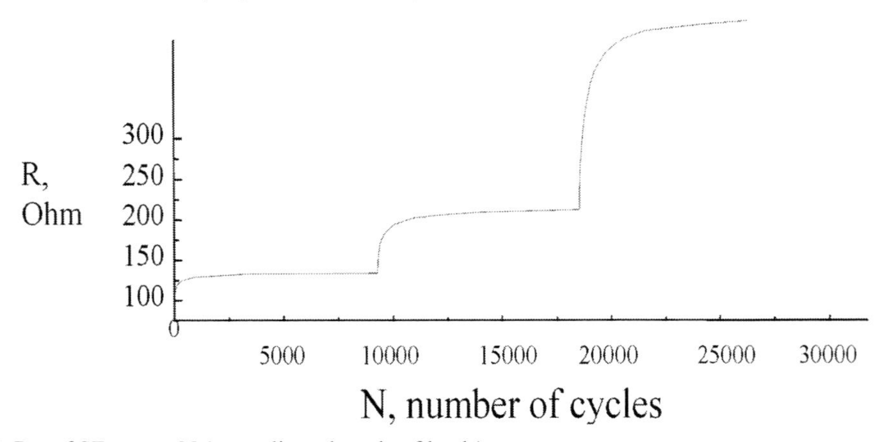

Figure 8. R_{eff} of SE versus N (complicated mode of loads).

The Original Technology to Forecast Fatigue Damage of an Aluminum Alloy

We shall consider both the fatigue curve of AA (Figure 9) and the equal change resistance curve (<E> - n curve for selected values of $r = (R_n – R_0) / R_0$) of SEs (Figure 10) were obtained at the same conditions of imposed spectrum of deformation.

Our description is very close to one, which was given in [14]. The difference is in the next. Instead of to select the SE's which ECRC is the "same" as investigated metal, we introduce the alarm state of an Aluminum Alloy Sample (AAS), It gives us the opportunity to escape the analysis of mechanical amplifier [14].

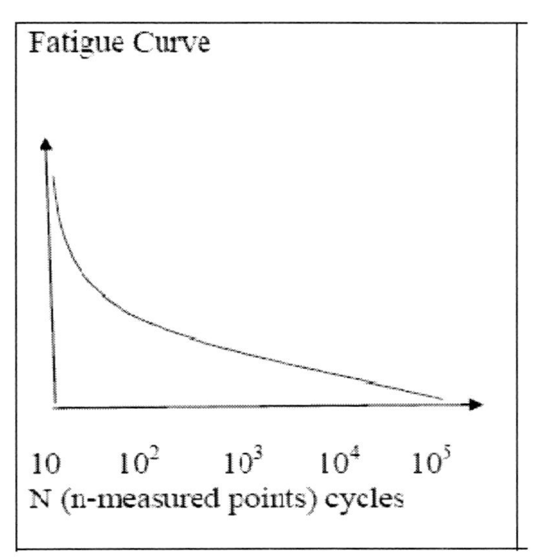

Figure 9. Fatigue curve of an Aluminum Alloy (D16-T) sample.

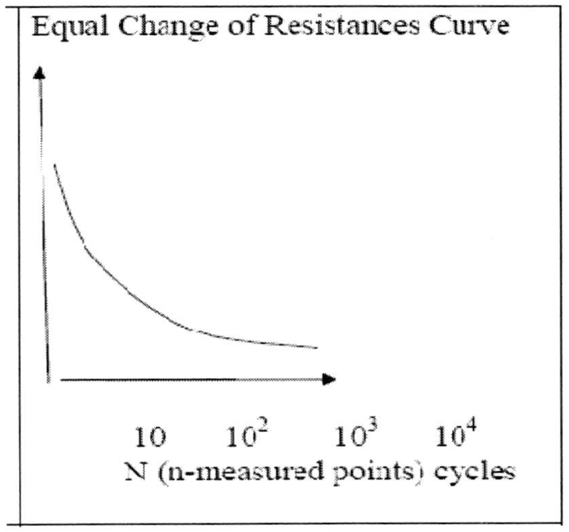

Figure 10. ECRC for SE, rigidly installed on the sample of AA.

Figure 10 shows the equal change of resistance curves {ECRCs) of FEs rigidly installed at the sample of AA. The rate of spectrum of deformation is the value <E>:

$$<E> = (<\varepsilon^2{}_\tau>)^{1/2} \; ; \quad <\varepsilon^2{}_\tau> = (1/\tau)^* \int_0^\tau \varepsilon^2 (t) * dt \tag{8}$$

ε (t) – value of imposed deformation; t – time ; τ – equal time interval at which the resistance's R_n were measured.

Figure 11 shows R_{eff} versus n for difference value of <E>. From these curves one can compose the Figure 10.

Figure 13 shows the operative curve for an AAS and to be suit for it the Alarm curve. A crossing point of these curves is the service AAS failure. One can adaptive forecast this point using the series (2) and formula (1).

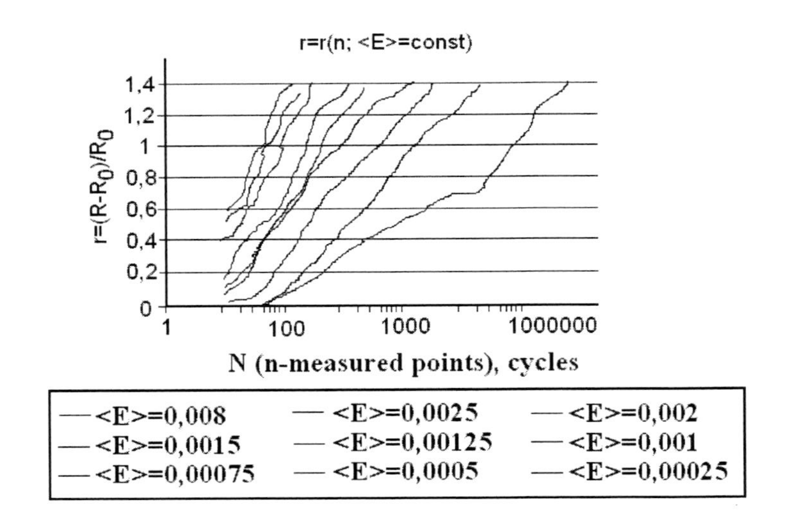

Figure 11. SE's r – n dependence for different value of <E.>.

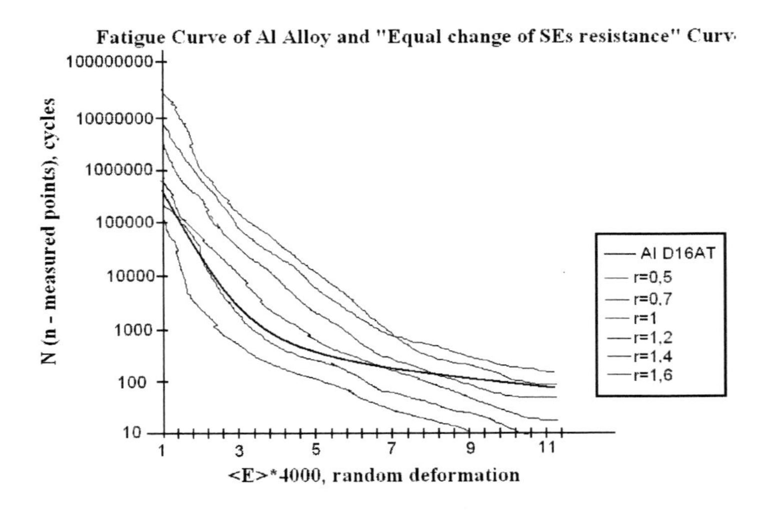

Figure 12. Crossing the Fatigue Curve with ECRCs forms the Alarm Curve for AAS.

Just bellow we present the methodical example helping to understand the main idea. (See Figure 14). At the Fig.1 Me there are two operative curve, which, for the simplicity, made as linear. The first have energy equal E1, the second have energy E2. (E2 > E1). The second Figure 2 Me is the same as Figure 1 Me. But instead of arbitrary fixed second line there one have a failure line. At the Figure 3 Me one can see fatigue curve for testing samples and a way haw it may be getting from Figure 2 Me. On the Figure 4 Me, one can see haw to get the Alarm curve for testing sample, using the data of Figure 3 and arbitrary fixed first and second line from Figure 1 Me.

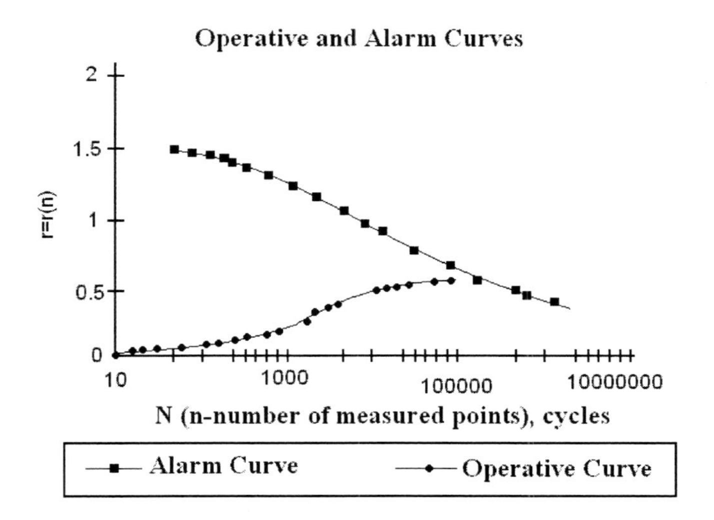

Figure 13. Alarm Curve for AAS and operative Curve for SEs , rigidly installed onto this AAS.

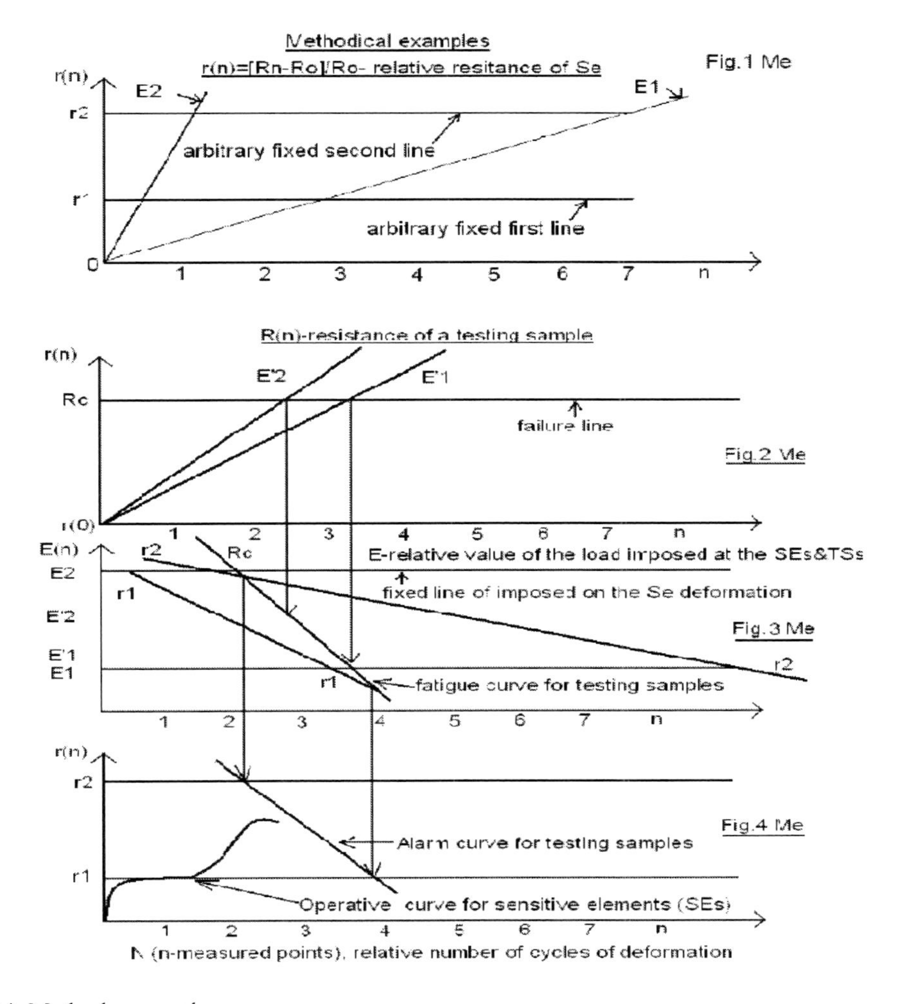

Figure 14. Methods examples.

REFERENCES

[1] Shamirzaev S. (1998) MHMs for Fatigue Gage. Published by The Iron and Steel Institute of Japan *Proceedings of the 7th International conference,* Steel Rolling'98, November, 9-11, 1998, Makuhari, Chibo, Japan, pp. 844-849.

[2] Shamirzaev S. Intern. J. of Fatigue (2002), no. 7/24,pp. 777-782

[3] Shamirzaev S. (1999) *Response of MHFMs at a Fast and a Slow Operational Loads.* A collection of technical papers of the 1999 AIAA/ASME/ASCE/AHS/ASC Structures, Structural Dynamics, and Materials Conference and Exhibit, St. Louis, MO, USA, 12-15 April,1999; volume 3, p.1717-1726.

[4] Rostoker W., McCaughey J.M., Markus H. (1960). *Embrittlement By Liqouid Metals.* Reinhold Publishing Corporation, New York, Chapman and Hall, LTD., London, 1960.

[5] Shamirzaev S. and Shamirzaeva G. (2000) The Rheological Model Of Fatigue Damage Of Cm. *Proceedings Of The Xiiith International Congress On Rheology*. Cambridge, United Kingdom, 20th to 25th August, 2000. Published by the British Society of Rheology. Volume 3, pp. 377-379.

[6] Shamirzaev S., Ganihanov Sh., Shamirzaeva G. (2001) The Monitoring Of Fatigue Features Of Cm For A Very High Cycle Of Loads. *Proceedings of the International Conference on FATIGUE in the Very High Cycle Regime*. 2-4 July, 2001. Vienna, Austria, pp. 245-252.

[7] Granato A.V. and Lucke K., (1956). J. Appl. Phys. 27 , (1956),563.

[8] Bratina W. J., (1966) in *Physical Acoustics* , Edited by W.P. MASON, Volume III, Part A, *The Effect Of Imperfections*, Paper #6, 1966, Academic Press, New York and London.

[9] Kuksenko V.S., Betechtin V.I., Ryskin V.S., Slutsker A.I. Intern. *J. Fracture,* 1975, 11, no.5,(1975) 829-840.

[10] Mason W. P.(1956). *Journal Acoust. Soc. Amer*., 28, (1956),1197 - 1219.

[11] Kardashev B.K., Nikanorov S.P. and Voinova O.A., (1972). *Phys. Stat. Sol.* (a) 12, (1972),375.

[12] Schenck H-., Schmidtmann E. , Kettler H., (1960) *Arch. Eisenhuettenw*., 31, (1960), 659 .

[13] S.Kh. Shamirzaev (2004)/ *Solid State Sciences* 6 (2004) 1125 – 1129.

[14] Troshchenko V.T., Boiko V. I. (1985) Fatigue Damage Sensor and Substatiation of its Application. Communications 1 and 2. *Problemi Prochnosti.* No.1, [187], Jun., 1985, pp.3-14.

In: Aluminum Alloys
Editor: Erik L. Persson
ISBN: 978-1-61122-311-8

Chapter 9

CATHODIC HYDROGEN CHARGING OF ALUMINIUM ALLOYS

C. N. Panagopoulos*, E. P. Georgiou and D. A. Lagaris
Laboratory of Physical Metallurgy,
National Technical University of Athens, Zografos,
15780, Athens, Greece

ABSTRACT

The growing importance of hydrogen among the alternative energy sources has given rise to new studies concerning the effect of hydrogen on the technological properties of metallic alloys. Among the most commonly used metallic materials are cast and wrought aluminium alloys, by virtue of their light weight, fabricability, physical properties, corrosion resistance and low cost. However, the interaction of hydrogen with aluminium alloys has not yet been fully understood.

In this research investigation hydrogen was introduced into the surface layers of important technological and industrial cast and wrought aluminium alloys via cathodic hydrogen charging technique. The effect of hydrogen absorption on the structure, surface microhardness and mechanical behavior of aluminium alloys was studied. In addition, the absorption of hydrogen was found to depend strongly on the charging conditions. The presence of aluminium hydride in the surface layers of hydrogen charged aluminium alloys, after intense charging conditions, was observed.

Hardening of the surface layers of aluminium alloys, due to hydrogen absorption and in some cases hydride formation, was also observed. Tensile experiments revealed that the ductility of aluminium alloys decreased with increasing hydrogen charging time, for a constant value of charging current density, and with increasing charging current density, for a constant value of charging time. However, their ultimate tensile strength was slightly affected by the hydrogen charging procedure. The cathodically charged aluminium alloys exhibited brittle transgranular fracture at the surface layers and ductile intergranular fracture at the deeper layers of the alloy. Finally, the effect of deformation treatment (cold rolling) on the hydrogen susceptibility of aluminium alloys was studied.

* Corresponding Author: Tel. ++3210 7722171, Fax: ++3210 7722119, e-mail: chpanag@metal.ntua.gr

Introduction

The growing importance of hydrogen among the alternative energy sources has given rise to new studies concerning the effect of hydrogen on the technological properties of metallic alloys [1-4]. However, the interaction of hydrogen with metallic materials has not yet been fully understood. Some of the proposed theories concerning hydrogen incorporation in metals are presented below.

The first theory suggested that molecular hydrogen under the influence of high pressures dissolved in the surface layers of a metallic material through micro-cracks and voids. The increased hydrogen concentration gradient caused increasing stresses in the material, which sometimes caused the formation of blisters [5]. Troiano [6] suggested that hydrogen diffuses mainly through easy paths, in the surface layers of the metallic materials. There the hydrogen in interstitial site reduces the atomic cohesive strength, creating localized microdefects and microcracks. Birnbaum and Sofronis [7] suggested that hydrogen atoms diffuse in the lattice of the metallic material and interact with the existing dislocations. This interaction enhances the dislocation mobility and causes plastic deformation and fracture (HELP theory). However, other investigators found that the concentrated hydrogen in the surface layers of the aluminium leads to the formation of a severely hardened region [8]. This increase of hardness in the surface layers is attributed to strain-hardening effects caused by the absorption of the hydrogen atoms, which reduce the dislocation mobility [8]. In any case, there is no unequivocal mechanism for hydrogen embrittlement of metallic materials.

The above complex phenomena are significantly affected by various factors during the electrolytic hydrogen absorption in metals, such as hydrogen charging conditions [9], heat treatments [10], surface treatments [11] etc. In this research paper the effect of chemical composition, hydrogen charging conditions and cold rolling on the hydrogen susceptibility of some significant technological aluminium alloy has been studied.

Experimental Procedure

The materials used in this study were commercially supplied aluminium alloys. From them a number of microhardness and standard tensile specimens were produced. In order to ensure that all the specimens had similar surface topography, all aluminium alloy specimens were polished by SiC papers with increasing finishes and 3μm diamond paste. The resulting roughness was found to range between 0.4 and 0.45μm. After the preparation of the specimens, a stress relief procedure was performed, by annealing all the specimens at 250 °C for 2 hrs and slowly cooling them at room temperature, in an automatic furnace with Ar atmosphere.

Hydrogen was introduced in the aluminium alloy specimens by electrolytic cathodic charging. This procedure was performed in a solution consisting of 75% vol. CH_3OH, 22.4% vol. H_2O and 2.6% vol. H_2SO_4, poisoned with As_2O_3 (0.1 gr/lt) as hydrogen recombination inhibitor, at room temperature. The charging currents employed were in the range of 15–90 mA/cm^2, while the charging time varied from 2 to 30 h, respectively. Graphite electrode was used as anode.

Mechanical testing was carried out immediately after charging, in order to minimize the loss of hydrogen. All the tension tests were performed at a strain rate of 3.3 x 10^{-4} s^{-1}, at room temperature. Microhardness testing was performed using a Shimadju Vickers indenter, imposing 0.15 N for 15 sec. The given values of the experimental parameters presented are the mean value of five independent experiments.

Metallographic study of the surface of both the cathodically hydrogen charged and the fractured specimens was performed with the aid of a Jeol 6100 scanning electron microscope, which was connected with an electron dispersive X-ray analyzer (EDAX). A Siemens D 5000 X-ray diffractometer, using a copper filter, was also used for the structural study of the cathodically charged aluminium alloy specimens.

RESULTS AND DISCUSSION

Structural characterization of cathodic hydrogen charged aluminium alloys, by means of X-ray diffraction, revealed the presence of aluminium hydride (AlH_3), figure 1(a,b).

The presence of aluminium hydride in the surface layers of hydrogen charged aluminium alloys was a result of increased hydrogen concentration, which excided the solubility limit of hydrogen in the aluminium matrix (5 x 10^{-6} % wt. at 660°C, under 0.1MPa hydrogen pressure). The detection of aluminium hydride on the surface layers of aluminium alloys, despite their different chemical composition, is considered extremely important as it is indicative of hydrogen "trapping" phenomena. Therefore aluminium alloys, under specific charging conditions, can be possibly used as hydrogen storage materials.

However, the absorption of hydrogen and the formation of hydrides were found to depend strongly on the charging conditions. As it can be seen from figure 2(a,b), during the cathodic hydrogen charging of 5083 aluminium alloy, aluminium hydrides form: (i) when charging time excided 6 hrs, for a constant charging current density of 30mA/cm^2 and (ii) when the applied charging current density excided 30mA/cm^2, for a constant charging time of 6 hrs.

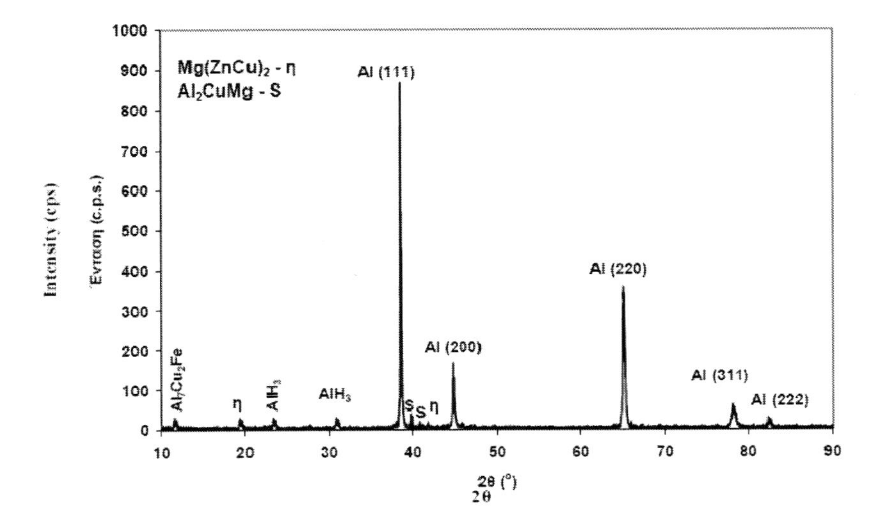

Figure 1. (Continued).

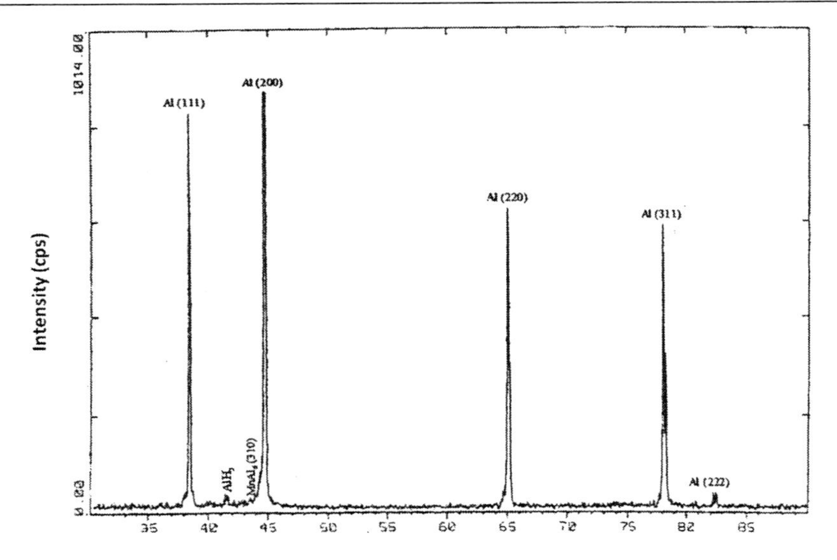

Figure 1. XRD patterns of: (a) 3109 aluminium alloy and (b) 7075 aluminium alloy.

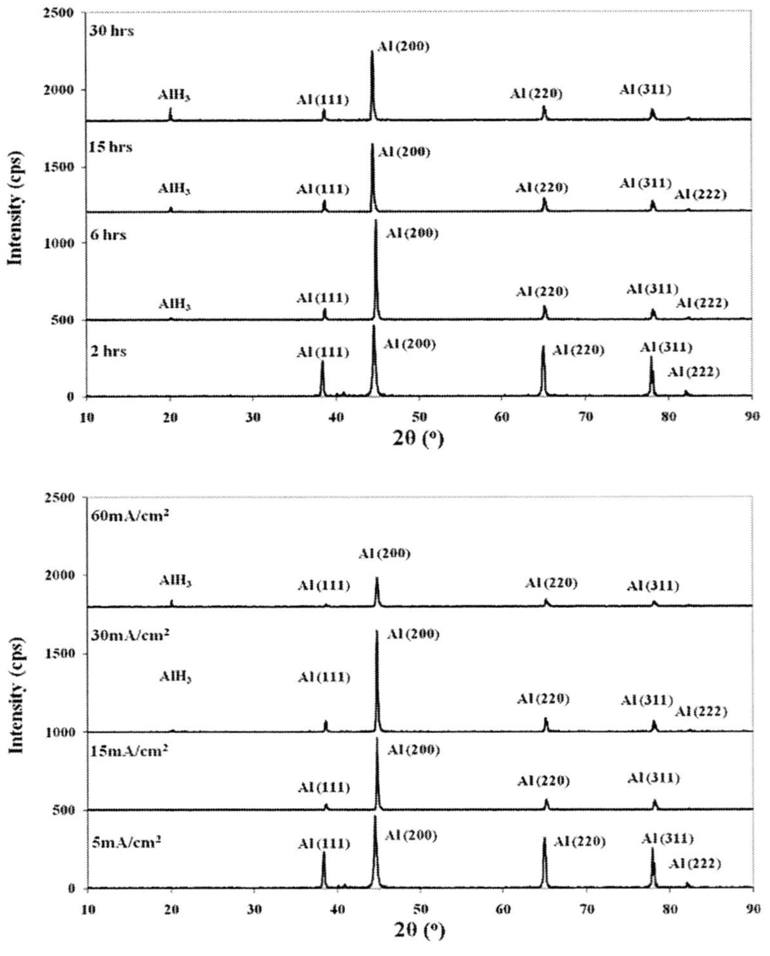

Figure 2. Comparative graph of the effect of: (a) charging time and (b) charging current density conditions on the XRD pattern of 5083 aluminium alloy.

The surface of the charged specimens was covered with a dark grey coloured film, Figure 3.

Figure 3. Surface metallography of hydrogen charged Al- 4%Zn-1%Mg aluminium alloy after cathodic hydrogen charging process.

The thickness of this film was observed to increase with increasing charging time, for a constant value of charging current density and with increasing charging current density, for a constant value of charging time. Analysis of the film with the aid of X-ray diffraction technique showed that it mainly consisted of As_2O_3, which was deposited from the solution to the surface of the specimens, during the cathodic hydrogen charging process. Thorough examination of the surface of cathodically charged aluminium alloys, after carefully removing the As_2O_3 thin film, revealed the formation of blisters, figure 4.

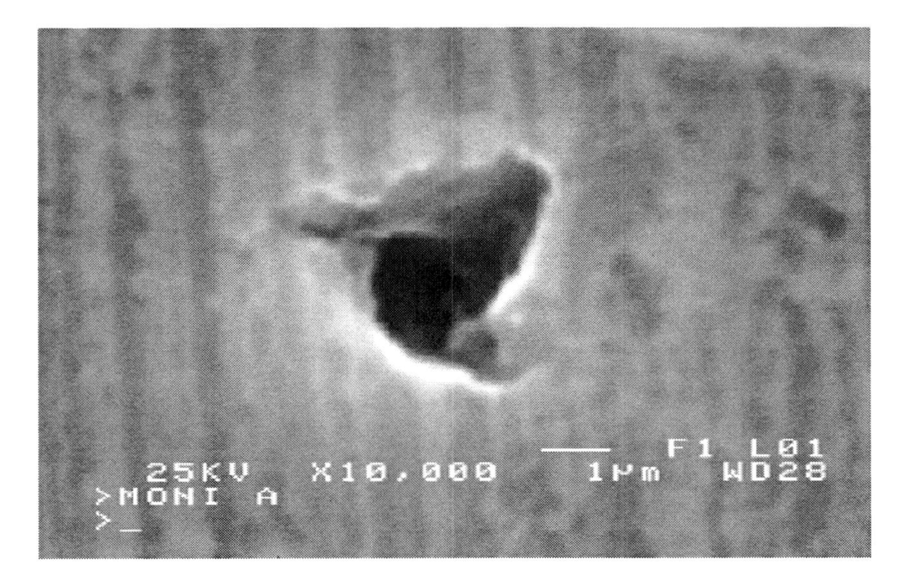

Figure 4. SEM micrograph of hydrogen-induced blistering of a charged aluminium alloy specimen at $30mA/cm^2$ for 6 hrs.

It is suggested that the following mechanism operates for the formation of blisters during the cathodic hydrogen charging of aluminium alloys [12]. The diffusion of hydrogen atoms in a metallic matrix, in supersaturated conditions, leads to the formation of molecular hydrogen which reacts with various microdefects such as voids, grain boundaries, cavities etc. to create the observed blisters.

Microhardness experiments on the surface layers of the cross section of charged aluminium alloys revealed increased surface hardening, due to hydrogen absorption. Figure 5 presents clear evidence that the microhardness of the surface layers of 3109 and 7075 aluminium alloys increased with increasing hydrogen charging time, for a constant value of charging current density, and with increasing charging current density, for a constant value of charging time respectively. The increase of microhardness of the surface layers due to hydrogen charging, can be explained in terms of dislocation pinning mechanisms.

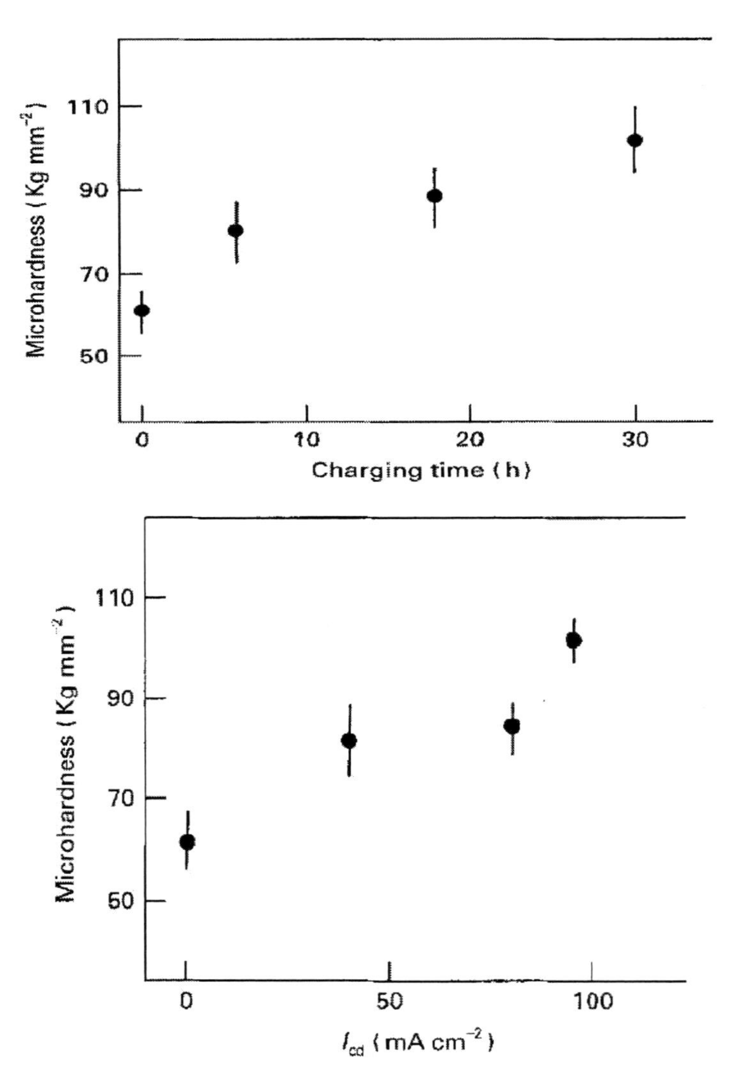

Figure 5. Comparative graph of the effect of charging time and charging current density conditions on the surface hardness of 5083 wrought aluminium alloy.

It has been proved that solute hydrogen atoms often act as dislocation pinning sites, which increase the surface hardness of the hydrogen charged alloy [13]. Dislocations may be generated by hydrogen concentration gradient, but with increasing difficulty, due to increased resistance to new dislocations production by the existing dislocations. Thus, higher charging current density and higher charging time lead to increased hydrogen fugacity in the surface layers of the alloy and consequently to increased surface hardness, due to decreased dislocation mobility.

In addition, Rodrigues and Kirchheim [14] proposed that the absorbed hydrogen atoms diffuse near dislocations, forming areas with increased concentration of hydrogen. The interaction of absorbed hydrogen and/or hydrides leads to the creation of dense Cottrell Clouds, which subsequently increase the hardness of the material. However, after a certain point (charging current- charging time), the microhardness of the surface layers tended to reach a saturation level. This phenomenon is probably attributed to the fact that the aluminium matrix became saturated and no more hydrogen could dissolve in the aluminium matrix.

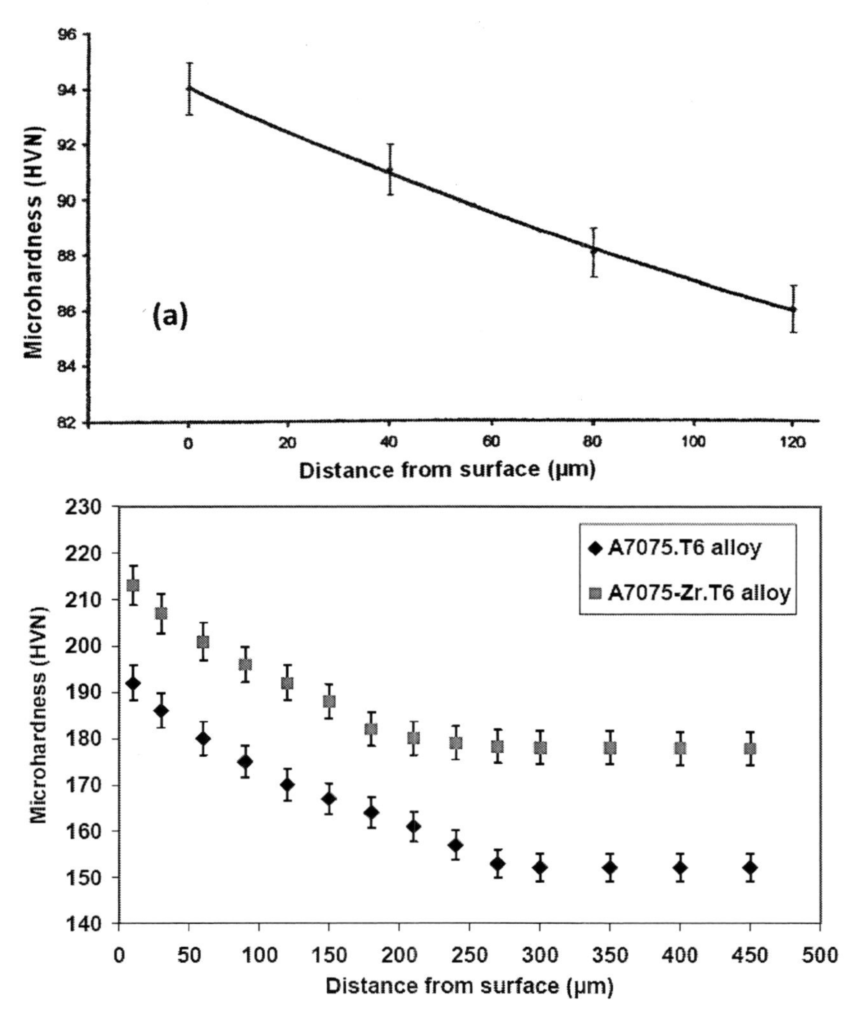

Figure 6. Hardness profile of: (a) 5083 and (b) 7075 hydrogen charged aluminium alloy specimens, at $30mA/cm^2$ for 6 hours, as a function diffusion depth.

A typical graph, connecting the specimens microhardness with the hydrogen depth in the surface layers of the alloy, is presented in figure 6(a,b). From the last figure, it can be said that the hardness in the surface layers of 3109 and 7075 aluminium alloys appeared to be higher, in comparison to the hardness of the as-received materials, due to the hardening phenomena explained previously. Taking into consideration that the distribution profile of the measured microhardness is possibly similar to the distribution profile of hydrogen, the diffusion profile can be seen as an approximate result of Fick diffusion mechanism [15] of hydrogen in the cathodically charged aluminium alloys. Therefore, the Diffusion Coefficient can be calculated, by the following mathematical equation:

$$D = x^2 / (4t)$$

where D is the diffusion coefficient (m^2s^{-1}), x is the diffusion length (m) of hydrogen in the surface layers of the examined aluminium alloy specimens, obtained from figure 6(a,b), and t is the charging time (sec). From the above equation the diffusion coefficient for 3019 aluminium alloy specimen was found to be 1.7 x 10^{-13} m^2s^{-1}, whereas for 7075 aluminium alloy specimen the hydrogen diffusion coefficient was 1 x 10^{-12} m^2s^{-1}. It should be noted that the calculated hydrogen diffusion coefficients were observed to be to in agreement with the diffusion coefficients of other similar aluminium alloys [16,17] obtained by other experimental techniques.

Typical engineering stress-strain curve of an uncharged and hydrogen charged 7075 aluminium alloy, are presented in figure 7. From figure 7, it could be said that hydrogen cathodic charging decreases slightly the ultimate tensile strength (UTS) and significantly the ductility of the examined alloy. The observed slight decrease, within experimental error, of the UTS of the hydrogen charged aluminium alloy could be mainly attributed to the presence of blisters and microcracks on the surface of the alloy [18], during the hydrogen cathodic charging, as it could be seen in figure 4.

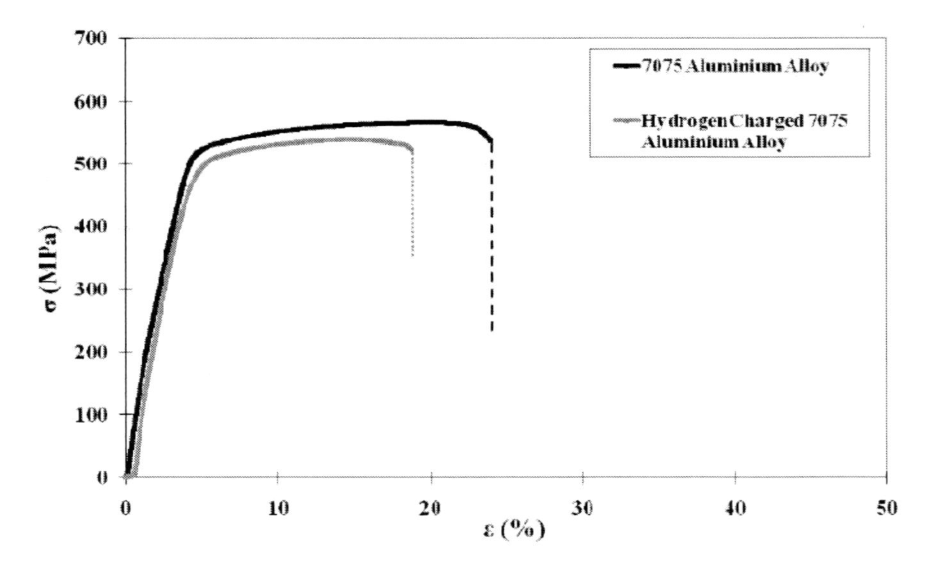

Figure 7. Typical Stess – Strain curves for as reveived and charged 7075 aluminium alloy, at 50mA/cm^2 for 6h.

On the other hand, the observed decrease of the ductility of this aluminium alloy could be attributed to the fact that the absorbed hydrogen atoms hinder the movement of dislocations and thus decrease the ductility of the alloy. In addition, the presence of blisters should also be taken in consideration.

The effect of hydrogen charging time on the UTS and ductility of 5083 wrought aluminium alloy, for a constant value of charging current density, was also studied. Figure 8(a) presents the effect of hydrogen charging time on the UTS and ductility of the hydrogen charged alloy respectively. From the above graph, it could be said that the increase in hydrogen charging time brought about an independence of the UTS of aluminium alloy, while the ductility of the alloy was substantially decreased. The decrease of the ductility of hydrogen charged alloy could be explained in terms of the dislocation pinning mechanism and blister formation, described in previous paragraphs.

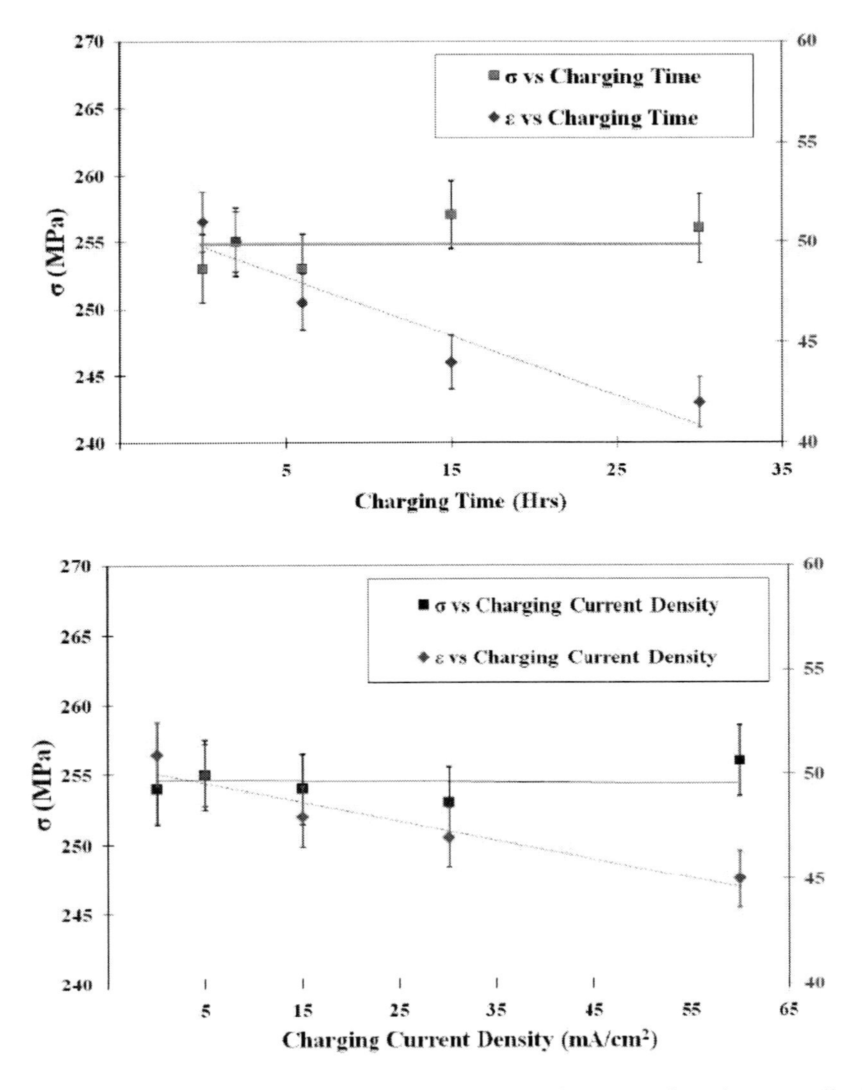

Figure 8. Graph presenting the effect of hydrogen charging time on the UTS and maximum tensile elongation of 5083 wrought aluminium alloy.

In a second series of tensile experiments, the effect of hydrogen charging current density on the UTS and ductility of 5083 wrought aluminium alloy, for a constant value of charging time, was studied. In figure 8b, is seen that the increase of hydrogen charging current density decreased the ductility of the examined alloy. However, the UTS was found to be independent, within experimental error, of the charging current density.

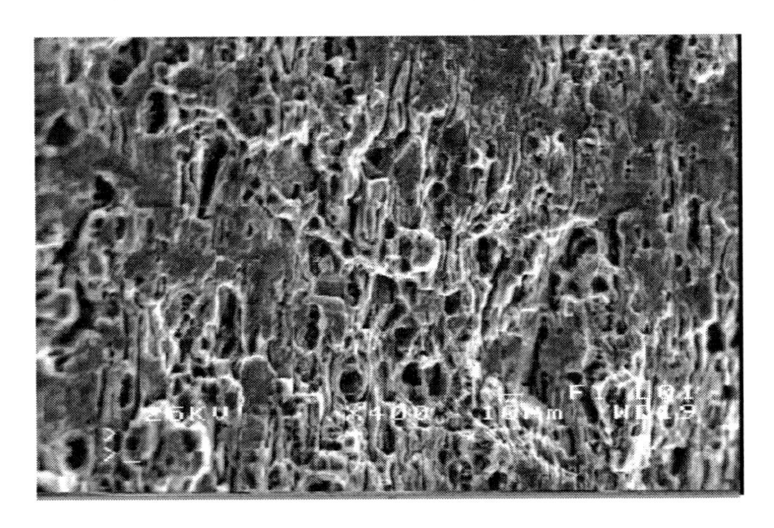

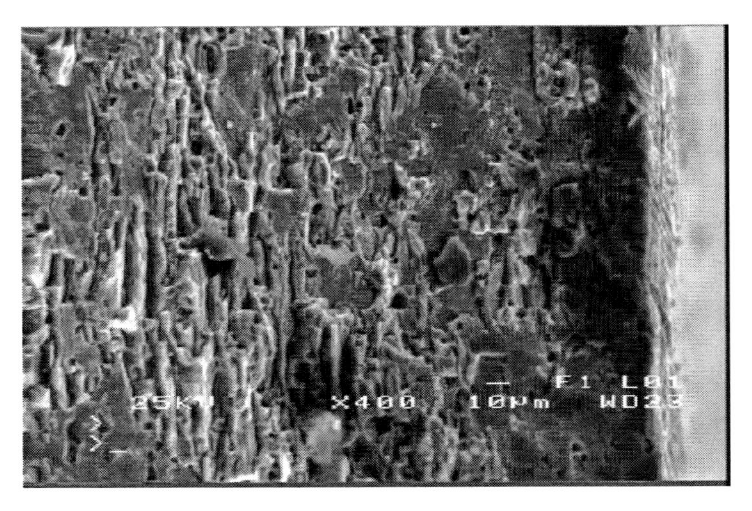

Figure 9. Graph presenting the effect of (a) charging time and (b) charging current density con the UTS and maximum tensile elongation of 5083 wrought aluminium alloy.

The fracture mechanism of the as received and hydrogen charged aluminium alloys, which were subjected to tensile testing experiments, were studied with the aid of a Scanning Electron Microscope. A typical morphology of the fractured surface of the as received and hydrogen charged aluminium specimens are shown in figures 9(a) and 9(b). As it can be seen from the fractograph presented in figure 9(a), the as received specimen exhibits ductile fracture, as it consists of dimples. On the other hand, the outer area of the charged specimen, figure 9(b), exhibits brittle interganular fracture. However, the inside area remains ductile, but the dimples are smaller and shallower when compared to the uncharged specimens of as received aluminium alloys. This intergranular cracking may be explained by the disordering and high-energy grain boundaries which act as strong traps for hydrogen. Therefore, the presence of hydrogen increases the susceptibility to cracking in the outer regions of hydrogen charged specimens, by intense mechanical stresses.

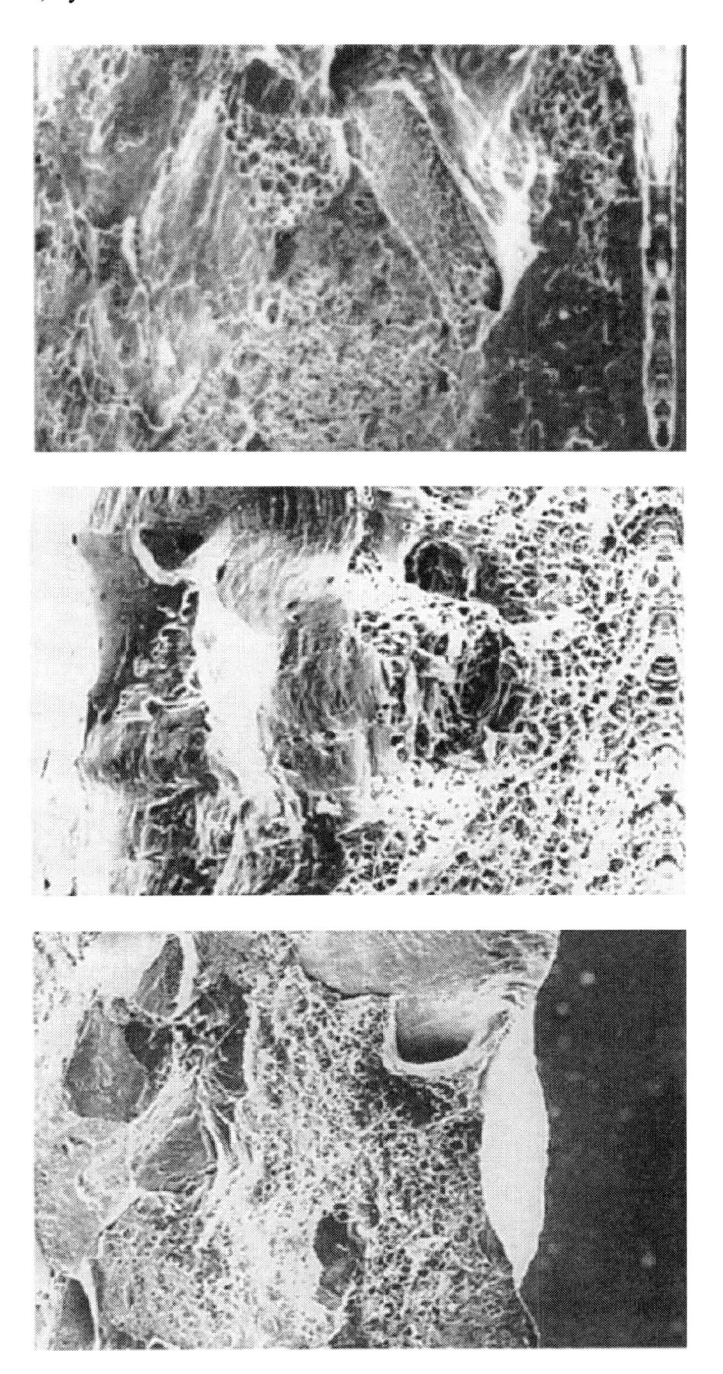

Figure 10. (Continued).

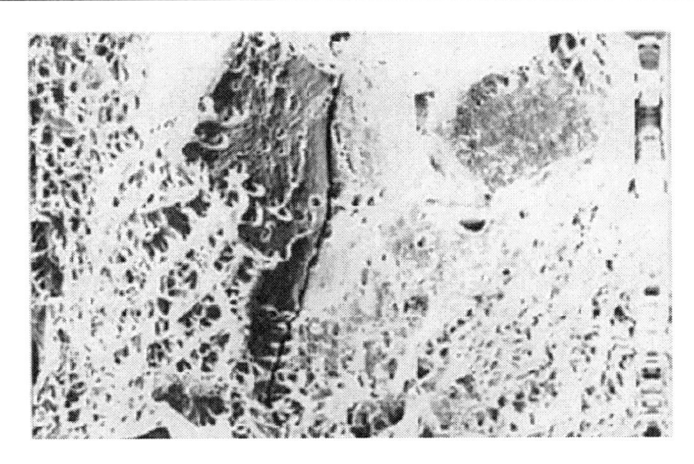

Figure 10. (a): SEM micrograph of the fractured surface of uncharged Al-4%Zn-1%Mg aluminium specimen. (x170). (b): SEM micrograph of the fractured surface of charged Al-4%Zn-1%Mg aluminium specimen at 25mA/cm^2 and for 3 hrs. (x170). (c): SEM micrograph of the fractured surface of charged Al-4%Zn-1%Mg aluminium specimen at 90mA/cm^2 and for 3 hrs. (x85). (d): Crack initiation and propagation in due to increased hydrogen diffusion. (x170).

In addition, the effect of hydrogen current density on the fracture mode of Al-4%Zn-1%Mg aluminium alloys was studied, for a constant charging time value of 3hrs. It was observed that the fracture mode of the uncharged specimen was completely transgranular ductile, figure 10(a), and results from internal nucleation, growth and linking up of voids. For values of the current density up to 25mAcm^{-2} the fracture mode of surface layer of the alloy was brittle transgranular, figure 10(b). This was possibly due to the hydrogen hardening mechanisms which were described previously. When the charging current density further increased to 40, 55 and 90mAcm^{-2} extensive areas of brittle fracture on the surface layers of the aluminium alloy where observed, figure 10(c). It has been proposed that these brittle areas can form from:

1. Crack initiation from blisters and microcracks. When the diffusion of hydrogen into the alloy increases, blisters and microcracks might be formed along precipitate free zones or grain boundaries, acting as crack initiation sites, figure 10(d).
2. Decohension of precipitates-intermetallic phases from the metallic matrix. It has been found that brittle fracture is greater when the precipitated phase is less coherent [19] and since η ($MgZn_2$) phase is incoherent, damage is expected to be significant.
3. Formation of brittle AlH_3 hydrides.

Finally, the effect of cold deformation on the cathodic hydrogen charging of 5083 aluminium alloy was investigated. The aluminium alloy was submitted to a cold rolling process, until the average thickness of the specimens was reduced by 7% and 15% respectively, with one pass.

Structural characterization by means of X-ray diffraction of the hydrogen charged 5083 wrought aluminium alloy, without and with 15% cold rolling, revealed that these specimens were consisted of: (i) aluminium solid solution and (ii) aluminium hydride (AlH_3), figure 11. The presence of aluminium hydride in the surface layers of hydrogen charged aluminium alloy was a result of increased hydrogen concentration.

In addition, a slight increase in aluminium hydride peak intensity was observed for the aluminium alloy with 15% cold rolling in contrast to the aluminium solid solution peaks. It has been observed that cold rolling processes can lead to the formation of various micro-defects, such as micro-pits, oil-pits, grooves and scratches [20].

However, these micro-defects can act as "easy paths", through which hydrogen can diffuse from the surface layers to the deeper layers of 5083 aluminium alloy, during the cathodic hydrogen process. A high concentration of diffusion pathways per material's volume is believed to increase the diffusity of hydrogen in the surface layers of a cold rolled material [21]. Thus, it can be observed that the cold rolling process led to an increase of the hydrogen susceptibility of 5083 aluminium alloy.

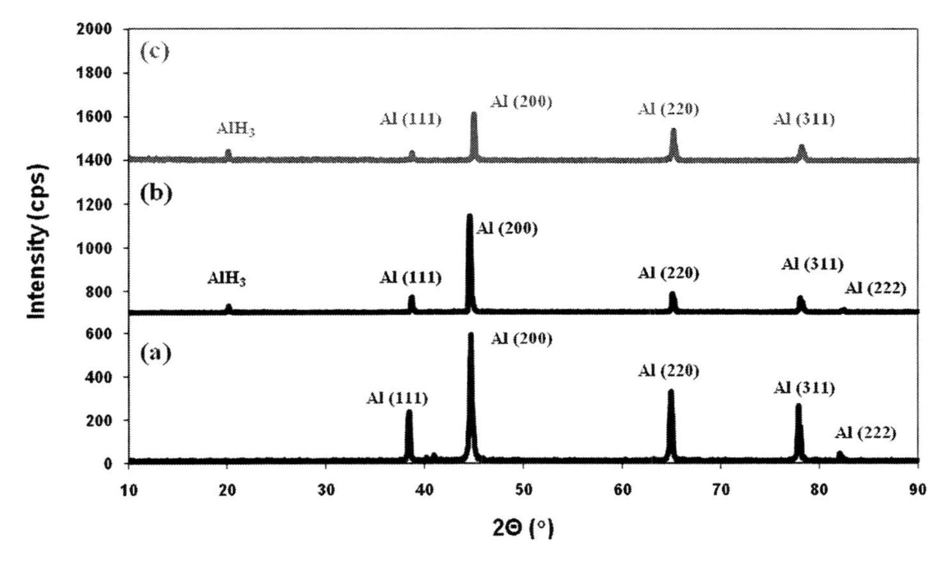

Figure 11. (a): XRD spectrogram of as received 5083 aluminium specimen. (b): XRD spectrogram of hydrogen charged aluminium specimen without cold rolling, under $30mA/cm^2$ current density and for 6hrs. (c): XRD spectrogram of hydrogen charged aluminium specimen with 15% cold rolling, under $30mA/cm^2$ current density and for 6hrs.

CONCLUSION

From the previous experimental investigation given, the following important deductions could be made:

1. As shown by X-ray diffraction results, AlH_3 formation was detected on the surface layers of all hydrogen charged aluminium alloy specimens, for specific charging conditions.
2. Cathodic hydrogen charging resulted in an increase of the microhardness of the surface layers of aluminium alloys.
3. The tensile experiments revealed that the ductility of aluminium alloys decreased with increasing hydrogen charging time, for a constant value of charging current density. The ductility also decreased with increasing charging current density, for a constant value of charging time.

4. The ultimate tensile strength of the examined alloy was approximately independent of the used hydrogen charging conditions.
5. The cathodically charged aluminium alloys exhibited brittle fracture at the surface layers, whereas ductile fracture was observed at the deeper layers of the same alloys.
6. Cold rolling process led to an increase of the hydrogen susceptibility of 5083 aluminium alloy.

REFERENCES

[1] A. Züttel, *Mater. Today* 6 (2003) 24-33.
[2] L. Schlapbach, A. Züttel, *Nature* 414 (2001) 353-358.
[3] A. Zaluska, L. Zaluski, J.O. Ström–Olsen, *J. Alloys Compd.* 288 (1999) 217-225.
[4] S.P. Cicconardi, E. Jannelli, G. Spazzafumo, *Int. J. Hydrogen Energy* 22 (1997) 897-902.
[5] M. Bernstein, A.W. Thompson, *Amer. Soc. For Metals, Metals Park*, Ohio, 1974.
[6] A. Troiano, *Trans. Am. Soc. Met.* 52 (1960) 54-80.
[7] H.K. Birnbaum, P. Sofronis, *Mater. Sci. Eng. A* 176 (1994) 191-202.
[8] J.W. Watson, Y.Z. Shen, M. Meshii, *Metall. Trans. A* 19 (1988) 2299-2304.
[9] C.N. Panagopoulos, P. Papapanayiotou, *J. Mater. Sci.* (1995) 3449-3456.
[10] D.A. Hardwick, A.W. Thompson, I.M. Bernstein, *Corros. Sci.* 28 (1988) 1127-1137.
[11] B.E. Wilde, M. Manohar, C.E. Albright, *Mater. Sci. Eng. A* 198 (1995) 43-49.
[12] C.N. Panagopoulos, A.S. El-Amoush, P.E. *Agathocleous Corros. Sci.* 40 (1998). 1837-1844.
[13] P. Rozenak, B. Ladna, H.K. Birnbaum, *J. Alloys Compd.* 415 (2006) 134-142.
[14] J.A. Rodrigues, R. Kirchheim, *Scripta Metall.* 17 (1983) 159-164.
[15] C.N. Panagopoulos, A.S. El-Amoush, K.G. Georgarakis, *J. Alloys Compd.* 392 (2005) 159-164.
[16] T.Ishikawa, R.B.Mclellan, *Acta Metall.* 34 (1986) 1091-1095.
[17] C. Thakur, R. Balasubramaniam, *J. Mater. Sci. Lett.* 15 (1996) 1397-1399.
[18] C.N. Panagopoulos, E.P. Georgiou, *Corros. Sci.* 49 (2007) 4443-4451.
[19] D. Kwon, S. Lee, R. J. Asaro, *Met. Trans.* 23A (1991) 1375.
[20] K. Kenmochi, I. Yarita, H. Abe, A. Fukuhara, T. Komatu, H. Kaito, *J. Mater. Proc. Technol.* 69 (1998) 106-101.
[21] R.D.K. Misra, D. Akhtar, *Mater. Res. Bull.* 21 (1986) 1473-1479.

In: Aluminum Alloys
Editor: Erik L. Persson

ISBN: 978-1-61122-311-8

Chapter 10

AlCu2,5Mg Aluminum Alloy Heat Treatment: Theory, Techniques and Applications

Alina-Adriana Minea and Dan Gelu Galusca
Technical University "Gh Asachi" lasi,
Materials Science and Engineering Faculty, 63,
B-dul D. Mangeron, 700050, Lasi, Romania

Abstract

This CHAPTER presents the results of an experimental study done on a wrought aluminum alloy (AlCu2,5Mg). This alloy has been heat treated in order to maximize its mechanical properties, especially its yield strength and its ductility. The experiment consisted in choosing 10 identical pieces of alloy which were to be differently heat treated regarding the quenching and artificial aging temperature. The methodology consists in the following steps: adopting the heat treatment technology for the proposed alloy; choosing the necessary hot working installations; choosing the equipments used for studying the yield strength; programming the experiment and the analytical interpretation of the results.

The present chapter presents the analysis of the yield strength and of the aging and quenching heating temperature. The equation for mechanical properties theoretical determination in standard or special heat treatment conditions has also been calculated. There have been carried out two nomograms for quick graphic determination of the necessary technology.

1. Introduction

Aluminum is a diverse material in its form and function. The wide range of products requires widely ranging mechanical properties. Furnaces are designed to accommodate material of all configurations requiring a variety of heat treatments.

Aluminum is supplied in almost an infinite variety of shapes and sizes for use in industries as diverse as transportation (automobiles, aircraft, railcars), packaging (foils, food, beverage cans), building products (siding, structural) and sports equipment (bats, shot-put circles, bleachers) to name a few [1]. Aluminum has an excellent strength-to-weight ratio and is easily cast, fabricated, formed and machined.

ALUMINUM HEAT TREATMENT

Heat treating is a critical step in the aluminum manufacturing process to achieve required end-use properties. The heat treatment of aluminum alloys requires precise control of the time-temperature profile, tight temperature uniformity and compliance with industry-wide specifications so as to achieve repeatable results and produce a high-quality, functional product. The most widely used specifications are AMS2770 (Heat Treatment of Wrought Aluminum Alloy Parts) and AMS2771 (Heat Treatment of Aluminum Alloy Castings)[2], which detail heat-treatment processes such as aging, annealing and solution heat treating in addition to parameters such as times, temperatures and quenchants. These specifications also provide information on necessary documentation for lot traceability and the quality-assurance provisions needed to ensure that a dependable product is produced.

Wrought aluminum alloys (Table 1) can be divided into two categories: non-heat treatable and heat treatable. Non-heat-treatable alloys, which include the 1xxx, 3xxx, 4xxx and 5xxx series alloys, derive their strength from solid solutioning and are further strengthened by strain hardening or, in limited cases, aging. Heat-treatable alloys include the 2xxx, 6xxx and 7xxx series alloys and are strengthened by solution heat treatment followed by precipitation hardening (aging).

Table 1. Wrought Alloy Designation System.

Alloy Series	Description or major alloying element
1XXX	Aluminum (99.00% minimum)
2XXX	Copper
3XXX	Manganese
4XXX	Silicon
5XXX	Magnesium
6XXX	Magnesium and silicon
7XXX	Zinc
8XXX	Other element
9XXX	Unused series

Cast aluminum alloys cannot be work hardened, so they are used in either the as-cast or heat-treated conditions. Common heat treatments include homogenization, annealing, solution treatment, aging and stress relief. Typical mechanical properties for commonly used casting alloys range from 20-50 Ksi (138-345 MPa) ultimate tensile strength and 15-40 Ksi (103-276 MPa) yield strength with up to 20% elongation [3].

Heat Treatment Processes

In general, the principles and procedures for heat treating wrought and cast alloys are similar. For cast alloys, however, soak times tend to be longer if the casting is allowed to cool below a process-critical temperature for the particular alloy. Solution soak times for castings can be significantly reduced to durations similar to that for wrought alloys if the castings are placed into the solution furnace while still hot (above the process-critical temperature) immediately following mold filling and solidification. The reduction of stress in complex cast shapes is achieved in large part by the control of quenching parameters such as agitation rate, quenchant temperature, rate of entry and part orientation in the quench.

Aging

The goal of aging is to cause precipitation dispersion of the alloy solute to occur (Figure 1). The degree of stable equilibrium achieved for a given grade is a function of both time and temperature. In order to achieve this, the microstructure must recover from an unstable or "metastable" condition produced by solution treating and quenching or by cold working.

Figure 1. Typical age hardening equipment.

The effects of age hardening or precipitation hardening on mechanical properties are greatly accelerated, and usually accentuated, by reheating the quenched material to about 212°F-424°F (100°C-200°C). A characteristic feature of elevated-temperature aging effects on tensile properties is that the increase in yield strength is more pronounced than the increase in tensile strength. Also ductility – as measured by percentage elongation – may decrease. Thus an alloy in the T6 temper has higher strength but lower ductility than the same alloy in the T4 temper.

In certain alloys, precipitation heat treating can occur without prior solution heat treatment since some alloys are relatively insensitive to cooling rate during quenching. Thus they can be either air cooled or water quenched. In either condition, these alloys will respond strongly to precipitation heat treatment.

In most precipitation-hardenable systems, a complex sequence of time-dependent and temperature-dependent changes is involved. The relative rates at which solution and precipitation reactions occur with different solutes depend upon the respective diffusion rates, in addition to solubility and alloy contents.

Annealing

Annealing is used for both heat-treatable and non-heat-treatable alloys to increase part ductility with a slight reduction in strength. There are several types of annealing treatments dependent to a large extent on the alloy type, initial and final microstructure and temper condition. In annealing it is important to ensure that the proper temperature is reached in all portions of the load (Figure 2). The maximum annealing temperature needs to be carefully controlled.

During annealing, the rate of softening is strongly temperature dependent – the time required can vary from a few hours at low temperature to a few seconds at high temperature. Full annealing (temper designation "O") produces the softest, most ductile and most versatile condition. Other forms of annealing include: stress-relief annealing, used to remove the effects of strain hardening in cold-worked alloys; partial annealing (or recovery annealing) done on non-heat-treatable wrought alloys to obtain intermediate mechanical properties; and recrystallization characterized by the gradual formation and appearance of a microscopically resolvable grain structure.

Homogenization (Ingot Preheating Treatments)

The initial thermal operation applied to castings or ingots (prior to hot working) is homogenization, which has one or more purposes depending upon the alloy, product and fabricating process involved. One of the principal objectives is improved workability since the microstructure of most alloys in the as-cast condition is quite heterogeneous. This is true for alloys that form solid solutions under equilibrium conditions and even for relatively dilute alloys.

Preheating Preheating of aluminum ingots prior to rolling, extruding, forming, forging or melting is used to reduce energy consumption by improved process efficiency, reducing cycle time and increasing safety.

Solution Heat Treatment The purpose of solution heat treatment is the dissolution of the maximum amount of soluble elements from the alloy into solid solution. The process consists of heating and holding the alloy at a temperature sufficiently high and for a long enough period of time to achieve a nearly homogenous solid solution in which all phases have dissolved.

Figure 2. Coil batch annealing furnace with nitrogen atmosphere and controlled cooling.

Care must be taken to avoid overheating or underheating. In the case of overheating, eutectic melting can occur with a corresponding degradation of properties such as tensile strength, ductility and fracture toughness. If underheated, solution treatment is incomplete and strength values lower than normal can be expected. In certain cases extreme property loss can occur. The solution soak times for castings can be reduced significantly by placing the casting directly into the solution furnace immediately following solidification. The casting is maintained at a temperature above a process-critical temperature (PCT), and the alloy solute is still in solution.

In general, a temperature variation of ±10°F (±5.5°C) from control set point is allowable, but certain alloys require even tighter tolerances. Tighter thermal variation (±5°F) allows the set point to be controlled closer to the eutectic, thus improving proportion and reducing required soak time. The time at temperature is a function of the solubility of the alloy solute and the temperature at which the aluminum casting or wrought alloy is removed from the mold and placed into the solution furnace. This time may vary from several minutes to many hours. The time required to heat a load to the treatment temperature increases with section thickness, air space around the casting for hot air to flow and the loading arrangement.

QUENCHING

Rapid and uninterrupted quenching in water or poly (alkylene) glycol in water is, in most instances, required to avoid precipitation detrimental to mechanical properties and corrosion resistance. The solid solution formed by solution heat treatment must be cooled rapidly enough to produce a supersaturated solution at room temperature that provides the optimal condition for subsequent age (precipitation) hardening. Quench types include hot water immersion, ambient water immersion, water spray, forced air, forced air with mist and poly (alkylene) glycol in water.

Quenching is, in many ways, the most critical step in the sequence of heat treating. In immersion quenching, cooling rates can be reduced by increasing the quenchant temperature. Conditions that increase the stability of a vapor film around the part decrease the cooling rate. Four factors that minimize distortion in the aluminum include:

- Temperature of the quenchant
- Agitation rate of the quenchant
- Speed of entry of casting into the quenchant
- Orientation of the aluminum part as it enters the quenchant

Stress Relief Stress-relief annealing can be used to remove the effects of strain hardening in cold-worked alloys. No appreciable holding time is required after the parts have reached temperature. Stress-relief annealing of castings provides maximum stability for service applications where elevated temperatures are involved.

TEMPERING

Tempering can be performed on heat-treatable aluminum alloys to provide the best combination of strength, ductility and toughness. These may be classified by the following designations:

- "F" - as fabricated
- "H" - strain hardened
- "O" - annealed
- "T" - thermally treated
- "W" - solution treated

The temper designation (Table 2) follows the alloy designation and consists of letters. Subdivisions, where required, are indicated by one or more digits following the letters.

Batch Installations

Batch designs are typically used for aluminum applications where throughput is not predictable or consistent, where the process varies from load to load or where a ramp-up in production favors sequential implementation. Batch units often run a variety of load configurations, so it is critical to have versatile airflow and tight temperature control. Typically, systems operate in temperature ranges of 350˚F, 500˚F, 650˚F and 850˚F. Heating systems can be provided in electric, direct gas or indirect gas heating. Loading can be accomplished by truck, rack/shelf, car, overhead trolley or monorail. Single and multiple chamber units provide maximum process flexibility. Common applications include aging, annealing, homogenizing and stress relief.

Table 2. "T" Temper designation.

First digit indicates sequence of treatments	
T1	Naturally aged after cooling from an elevated-temperature shaping process
T2	Cold worked after cooling from an elevated-temperature shaping process and naturally aged
T3	Solution heat treated, cold worked and naturally aged
T4	Solution heat treated and naturally aged
T5	Artificially aged after cooling from an elevated-temperature shaping process
T6	Solution heat treated and artificially aged
T7	Solution heat treated and stabilized (overaged)
T8	Solution heat treated, cold worked and artificially aged
T9	Solution heat treated, artificially aged and cold worked
T10	Cold worked after cooling from an elevated-temperature shaping process and artificially aged
Second digit indicates variation in the basic treatment	
Examples:	T42 or T62 – Heat treated to temper by user
Additional digits indicate stress relief	
Examples:	TX51 – Stress relieved by stretching
	TX52 – Stress relieved by compressing
	TX54 – Stress relieved by stretching and compressing

Continuous Installations

Continuous designs are typically used for automated production and include mesh-belt conveyors, roller hearths, roller rails, rotary hearth, pusher and walking-beam units. Capacities ranging from several hundred to several thousand pounds per hour are typical. Parts are loaded consistently and uniformly for continuous process flow, so minimum labor is required. Continuous furnace systems significantly reduce operating costs. Typically, systems operate in temperature ranges up to 1100˚F. Heating systems can be provided in electric, direct gas or indirect gas heating. If desired, loading can be automated in a number of different ways. Common heat treatments include aging, annealing and solutioning for products as diverse as castings, forgings, plate, bar and tube products.

INTEGRATION WITH LEAN AND AGILE MANUFACTURING

In order to integrate aluminum heat-treating systems into lean and green manufacturing strategies, system automation and control for loading, unloading and transferring workloads is required. Robotics, roller conveyor systems, manipulators and charging cars are typical examples of equipment supplied to increase production efficiency while reducing manpower requirements.

The processing of aluminum requires a combination of rapid heating and close temperature uniformity throughout the entire load. Many components are safety-critical and must be heat treated with high precision and repeatability. Often plant floor space is important and compact designs are highly desirable. Integrating with the SCADA systems for real-time data acquisition and integration with upstream and downstream processes is essential. The heat treatment of aluminum demands that all aspects of the process are monitored and controlled.

The evolution of heating methods has an important role on the quantitative and qualitative development of heat treatment technologies. This evolution is decisive for the charge heating rate, for the value of energetic consumption of the process and for the technological effects subsequently obtained [4]. It is well-known the influence of heat treatment on mechanical properties and the corrosion resistance of various alloys [5].

Heat treating on its broadest sense, refers to any of the heating and cooling operations that are performed for the purpose of changing the mechanical properties, the metallurgical structure, or the residual stress state of a metal product. When the term is applied to aluminum alloys, however, its use frequently is restricted to the specific operations employed to increase strength and hardness of the precipitation – hardenable wrought and cast alloys.

These usually are referred to as the "heat-treatable" alloys to distinguish them from those alloys in which no significant strengthening can be achieved by heating and cooling. The latter, generally referred to as "non heat-treatable" alloys depend primarily on cold work to increase strength. Heating to decrease strength and increase ductility (annealing) is used with alloys of both types; metallurgical reactions may vary with type of alloy and with degree of softening desired.

One essential attribute of a precipitation-hardening alloy system is a temperature-dependent equilibrium solid solubility characterized by increasing solubility with increasing temperature. The mayor aluminum alloy systems with precipitation hardening include:

- Aluminum-copper systems with strengthening from $CuAl_2$
- Aluminum-copper-magnesium systems (magnesium intensifies precipitation)
- Aluminum-magnesium-silicon systems with strengthening from Mg_2Si
- Aluminum-zinc-magnesium systems with strengthening from $MgZn_2$
- Aluminum-zinc-magnesium-copper systems

The general requirement for precipitation strengthening of supersaturated solid solutions involves the formation of finely dispersed precipitates during aging heat treatment (which may include either natural aging or artificial aging). The aging must be accomplished not only below the equilibrium solvus temperature, but below a metastable miscibility gap called the Guinier-Preston (GP) zone solvus line.

The commercial heat-treatable alloys are, with few exceptions, based on ternary or quaternary systems with respect to the solutes involved in developing strength by precipitation. Commercial alloys whose strength and hardness can be significantly increased by heat treatment include 2xxx, 6xxx, and 7xxx series wrought alloys and 2xx.0, 3xx.0 and 7xx.0 series casting alloys.

Some of these contain only copper, or copper and silicon as the primary strengthening alloy addition. Most of the heat-treatable alloys, however, contain combinations of magnesium with one or more of the elements, copper, silicon and zinc. Characteristically, even small amounts of magnesium in concert with these elements accelerate and accentuate precipitation hardening, while alloys in the 6xxx series contain silicon and magnesium approximately in the proportions required for formulation of magnesium silicide (Mg_2Si). Although not as strong as most 2xxx and 7xxx alloys, 6xxx alloys have good formability, weldability, machinability, and corrosion resistance, with medium strength.

In the heat-treatable wrought alloys, with some notable exceptions (2024, 2219, and 7178), such solute elements are present in amounts that are within the limits of mutual solid solubility at temperatures below the eutectic temperature (lowest melting temperature).

In contrast, some of the casting alloys of the 2xx.0 series and all of the 3xx.0 series alloys contain amounts of soluble elements that far exceed solid-solubility limits. In these alloys, the phase formed by combination of the excess soluble elements with the aluminum will never be dissolved, although the shapes of the undissolved particles may be changed by partial solution.

Heat treating to increase strength of aluminum alloys is a two – step process:

- quenching heat treating: dissolution of soluble phases and development of supersaturating;
- aging (age hardening) heat treating: precipitation of solute atoms either at room temperature (natural aging) or elevated temperatures (artificial aging or precipitation heat treating).

Quality-assurance criteria that heat-treated materials must meet always include minimum tensile properties and, for certain alloys and tempers, adequate fracture toughness and resistance to detrimental forms of corrosion (such as intergranular or exfoliation attack) or to stress-corrosion cracking. All processing steps through heat treatment must be carefully controlled to ensure high and reliable performance.

In general, the relatively constant relationships among various properties allow the use of tensile properties alone as acceptance criteria. The minimum guaranteed strength is ordinarily that value above which it has been statistically predicted with 95% probability that 99% or more of the material will pass. Testing provides a check for evidence of conformance; process capability and process control are the foundations for guaranteed values.

Published minimum guaranteed values are applicable only to specimens cut from a specific location in the product, with their axes oriented at a specific angle to the direction of working as defined in the applicable procurement specification. In thick plate, for example, the guaranteed values apply to specimens taken from a plane midway between the center and the surface, and their axes parallel to the width dimension (long transverse). Different properties should be expected in specimens taken from other locations or in specimens whose axes were parallel to thickness dimension (small transverse) [3]. However, the specified "referee" locations and orientations do provide a useful basis for lot-to-lot comparisons, and constitute a valuable adjunct to other process-control measures.

Tensile tests can be used to evaluate the effects of changes in the process, provide specimens are carefully selected. A variation in process that produces above-minimum properties on test specimens, is not necessarily satisfactory-acceptability can be judged only by comparing the resulting properties with those developed by the standard process on similarly located specimens. Finally, variations in heat-treating procedure are likely to affect the relationships among tensile properties and other mechanical properties. In applications where other properties are more important than tensile properties, the other properties should be checked also [4, 7, 8].

Elasticity tests are less valuable for acceptance and rejection of heat-treated aluminum alloys than they are for steel. Nevertheless, elasticity tests have some utility for process control [9, 10, 11].

The objective is to select the cycle that produces optimum precipitate size and distribution pattern. Unfortunately, the cycle required to maximize one property, such as tensile strength, is usually different from that required to maximize others, such as yield stress and corrosion resistance. Consequently, the cycles used represent compromises that provide the best combination of mechanical properties (it can be refereed at a combination of hardness and mechanical stress).

Good temperature control and uniformity throughout the furnace and load are required for all precipitation heat treating. Recommended temperatures are generally those that are least critical and that can be used with practical time cycles.

The general methods for heat treating aluminum alloys include the use of molten salt – baths, air – chamber furnaces and induction heaters. Air furnaces are used more widely because they permit greater flexibility in operating temperature. Air furnaces are also more economical when the product mix includes a few parts; holding temperature of a large volume of salt in readiness for an occasional part is far more expensive than heating an equal volume of air. Also, induction methods can provide high heating rates, which affects transformation behavior.

This chapter presents a wrought aluminum alloy, AlCu2,5Mg that are used for aeronautical parts. These alloys are heat treated in order to establish the optimum mechanical properties, and I am referring especially at micro hardness, tensile strength and elasticity. So, were took 10 parts, of identical measures, from this alloy and was applied the final heat treating. At the end, it was studied the elasticity, micro-hardness and tensile strength, in order to establish the optimum technology.

Establishing the Experimental Technological Conditions. Programming the Experiment

Establishing the Preliminary Experimental Conditions

The proposed working methodology for improving furnace heat transfer has to follow further steps:

- adopting the correct heat treatment technology
- choosing the most suitable testing equipments
- choosing the most suitable heating furnaces for reaching the technological objectives
- conducting the experimental work and results interpretation.

Adopting the Correct Heat Treatment Technology for Studied Aluminum Alloys

In this chapter, the studied aluminum alloys will be presented along with their technological parameters for heat treatment. This is a widely used wrought alloy and the chemical composition is according to SR EN 573-3.

The parts have the geometrical configuration like in figure 3 and the dimensions are in table 3.

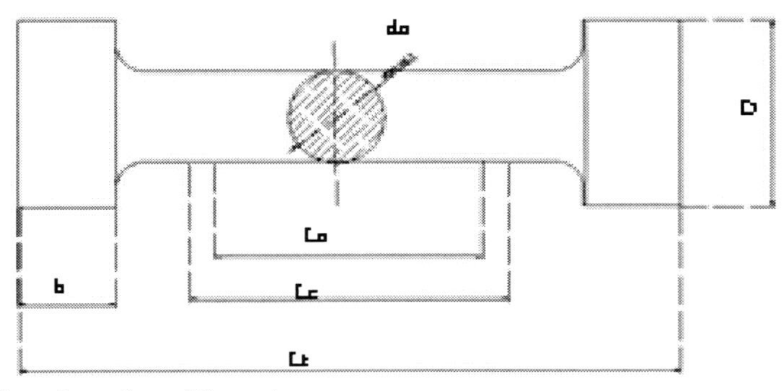

Figure 3. Geometrical configuration of the parts.

Table 3. Part dimensions, in mm

	d_0	S_0	D	h	L_t	L_0	L_c
AlCu2,5Mg	16	200,96	20	50	205	80	96

The final heat treating applied to each charge is quenching followed by artificial aging.

Chemical Composition Tests

It is well known that little variation in composition can affect the alloy properties. So, all the tests have to start with a precise chemical composition test. In this idea, the experimental work started with some chemical tests based on optical spectrometry for each alloy.

Table 4. Chemical composition tests

	probe	chemical composition, %									
		Mg	Cu	Fe	Ni	Si	Sn	Mn	Zn	Ti	Cr
AlCu2,5Mg	1	0,33	2,00	0,49	0,02	1.14	0.02	0,70	0,23	0,08	0,01
	2	0,34	2,01	0,49	0,02	1,13	0,01	0,70	0,20	0,08	0,01
	media	0,33	2,01	0,49	0,02	1,14	0,01	0,70	0,22	0,08	0,01

Further there are presented the results for each determination and a media of the obtained values.

In table 4. are AlCu2,5Mg alloy tests.

Programming the Experiment

In order to accomplish the desired mathematical modeling, the experiment has to be programmed and this implies the following:

- Establishing the necessary and sufficient number of experiences and the necessary conditions in order to accomplish them;
- Establishing the regression equation which represent the process model;
- Establishing the conditions necessary to accomplish the optimum value of the process performance fulfilled.

In this context, for each variable, there have been established the basic levels as well as the variation intervals. By adding the variation level to the basic level there has been obtained the superior level, and by decreasing it the inferior level of the variable. Choosing the variation interval must offer the most accurate values from the functional point of view. A first step is to establish the basic levels and the variation intervals. In table 5, the variation interval and the basic level for programming the experiment are presented.

The interpretation of the experimental results consists of establishing the experimental variation curves of the specific mass of each charge according to energetic consumption of the equipment; the interpretation of the experimental results shall be finalized by determining the analytical equations that describe the experimental curves obtained.

Table 5.The matrix bases for k=2

Quenching temperature, Tc,°C		Aging temperature, Ta, °C	
base level	515°C	base level	160°C
variation	10°C	variation	10°C
calculation method	$x_1 = \frac{Tc-515}{10}$	calculation method	$x_2 = \frac{Ta-160}{10}$

Final heat treating process optimization consists in establishing the minimum number of experiments needed for the correct process description. In this case, using the 3^k factorial experiment model, are sufficient a number of 3 experiments (k = 1 for one variable: mass).

Therefore, for each process variable it will be establish a base level and a variation area for a correct description of the heat treating technology.

Heat Treatment Optimization on an AlCu2,5Mg Aluminum Alloy

In order to realize this study, a process optimization has been done by establishing some various heating temperatures, in standard limits and the mechanical properties obtained in every case and I am referring at hardness and mechanical stress were measured.

It have made experiments in the same conditions of preheating for the furnace, and was used the same equipment for every stress and hardness determinations. The experiments are illustrated in Table 6.

Final heat treating parameters, from aluminum alloy standards, are:

- For quenching heat treating, the heating temperatures are 505 – 525°C, for all types of parts. The maintaining time at the heating temperature is 20 minutes. The cooling is at high rate, in water.
- For artificial aging heat treating, the standards for AlCu2,5Mg recommend a 150 – 170°C heating temperature, maintained for 6 hours.

Table 6. The programmed experiment for AlCu2,5Mg aluminum alloy

experiment	variation factor, x_1	T_c, °C	variation factor, x_2	T_a, °C
0.	-	-	-	-
1.	-1	505	-1	150
2.	-1	505	0	160
3.	-1	505	1	170
4.	0	515	-1	150
5.	0	515	0	160
6.	0	515	1	170
7.	1	525	-1	150
8.	1	525	0	160
9.	1	525	1	170

Results and Discussions

In order to realize the proposed technology, I used an electrical furnace with forced air circulation. For heat treatment I chose a SUPERTHERM furnace. This equipment is for laboratory use and can obtain any proposed heating diagrams. Mechanical stress curves and values were obtained on a servo - hydraulically MTS 824.10 Test Star II machine, which have a computer with TestStar System soft, digital controller and a force unit. Hardness was obtained on a Neophot microscope with Hanemann installation.

After experiments and the mechanical characteristics determinations I obtained a mathematical model which will describe the final heat treating process. The mathematical model is based on a programmed experiment that consists of:

- establish the necessary and sufficient number of experiments for the studied alloy;
- establish the regression equation which represents the model of the process;
- establish the conditions to obtain an optimum value of the process performance;
- determine the mathematical model of final heat treating of the AlCu2,5Mg alloy.

Final heat treating process optimization consists in establishing the minimum number of experiments needed for the correct process description. In this case, using the 3^k factorial experiment model, are sufficient a number of 9 experiments ($k = 2$ for two variables: artificial aging heat treating temperature and quenching heat treating temperature). Therefore, for each process variable it will be establish a base level and a variation area for a correct description of the heat treating technology. In order to obtain the proposed mathematical model, I have made the experiments that are illustrated in Table 4. Also, mechanical stress curves are illustrated in Figures 4-13.

During the experiment, every part had a particular code. The experimental results are in Table 7, Table 8 and Table 9.

Table 7. AlCu2,5Mg mechanical stress experiment

part code	mechanical stress, Rm	elongation at Rm	specific deformation at Rm	fracture stress	fracture elongation	specific deformation at fracture	stress Rp
	MPa	mm	%	MPa	mm	%	MPa
0	243,426	14,5	14,20447	243,4261	14	14,2	120,4240
1	374,901	13,8	13,78661	374,7300	14	14,2	264,3085
2	355,878	19,0	18,75295	354,7175	21	20,3	223,1074
3	411,619	9,4	9,10137	409,6451	10	9,9	359,8068
4	345,944	18,8	18,64251	342,0597	23	22,5	221,9793
5	350,572	18,2	18,03330	348,5886	20	19,6	219,4174
6	396,037	9,5	9,07510	394,8653	11	10,3	337,2006
7	351,561	22,8	22,58176	348,1774	24	24,1	224,2233
8	356,100	18,0	17,86037	350,3231	22	21,8	219,0241
9	363,547	8,8	8,6197	363,5469	9	8,6	304,622

Table 8. AlCu2,5Mg elasticity tests

part code	Elasticity	Elongation, %
0.	24895,50	-
1.	72724,54	18,37
2.	70714,14	24,12
3.	72556,72	13,37
4.	68526,10	26,62
5.	69234,11	23,75
6.	72313,10	14,00
7.	70661,72	28,37
8.	71811,01	24,00
9.	70301,95	11,87

With the results illustrated in Table 7, 8 and 9, using 3^k factorial experiment model and with a specific computer program I obtained the curves that are described by the following equations. In these cases, the needed function is a degree m polynomial and the variables are:

- mechanical stress;
- hardness;

- elasticity modulus;
- elongation.

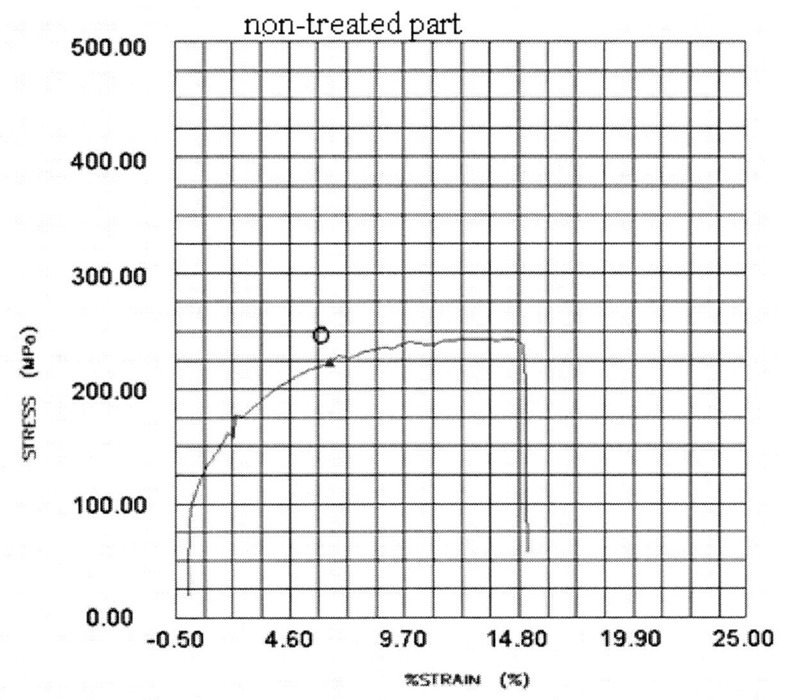

Figure 4. Stress diagram – test part, non-treated.

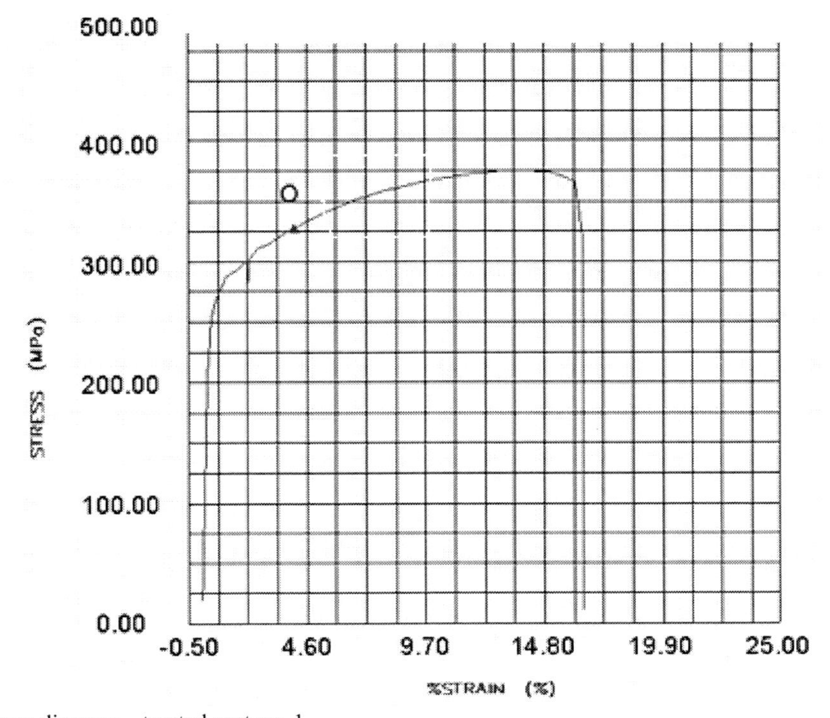

Figure 5. Stress diagram – treated part no. 1.

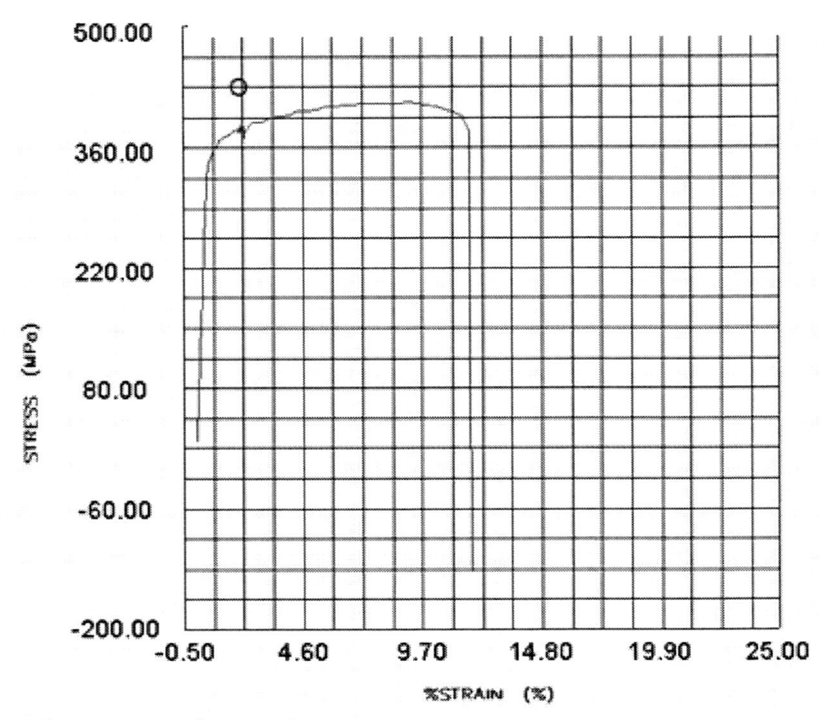

Figure 6. Stress diagram – treated part no. 2.

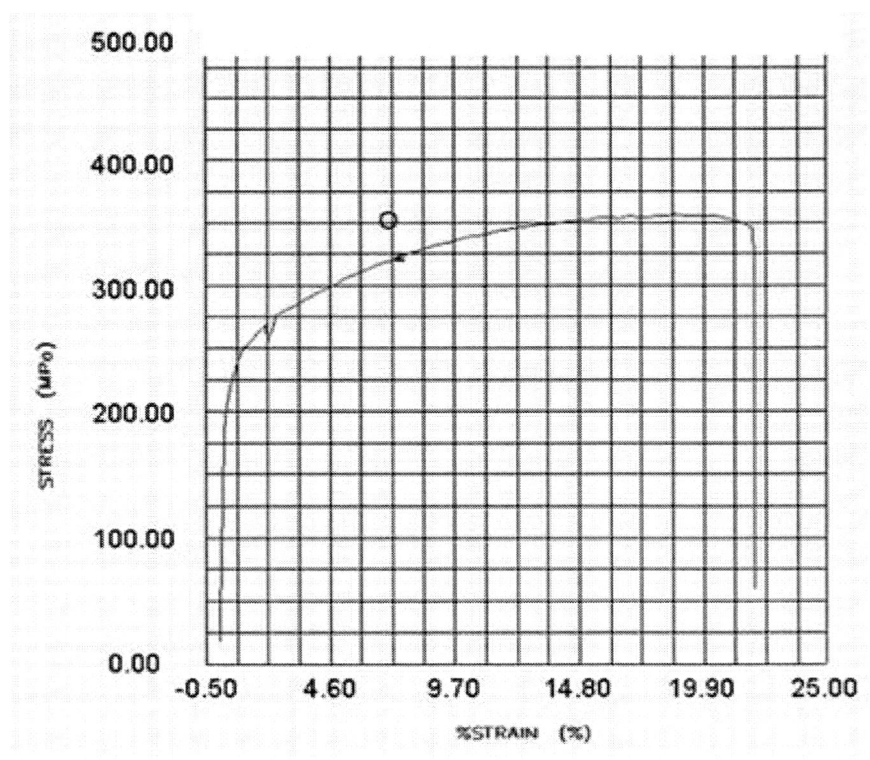

Figure 7. Stress diagram – treated part no. 3.

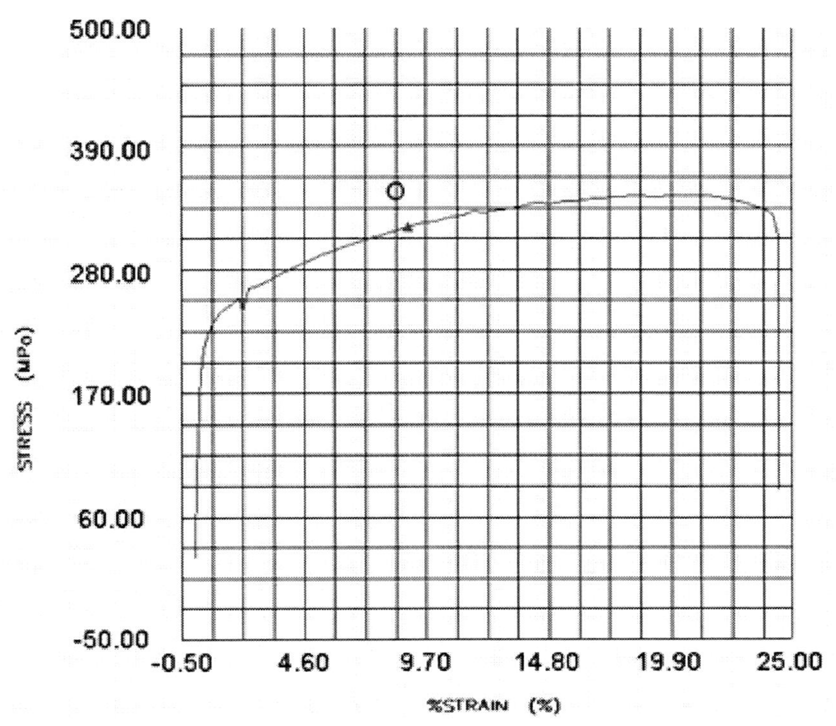

Figure 8. Stress diagram – treated part no.4.

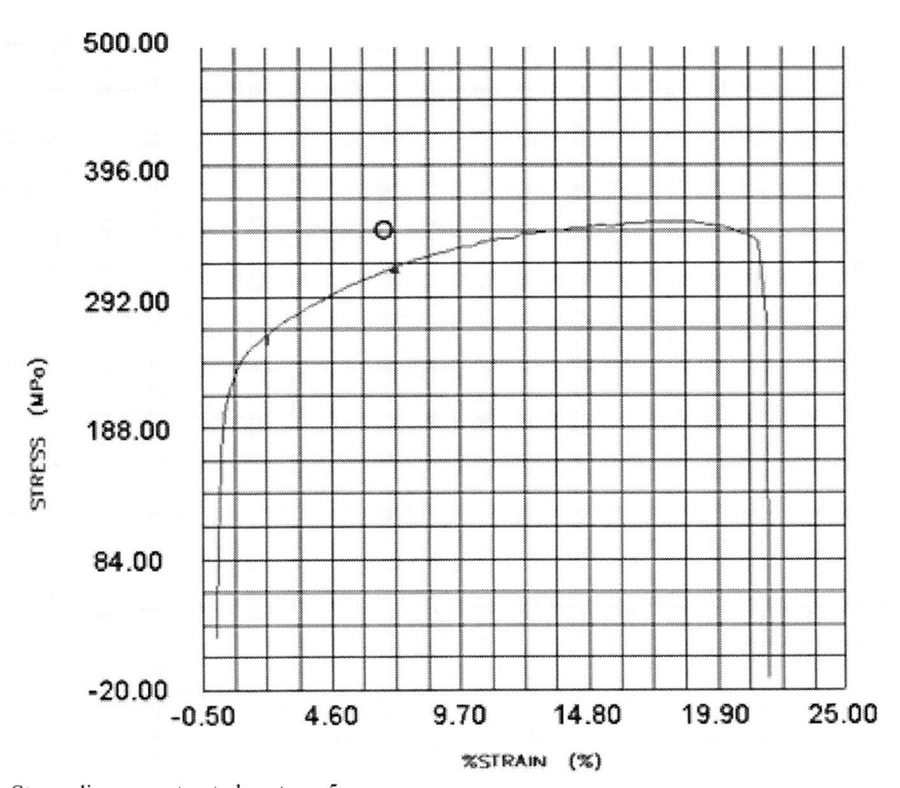

Figure 9. Stress diagram – treated part no. 5.

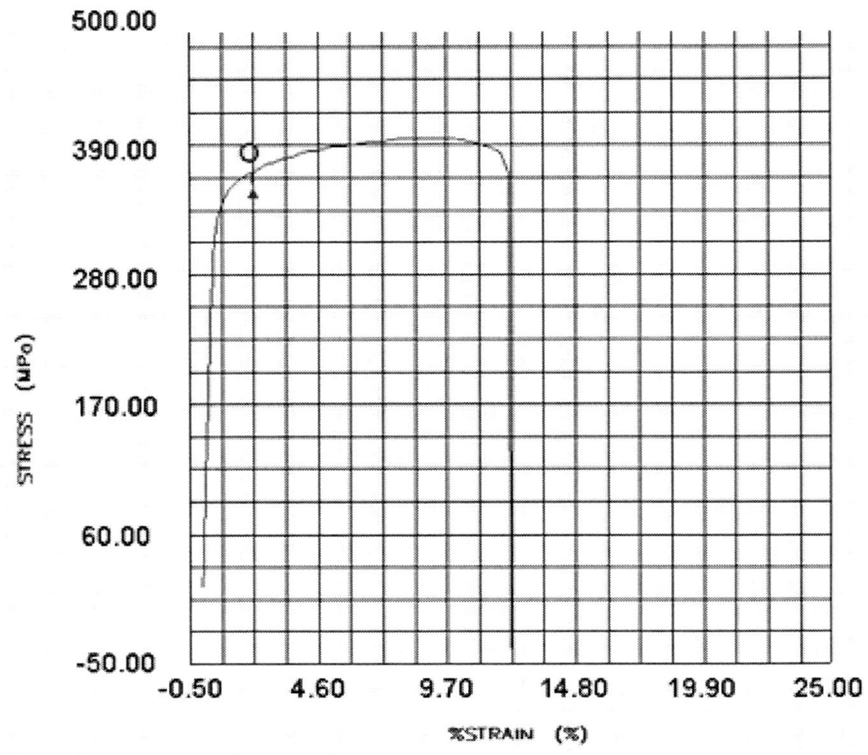

Figure 10. Stress diagram – treated part no. 6.

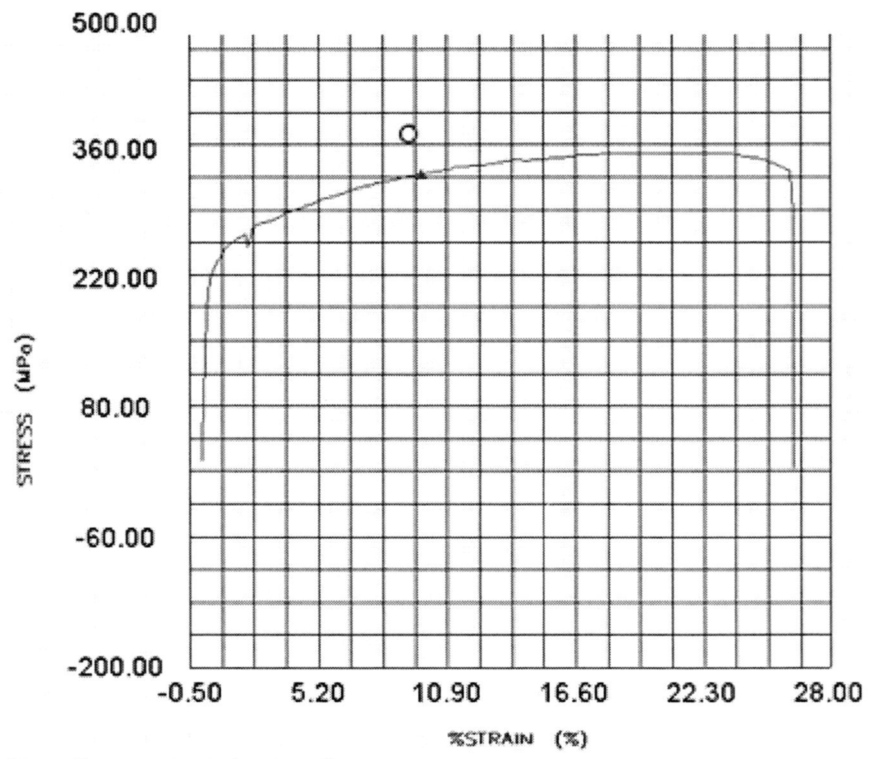

Figure 11. Stress diagram – treated part no. 7.

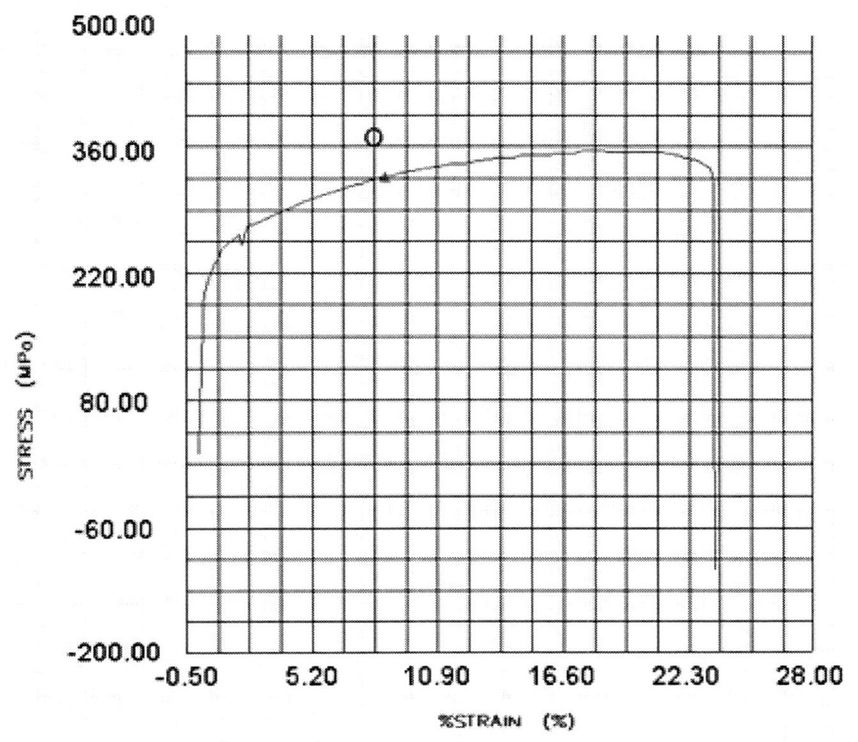

Figure 12. Stress diagram – treated part no. 8.

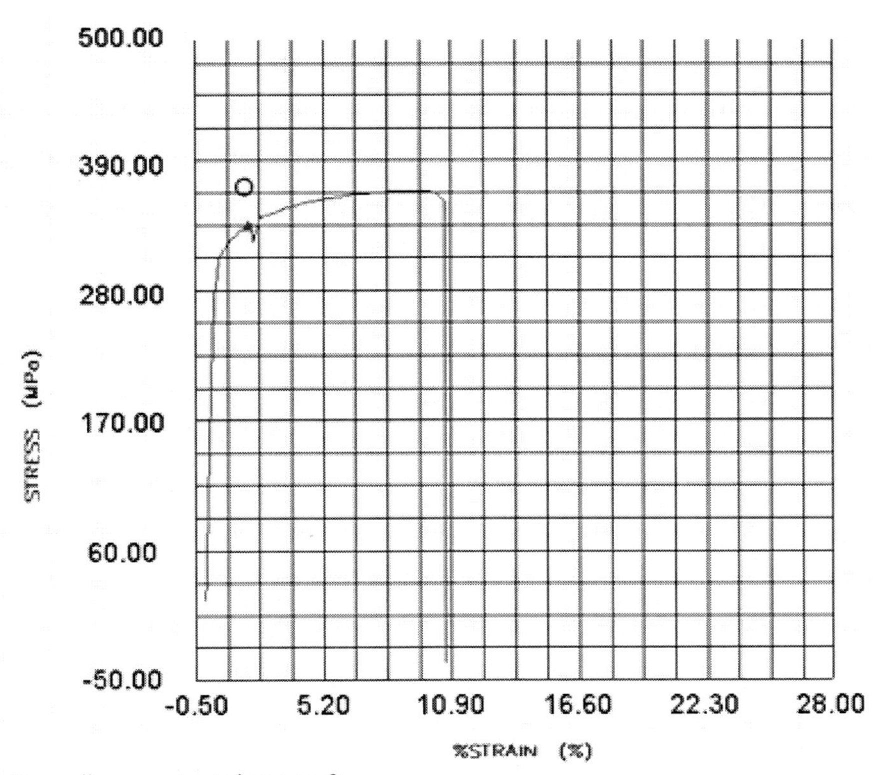

Figure 13. Stress diagram – treated part no. 9.

The regression, which describes the process, is:

$$R = 351{,}017 - 11{,}865x_1 + 16{,}466x_2 - 6{,}183x_1x_2 + 4{,}75x_1^2 + 19{,}751x_2^2 \quad (1)$$

$$HV = 151{,}6 - 1{,}35x_1 + 2{,}4x_2 + 5{,}5x_2^2 \quad (2)$$

$$M = 69628{,}271 - 536{,}787x_1 + 543{,}235x_2 - 47{,}988x_1x_2 + 1437{,}229x_1^2 + 594{,}262x_2^2 \quad (3)$$

$$A = 23{,}957 - 1{,}397x_1 - 5{,}687x_2 - 2{,}785x_1x_2 - 5{,}19x_2^2 \quad (4)$$

Table 9. AlCu2,5Mg hardness experiment

part code	experiment number										hardness average, HV
	1	2	3	4	5	6	7	8	9	10	
0	95,2	83,3	84,6	85,3	79,7	79,7	79,1	79,1	97,1	81,5	84,46
1	139	156	162	146	146	152	148	177	169	157	155,2
2	152	159	156	154	154	152	156	156	159	156	155,4
3	164	160	154	162	159	157	160	166	160	160	160,2
4	152	157	159	156	157	154	157	157	159	157	156,5
5	150	159	149	148	148	145	145	143	145	145	147,7
6	159	160	161	158	160	160	161	161	160	160	159,7
7	148	149	151	154	152	156	156	152	154	152	152,4
8	149	152	148	146	146	152	154	156	158	156	151,7
9	149	151	152	154	154	152	156	154	154	152	152,8

In these equations, x_1 and x_2 (as seen in Table 3) are the factors that describe the variation of quenching heat treating temperature and artificial aging heat treating temperature. For the two parameters, the values are:

- x_1 describes the variation of quenching heat treating temperature and can be determined with the relation: $T_c = T_{c\,0} + \Delta T_c\, x_1$, where T_{c0} is the base level for the quenching temperature (515°C) and ΔT_c is the proposed variation area(10°C).
- x_2 describes the variation of artificial aging heat treating temperature and can be determined with the relation: $T_a = T_{a\,0} + \Delta T_a\, x_1$, where T_{i0} is the base level for the artificial aging temperature (160°C) and ΔT_c is the proposed variation area(10°C).
- R describes the variation of the mechanical stress.
- HV is the variation of the part Vickers micro hardness
- M is elasticity modulus
- A is elongation.

Also, I made a study regarding the dual influence of the mechanical stress and hardness on heat treating parameters. So, two cases were considered:

- case 1: quenching heat treating temperature is constant.
- case 2: artificial aging heat treating temperature is constant.

Case 1 In this situation, according to Table 2, I have three equations, for the three values of the quenching temperature:

$$T_c = 505°C \qquad T_a = 129,06 - 0,17\ R + 0,61\ HV \qquad (5)$$

$$T_c = 515°C \qquad T_a = 406,422 + 1,27\ R - 4,57\ HV \qquad (6)$$

$$T_c = 525°C \qquad T_a = 1335,12 + 6\ R - 21,56\ HV \qquad (7)$$

Equations (5), (6) and (7) represent the variation equations between artificial aging heat treating temperature of the AlCu2,5Mg aluminum alloy and the hardness and mechanical stress.

Case 2 In this situation, according to table 2 I have three equations, for the three values of the artificial aging heat treating temperature:

$$T_a = 150°C \qquad T_c = 515 \pm \sqrt{21R - 88,5HV + 6248,6} \qquad (8)$$

$$T_a = 160°C \qquad T_c = 515 \pm \sqrt{21R - 185HV + 20660,5} \qquad (9)$$

$$T_a = 170°C \qquad T_c = 515 \pm \sqrt{5R - 21HV + 1568,9} \qquad (10)$$

Equatons (8), (9) and (10) represent the variation equations between quenching heat treating temperature of the AlCu2,5Mg aluminum alloy and the hardness and mechanical stress.

Analyzing the determined mathematical models it reveals a conclusion about the existing area for mechanical characteristics, after final heat treating:

- for mechanical stress, AlCu2,5Mg alloy have 340 – 450 MPa;
- for hardness, AlCu2,5Mg alloy have 150 – 170HV.

Moreover, if it consider a more specialized study, it can detail the results, as shown in the next sections.

Theoretical Contributions for Choosing Heat Treatment Parameters According to Stress Conditions

For the study equation (1) was interpretated:

$$R = 351,017 - 11,865x_1 + 16,466x_2 - 6,183x_1x_2 + 4,75x_1^2 + 19,751x_2^2 \qquad (1)$$

According to table 4, the equation (1) can be written as:

$$R = 16968{,}475 - 40{,}223T_c - 29{,}727T_a - 0{,}0618T_cT_a + 0{,}0475T_c^2 + 0{,}1975T_a^2 \quad (11)$$

For studying the heat treatment parameters variation according to mechanical stress two cases were considered:

- Case I: determining the aging temperature when mechanical stress and quenching temperature are fixed for the alloy;
- Case II: determining the quenching temperature when mechanical stress and aging temperature are fixed for the alloy;

Case I:it will determine the aging temperature when mechanical stress and quenching temperature are fixed for the alloy, and it must consider three situations:

$T_c = 505°C$

$$T_a = 154{,}25 + \sqrt{5R - 1828{,}4} \quad (12)$$

This equation has a restriction for mechanical stress: $R > 365{,}68$ MPa.

$T_c = 515°C$

$$T_a = 158{,}27 + \sqrt{5R - 1759{,}9} \quad (13)$$

This equation has a restriction for mechanical stress: $R > 351{,}98$ MPa.

$T_c = 525°C$

$$T_a = 157{,}4 + \sqrt{5R - 1715{,}2} \quad (14)$$

This equation has a restriction for mechanical stress: $R > 343{,}04$ MPa.

Equations (12), (13) and (14) are graphically represented in figure 14 and represent all the equations for aging temperature variation for AlCu2,5Mg alloy. Also figure 14 can be used as a nomogram for temperature determination.

Case II:it will determine the quenching temperature when mechanical stress and aging temperature are fixed for the alloy, and it must consider three situations:

$T_a = 150°C$

$$T_c = 520{,}98 + \sqrt{21R - 7423} \quad (15)$$

This equation has a restriction for mechanical stress: $R > 353{,}476$ MPa.

$T_a = 160°C$

$$T_c = 527{,}49 - \sqrt{21R - 7264{,}9} \quad (16)$$

This equation has a restriction for mechanical stress: R > 345,947 MPa.

$T_a = 170°C$

$$T_c = 534 - \sqrt{21R - 7791{,}2} \quad (17)$$

This equation has a restriction for mechanical stress: R > 371,009 MPa.

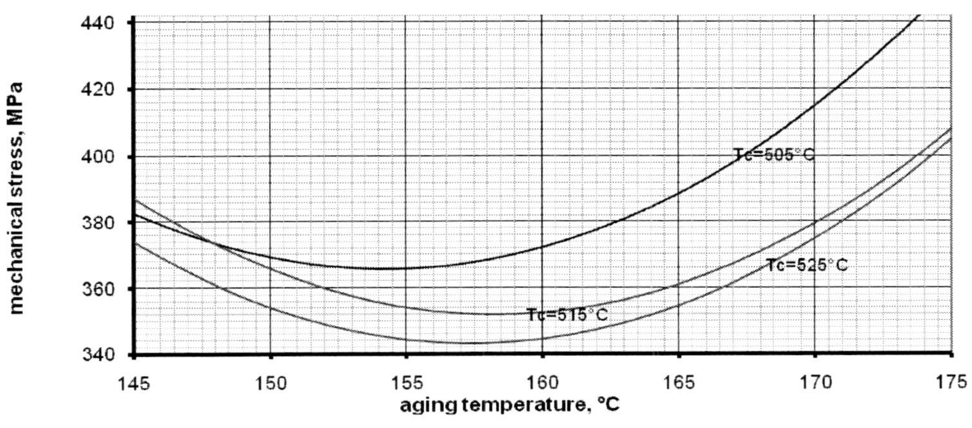

Figure 14. Nomogram for aging temperature determination for an AlCu2,5Mg aluminum alloy.

Equations (15), (16) and (17) are graphically represented in figure 15 and represent all the equations for quenching temperature variation for AlCu2,5Mg alloy. Also figure 15 can be used as a nomogram for temperature determination.

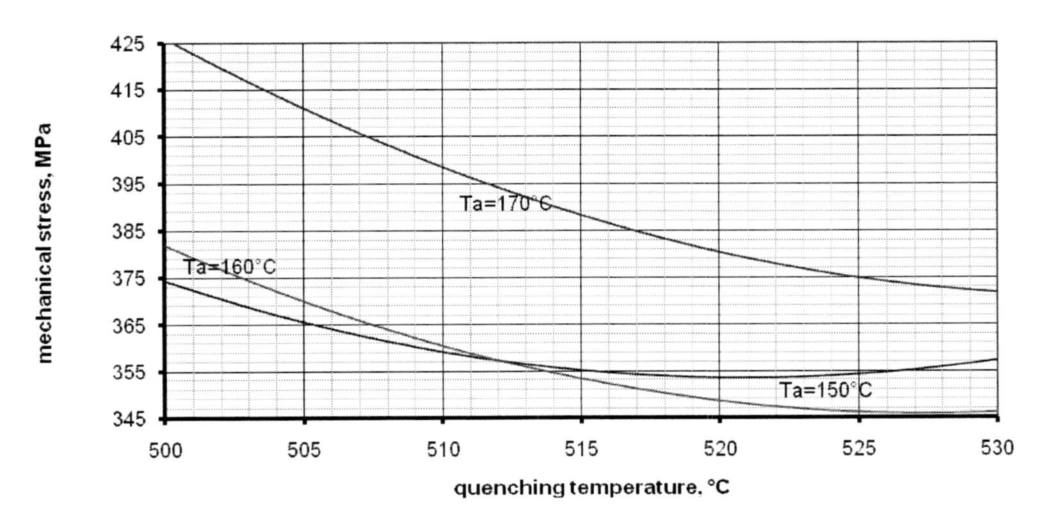

Figure 15. Nomogram for quenching temperature determination for an AlCu2,5Mg aluminum alloy.

3.1.2. Theoretical Contributions for Choosing Heat Treatment Parameters According to Stress Conditions

For the study equation (2) was considered:

$$HV = 151{,}6 - 1{,}35x_1 + 2{,}4x_2 + 5{,}5x_2^2 \quad (2)$$

Also, this equation according to table 4 can be written as:

$$HV = 1590{,}725 - 0{,}135T_c - 17{,}36T_a + 0{,}055T_a^2 \quad (18)$$

For studying the heat treatment parameters variation according to hardness two cases were considered:

- Case I: determining the aging temperature when hardness and quenching temperature are fixed for the alloy;
- Case II: determining the quenching temperature when hardness and aging temperature are fixed for the alloy;

Case I:it will determine the aging temperature when hardness and quenching temperature are fixed for the alloy, and it must consider three situations:

$T_c = 505°C$

$$T_a = 157{,}82 \pm \sqrt{18{,}1HV - 2776{,}1} \quad (19)$$

This equation has a restriction: HV>153,38.

$T_c = 515°C$

$$T_a = 157{,}82 \pm \sqrt{18{,}1HV - 2751{,}5} \quad (20)$$

This equation has a restriction for hardness: HV>152,019.

$T_c = 525°C$

$$T_a = 157{,}82 \pm \sqrt{18{,}1HV - 2731{,}8} \quad (21)$$

This equation has a restriction for hardness: HV>150,929.

Equations (19), (20) and (21) are graphically represented in figure 16 and represent all the equations for aging temperature variation for AlCu2,5Mg alloy. Also figure 16 can be used as a nomogram for temperature determination when hardness is imposed.

Case II:it will determine the quenching temperature when hardness and aging temperature are fixed for the alloy, and it must consider three situations:

$T_a = 150°C$

$$T_c = 1660{,}92 - 7{,}4HV \tag{22}$$

This equation has a restriction for hardness: 153,5 < HV < 156,2.

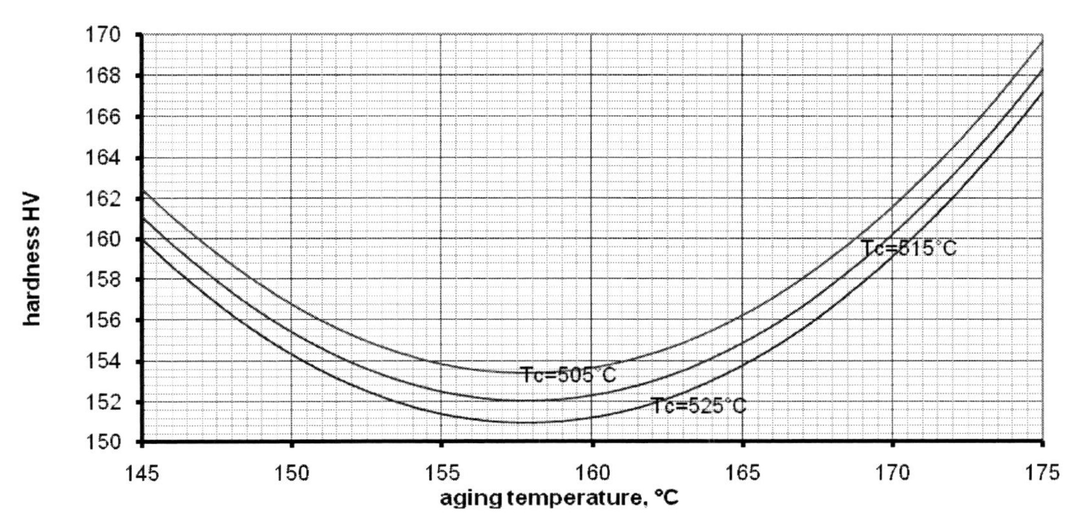

Figure 16. Nomogram for aging temperature determination for an AlCu2,5Mg aluminum alloy, when hardness is imposed.

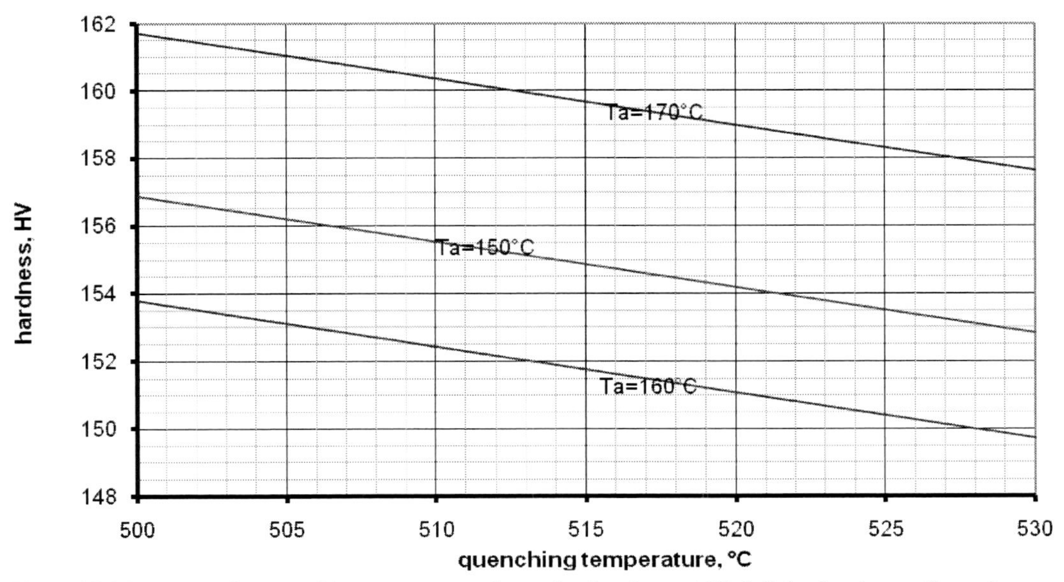

Figure 17. Nomogram for quenching temperature determination for an AlCu2,5Mg aluminum alloy, when hardness is imposed.

$T_a = 160°C$

$$T_c = 1637{,}96 - 7{,}4HV \tag{23}$$

This equation has a restriction for hardness: $150{,}4 < HV < 153{,}102$.

$T_a = 170°C$

$$T_c = 1696{,}48 - 7{,}4HV \quad (24)$$

This equation has a restriction for hardness: $158{,}3 < HV < 161{,}01$.

Equations (22), (23) and (24) are graphically represented in figure 17 and represent all the equations for quenching temperature variation for AlCu2,5Mg alloy. Also figure 17 can be used as a nomogram for temperature determination when hardness is imposed.

Conclusion

This particular study helps on predicting some heat treatment parameters in order to help the technology designers. Thus, equations (1) – (24) are theoretical obtained on experimental bases and helps on heat treatment temperatures estimation as well as mechanical stress, hardness, elasticity modulus and elongation. Also, the determined temperatures have to be in standard limits recommendations for this particular alloy: AlCu2,5Mg. In those situations when the calculated temperature it is not in the standard limits, it have to give up on imposing many characteristics and start imposing just one, for example mechanical stress or hardness. The other mechanical characteristics will be designed further, on the bases of heating temperatures.

Anyway, a final conclusion can be outlined along with some limits for hardness and mechanical stress values as:

- $340 < R < 450$ MPa for mechanical stress
- $150 < HV < 170$ for hardness.
-

Besides those limits, it cannot be possible to obtain certain mechanical characteristics through regular heat treatment and it have to consider specialized heating and cooling cycles.

MICROSTRUCTURAL CONSIDERATIONS OF A HEAT TREATED ALCU2,5MG ALLOY

This chapter presents a wrought aluminum alloy, AlCu2,5Mg that is studied in order to establish microstructure profile, after diverse heat treatments.

On the basis of these experiments a theoretical study was done, regarding to establish of the heating parameters for quenching and aging in order to obtain certain properties, needed in service for this alloy.

Also, it has to mention that the determined temperatures must be in the limits that are recommended in the material standards.

For the microstructure study, few parts were chosen: test part along with parts no. 1, 3 and 5. The characteristics are given in the table 10:

Table 10

experiment	T_c, °C	T_a, °C	mechanical stress, R	Elasticity	Elongation, %	Hardness, HV
test part	-	-	243,426	24895,50	-	84,46
1.	505	150	374,901	72724,54	18,37	155,2
3.	505	170	411,619	72556,72	13,37	160,2
5.	515	160	350,572	69234,11	23,75	147,7

In figure 18 is AlCu2,5Mg alloy microstructure for the test part. Further Figures 19, 20 and 21 are the microstructures for parts 1, 3 and 5, respectively.

In all the microstructures can be noticed the α solid solution as a base compound along with chemical compounds Al_2Cu (with black) and Cu oxides (little dark points). These oxides appear in the structure mainly because of forming procedures.

Further the influence of Al_2Cu was studied. Theoretically an increased quantity of this compound will determine a higher alloy hardness and good mechanical behavior. Comparing the data-s from Figures 18-21 and from table 10 it remarks that part 3, that have the highest hardness corresponds to the most increased bitplane: 16.658 %

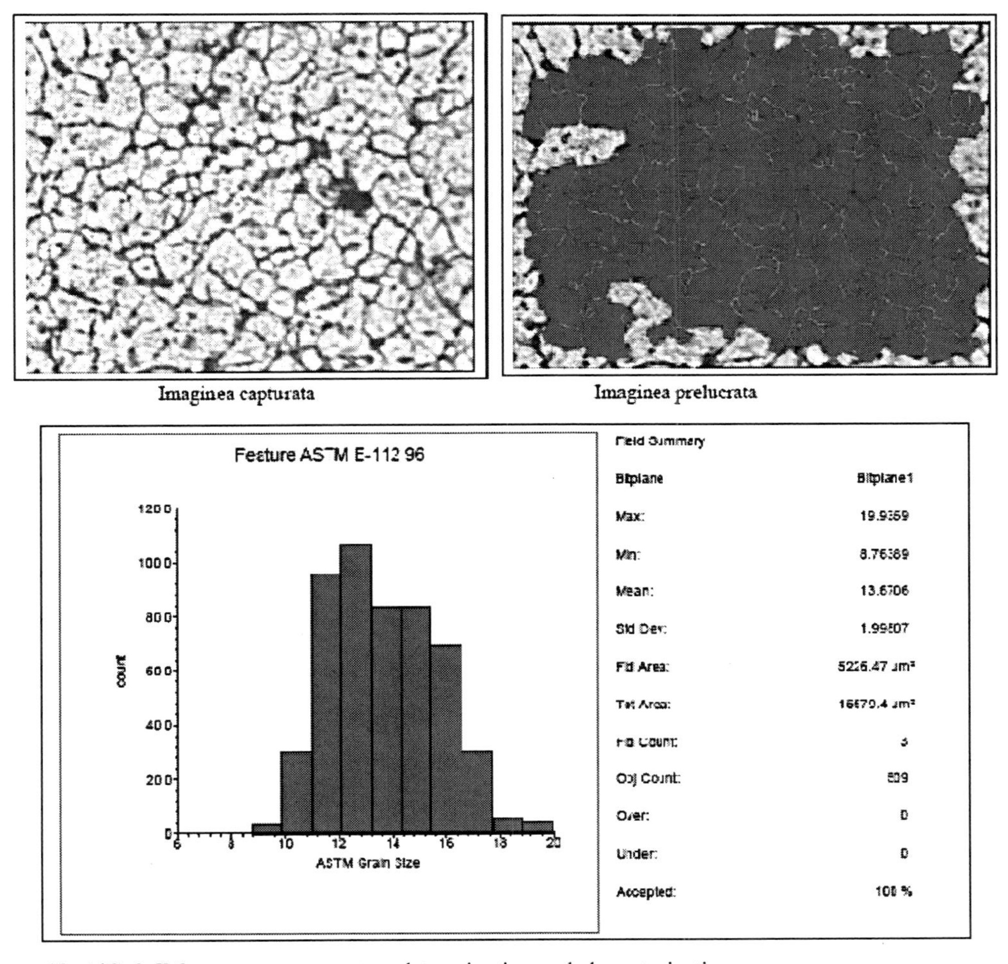

Figure 18. AlCu2,5Mg, test part – structure determination and characterization.

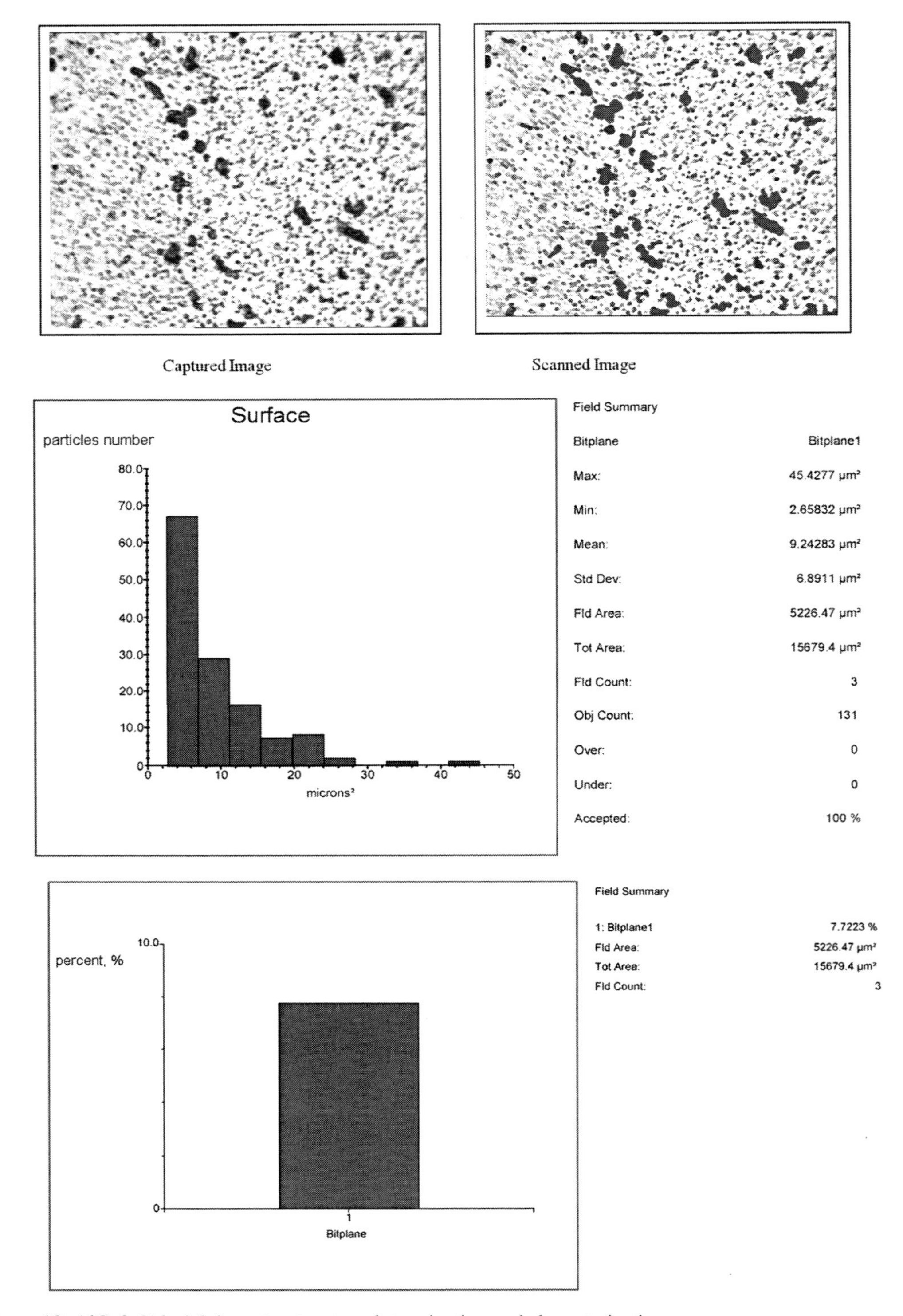

Figure 19. AlCu2,5Mg,1.1.1. part – structure determination and characterization.

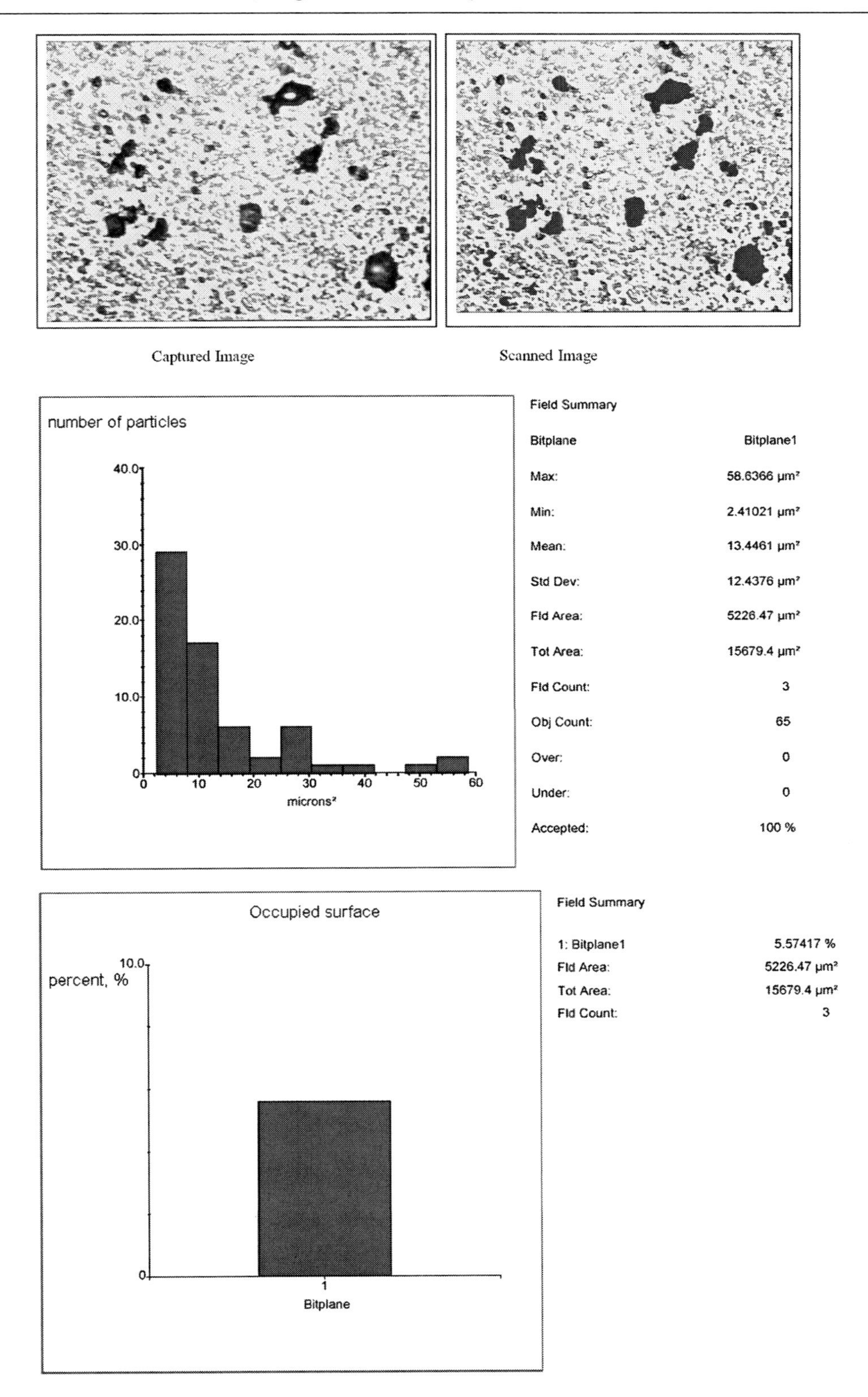

Figure 20. AlCu2,5Mg,2.2.1. part – structure determination and characterization.

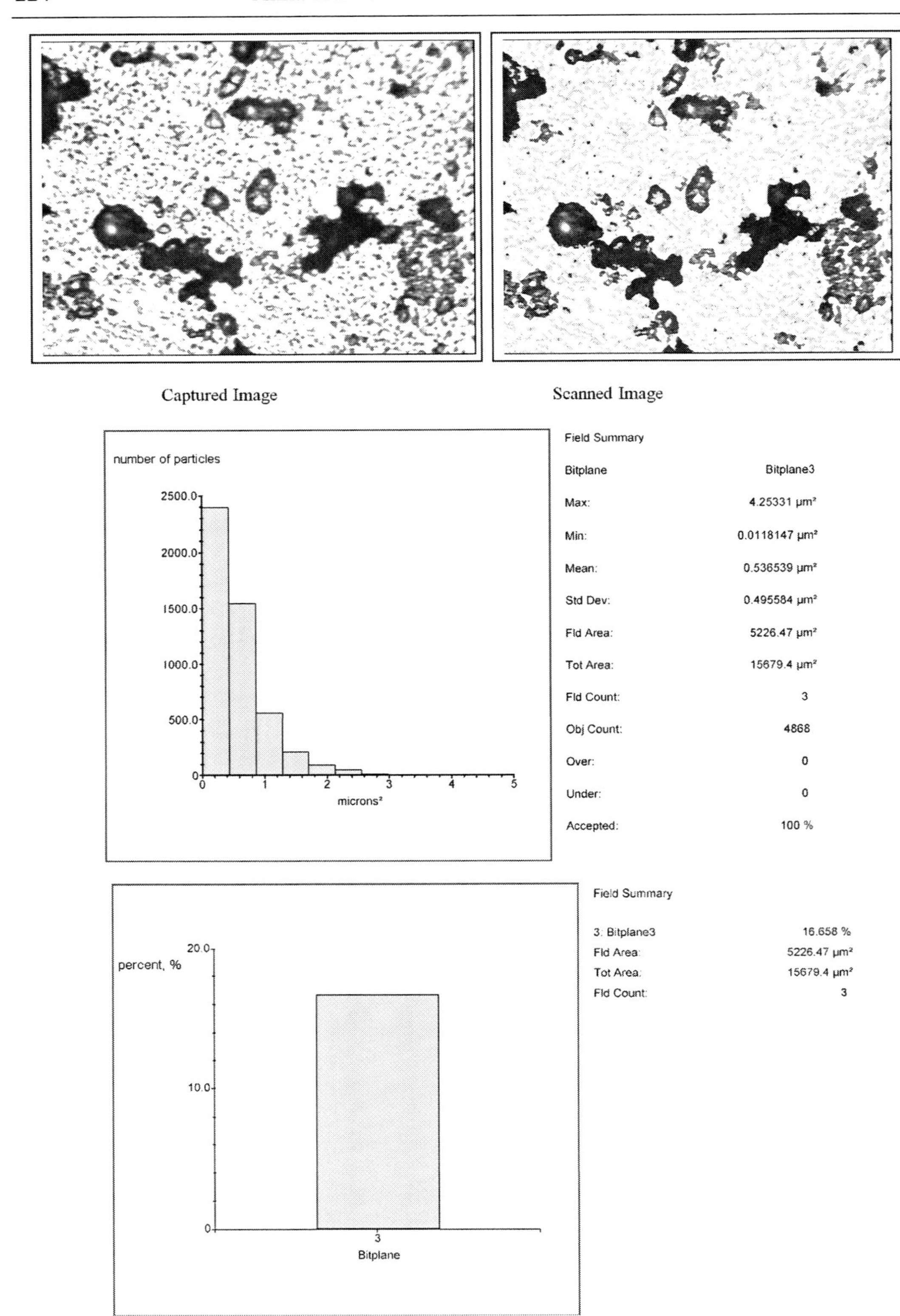

Figure 21. AlCu2,5Mg,1.1.2. part – structure determination and characterization.

Conclusion

This chapter contains a study regarding the improvement of heating parameters, which determine a better technology for aluminum alloys and low energy consumption. I consider that the researching methodology chose is original and allow realizing the experiments to reveal improvements of the final heat treating technology for an AlCu2,5Mg alloy:

- correct calculation of the final heat treating technology for the studied alloy;
- using modern equipment for heating and testing the final heat treated parts, and I are referring at mechanical stress and hardness;
- correct programmed experiments and interpretations.

Further, an experimental study interpretation was made and the conclusion will have the test part as a base of comparison.

a) Hardness. The experimental results are in table 9. The minimum value is for the test part and for the treated parts the maximum was obtained for part 3 and the minimum for part 5. In this case hardness was increased by 89.67% comparing with the non-treated part. The heat treatment applied for part 3 was a quenching at 505°C and an artificial aging at 170°C. Also, applying the quenching at 515°C and the artificial aging at 160°C the minimum hardness was obtained.
b) Mechanical stress. The experimental results are in table 7. The minimum value is for the test part and for the treated parts the maximum was obtained for part 3 and the minimum for part 4. In this case the mechanical stress was increased by 42.11 – 69.09% comparing with the non-treated part. The heat treatment applied for part 3 was a quenching at 505°C and an artificial aging at 170°C.
c) Elongation is varying from 11.87 % for part no 9 to28.37% for part 7.
d) Elasticity modulus. The experimental results are in table 8. The minimum value is for the test part and for the treated parts the maximum was obtained for part 1 and the minimum for part 4. In this case the elasticity was increased by 292.12 % comparing with the non-treated part. The heat treatment applied for part 1 was a quenching at 505°C and an artificial aging at 150°C.

In this context, final heat treating technology optimization for the AlCu2,5Mg aluminum alloys, with a high degree of predicting the mechanical characteristics of the working parts is an important contribution in the industry field:

- reduced energy consumption by optimization the technology described;
- low cost for the heat treated parts, by anticipation of the mechanical characteristics;
- reduced number of persons involved in heat treating technology designing, by obtaining mathematical models, which describe the variation of the characteristics.
- Some contributions in the chapter area are:
- the possibility of choosing a heat treating technology based on a certain mechanical stress needed in service, for an AlCu2,5Mg aluminum alloy;

- the possibility of choosing a heat treating technology based on a certain hardness or elasticity;
- the possibility of predicting the heat treating temperatures based on a certain mechanical stress, elasticity and hardness needed in service;
- analytical calculation of hardness, elasticity and mechanical stress for an AlCu2,5Mg alloy.

With these experimental results, using the specified experiment model we made a study regarding the necessity of heat treatment.

This study helps to determine certain material characteristics that are needed in the service of the parts and also to determine exactly the optimum final heat treatment.

As a conclusion, this chapter presents the algorithm for applying the optimum heat treatment in order to obtain the necessary properties for the working parts.

REFERENCES

[1] Banno, T., Heat Treatment Technology - Present Status and Challenges, *Heat Treatment of Metals*, 1994..

[2] A.A. Minea, A. Dima, Influence of the heating process on aluminum aloys failure, Materials Engineering, Italy, 2002, vol. 13, n.2, pp. 187-190, ISSN 1120-7302.

[3] Dima, A.A. Minea, Theoretical contributions regarding atmosphere circulation in a radiative furnace, *Materials Engineering*, Italy, 2002, vol. 13, n.3, pp.265-267, ISSN 1120-7302.

[4] A.A. Minea, O. Minea, P. Dumitrash, Studies about AlCu2Mg1,5Ni behavior at heat treatment, Elektronaia obrabotka materialov, *Moldavian Academy Journal*, Republic of Moldova, no.6, 2003, pp.82-84, ISSN 0013-5739.

[5] Minea, A.A., Minea O., Dima A., Experimental and theoretical contributions on hardness profile of an AlCu2Mg1,5Ni alloy, International Conference "Advanced materials and technologies" Galati, Romania, 20-22 nov., 2003, pp. 187-189, ISBN 973-627-066-1.

[6] Minea, A.A., Dima A, Experimental studies on microstructure profile of an AlCu2Mg1,5Ni alloy, International Conference "Advanced materials and technologies" Galati, Romania, 20-22 nov., 2003 pp. 419-421, ISBN 973-627-066-1.

[7] Hatch J.E, (1984), Aluminum Properties and Physical Metallurgy, *American Society for Metals*, p. 165-166.

[8] Manzini S.G., (1994), Influence of deformation before artificial aging on properties of Al-Cu-Mg aluminum alloy, Scripta, Met. Et Materialia, 31, p. 1127 – 1130.

[9] Minea A.A, (2005), Mass and Energy Transfer, Editura Cermi, Iaşi, p.122.

[10] Minea A.A.,. Sandu I.G, (2006), Heat treatment optimization on AlCuMg aluminum alloy, *Revista de Chimie* 6, vol.57, p. 586-590.

[11] Minea A. A., (2006), Metallurgical implications of heat treatment of aluminum alloys in electrical furnaces, Conferința Națională De Metalurgie Şi Ştiința Materialelor, Romat, p. 311-315.

[12] Schmidt U., Unger R., (1993), Correlation between microstructure and thermomechanical properties studied on AlCuMg-based alloys, *Journal De Physique 3*, p. 197-205.

[13] Minea,A.A.,Transfer de căldură si instalații termice, Editura Tehnica,Stiintifica si Didactica Cermi Iasi, 2003.

[14] Minea, A.A., Minea, O. Metode de protecție şi tratamente termice, Ed. Cermi, Iaşi, 1999.

INDEX

D

E

F

G

H

I

N

O